W0051135

ISBN 978-3-662-16020-6   ISBN 978-3-7985-1797-4 (eBook)

DOI 10.1007/ 978-3-7985-1797-4

CONTENTS · INHALT

(Fortsetzung Seite III)

PROGRESS IN COLLOID AND POLYMER SCIENCE

Fortschrittsberichte über Kolloide und Polymere

Supplements to "Colloid and Polymer Science" · Continuation of „Kolloid-Beihefte"

Vol. 58   1975

Progr. Colloid & Polymer Sci. **58**, 1—18 (1975)

Meß- und Prüflaboratorium der BASF Aktiengesellschaft Ludwigshafen am Rhein

Zur Defekttheorie des ein-phasigen Vorschmelzens von n-Paraffinen
Teil II: Quasi-chemische Näherung; Lagaly-Fitz-Weiss-Effekt

H. Baur

Mit 13 Abbildungen und 3 Tabellen

(Eingegangen am 6. August 1974)

1. Modell und Variable

Wie im ersten Teil dieser Arbeit (1) betrachten wir eine Lamelle aus N parallel angeordneten Paraffin-Molekülen gleicher Länge, die mit Konformationsfehlern behaftet sein können. Die Konformationsfehler [Abweichungen von der gestreckten ebenen Zickzack-Kette, z.B. Kinken im Sinne von *Pechhold* und *Blasenbrey* (2)] sollen die Lamellenstruktur nicht wesentlich stören.

Mit $x_j = N_j/N$ bezeichnen wir die Konzentration der Ketten, die mit j Konformationsfehlern behaftet sind (N_j: Anzahl dieser Ketten), mit $x_{jk} = 2 N_{jk}/qN$ die Konzentration der Paare benachbarter Ketten, von denen die eine Kette j und die andere Kette k Konformationsfehler enthält (N_{jk}: Anzahl dieser Paare; q: Koordinationszahl). L sei die maximal mögliche Anzahl der Konformationsfehler in einer Kette.

Die Konzentrationen genügen den Nebenbedingungen[1])

$$\sum_j x_j = 1 \,, \qquad [1]$$

$$x_{jk} = x_{kj} \quad (j, k = 0 \ldots L) \,, \qquad [2]$$

$$x_{jj} + \tfrac{1}{2} \sum_{k \neq j} x_{jk} = x_j \quad (j = 0 \ldots L) \,. \qquad [3]$$

---

[1]) Zur Abkürzung schreiben wir wieder
$$\sum_{j=0}^{L} \equiv \sum_j \quad \text{und} \quad \sum_{j=1}^{L} \equiv \sum_j{}' \,.$$

Von den $(L+1)$ Variablen x_j sind also nur L und von den $(L+1)^2$ Variablen x_{jk} nur $L(L+1)/2$ voneinander unabhängig.

Zur energetischen Kennzeichnung des Systems führen wir die folgenden Größen ein:

u_0:   innere Energie einer ungestörten Kette,

Δu: Exzeßenergie eines einzelnen Konformationsfehlers,

$w_{jk} < 0$:   Bindungsenergie zweier benachbarter Ketten, von denen die eine mit j und die andere mit k Konformationsfehlern behaftet ist.

u_0 und Δu seien unabhängig von den zwischenmolekularen Wechselwirkungskräften.

Die w_{jk} hängen natürlich von der Anordnung der Konformationsfehler in den Ketten ab. In der *Bragg-Williams*-Näherung (1) haben wir diese Abhängigkeit vernachlässigt. In der quasichemischen Näherung, die im Gegensatz zur *Bragg-Williams*-Näherung auch Aussagen über die Nahordnung enthält, ist dagegen eine solche Vernachlässigung nicht mehr erlaubt. Wir unterscheiden daher im folgenden zwischen $(j; k)$-Paaren $(j; k)'$ mit der Wechselwirkungsenergie w'_{jk}, bei denen die Konformationsfehler gegenseitig eine energetisch günstige Lage einnehmen, und $(j; k)$-Paaren $(j; k)''$ mit der Wechselwirkungsenergie $w''_{jk} > w'_{jk}$, bei denen die Konformationsfehler gegenseitig eine energetisch ungünstige Lage einnehmen. Entsprechend haben wir dann auch die Konzentrationen x'_{jk} und x''_{jk} zu unterscheiden, wobei die Nebenbedingungen

$$x'_{jk} + x''_{jk} = x_{jk} \quad (j, k = 1 \dots L) \tag{4}$$

gelten.

Die Gesamtzahl der voneinander unabhängigen Variablen erhöht sich damit auf $L^2 + 2L$. Als unabhängige Variable betrachten wir im folgenden

$$
\begin{array}{lll}
x_k & (k > 0); & \text{Anzahl: } L, \\
x_{k0} & (k > 0); & \text{Anzahl: } L, \\
x_{jk} & (j > k > 0); & \text{Anzahl: } L(L-1)/2, \\
x'_{kk} & (k > 0); & \text{Anzahl: } L, \\
x'_{jk} & (j > k > 0); & \text{Anzahl: } L(L-1)/2.
\end{array}
$$

2. Die Gleichungen des quasi-chemischen Gleichgewichtes

Bei vorgegebener Temperatur T und vorgegebenem Deformationszustand und nach Abspaltung des Beitrages der Phononen und der inneren Freiheitsgrade der Ketten wird das thermodynamische Gleichgewicht der Lamelle bestimmt durch die Zustandssumme

$$Q = \sum_{z} g(N; z) \exp\left\{ -\frac{E(N; z)}{kT} \right\}, \tag{6}$$

in der $E(N; z)$ die möglichen Energiezustände des Systems, $g(N; z) - 1$ deren Entartungsgrad und k die *Boltzmann*sche Konstante bezeichnen. Zu summieren wäre eigentlich über alle möglichen $(L^2 + 2L)$-tupel

$$z = \{x_k; x_{k0}; x_{jk}; x'_{kk}; x'_{jk}\} \tag{7}$$

mit $j > k > 0$. Wir beschränken uns jedoch darauf, die Summe durch ihr maximales Glied zu ersetzen.

Die Energie ist durch

$$\frac{E}{N} = u_0 + \varDelta u \sum_{j}' j x_j + \frac{q}{2}\left(w_{00} x_{00} + \sum_{j}' w_{j0} x_{j0} \right.$$

$$
\left. + \sum_{j}' w'_{jj} x'_{jj} + \sum_{j}' w''_{jj} x''_{jj} \right.
$$
$$
\left. + \sum_{j>k}' \sum w'_{jk} x'_{jk} + \sum_{j>k}' \sum w''_{jk} x''_{jk} \right) \tag{8}
$$

gegeben [vgl. (1)].

Zur Ermittlung des Faktors g betrachten wir die Paare $(j; k)$ als voneinander unabhängige statistische Elemente, die quasichemischen Reaktionsgleichungen z. B. der Form

$$(l; m)' \rightleftharpoons (l; m)''$$

oder

$$(l; l)'' + (m; m)'' \rightleftharpoons 2 (l; m)''$$

genügen (vgl. weiter unten die Gln. [17] und [16]). In dieser sog. quasi-chemischen Näherung von *Fowler* und *Guggenheim* (3) ist der Faktor $g(N; z)$ aufzuspalten in ein Produkt

$$g = h g', \tag{9}$$

in dem $g'(N; z)$ die Anzahl der unterscheidbaren Anordnungen bei statistischer Verteilung der Kettenpaare $(j; k)$ angibt und

$$h = g''/g'(N; z^*) \tag{10}$$

ein Normierungsfaktor ist, der für die Widerspruchsfreiheit des quasi-chemischen Ansatzes beim Grenzübergang $T \to \infty$ sorgt. g'' bezeichnet die Anzahl der unterscheidbaren Anordnungen bei statistischer Verteilung der Ketten (j) (*Bragg-Williams*-Näherung), z^* die dazugehörenden Werte von [7].

Ist γ_j die Anzahl der unterscheidbaren Konformationsisomeren einer (j)-Kette, $\varPhi_{jj}$ der Bruchteil der γ_j^2 $(j; j)$-Paare, die zur Wechselwirkungsenergie w_{jj} führen und $\varPhi_{jk}$ $(j > k > 0)$ der Bruchteil der $2 \gamma_j \gamma_k$ $(j; k)$-Paare, die zur Wechselwirkungsenergie w'_{jk} führen, so gilt

$$g' = \frac{(qN/2)!}{\prod\limits_{j \geq k} P_{jk}}$$

mit

$$P_{00} \equiv N_{00}!,$$

$$P_{j0} \equiv \left(\frac{N_{j0}}{2 \gamma_j} \right)!^{2\gamma_j}; \quad j > 0,$$

$$P_{jj} \equiv \left(\frac{N'_{jj}}{\varPhi_{jj} \gamma_j^2} \right)!^{\varPhi_{jj} \gamma_j^2} \left(\frac{N''_{jj}}{(1 - \varPhi_{jj}) \gamma_j^2} \right)!^{(1 - \varPhi_{jj}) \gamma_j^2}; \quad j > 0,$$

$$P_{jk} \equiv \left(\frac{N'_{jk}}{2 \varPhi_{jk} \gamma_j \gamma_k} \right)!^{2\varPhi_{jk} \gamma_j \gamma_k} \left(\frac{N''_{jk}}{2 (1 - \varPhi_{jk}) \gamma_j \gamma_k} \right)!^{2(1 - \varPhi_{jk}) \gamma_j \gamma_k}; \quad j > k > 0.$$

Unter Verwendung der *Stirling*schen Formel (Voraussetzung: große laterale Ausdehnung der Lamelle) folgt daraus

$$\frac{2}{qN} \ln g' = - x_{00} \ln x_{00} - {\sum_j}' x_{j0} \ln \left(\frac{x_{j0}}{2\,\gamma_j} \right) - {\sum_j}' x'_{jj} \ln \left(\frac{x'_{jj}}{\Phi_{jj}\,\gamma_j^2} \right) - {\sum_j}' x''_{jj} \ln \left[\frac{x''_{jj}}{(1-\Phi_{jj})\,\gamma_j^2} \right]$$

$$- {\sum_{j>k}}' x'_{jk} \ln \left(\frac{x'_{jk}}{2\,\Phi_{jk}\,\gamma_j\,\gamma_k} \right) - {\sum_{j>k}}' x''_{jk} \ln \left[\frac{x''_{jk}}{2\,(1-\Phi_{jk})\,\gamma_j\,\gamma_k} \right]. \qquad [11]$$

Ganz entsprechend folgt ferner [vgl. (1)]

$$\frac{1}{N} \ln g'' = - x_0 \ln x_0 - {\sum_j}' x_j \ln \left(\frac{x_j}{\gamma_j} \right). \qquad [12]$$

Bei statistischer Verteilung der einzelnen Ketten ist

$$x_{00}^* = x_0^2$$
$$x_{j0}^* = 2\,x_j\,x_0 \qquad (j>0),$$
$$x_{jj}'^* = \Phi_{jj}\,x_j^2 \qquad (j>0),$$
$$x_{jj}''^* = (1-\Phi_{jj})\,x_j^2 \qquad (j>0),$$
$$x_{jk}'^* = 2\,\Phi_{jk}\,x_j\,x_k \qquad (j>k>0),$$
$$x_{jk}''^* = 2\,(1-\Phi_{jk})\,x_j\,x_k \qquad (j>k>0).$$

Setzen wir diese Werte in [11] ein, so erhalten wir für den Normierungsfaktor [10]

$$\frac{2}{qN} \ln h = \frac{2}{qN} \ln g'' - \frac{2}{qN} \ln g'(N; z^*)$$

$$= 2\,\frac{q-1}{q} \left[x_0 \ln x_0 + {\sum_j}' x_j \ln \left(\frac{x_j}{\gamma_j} \right) \right],$$

d.h.

$$\ln h = (1-q) \ln g'',$$

und an Stelle von [9]

$$g = (g'')^{1-q}\,g'.$$

Bei Beschränkung auf das maximale Glied $\hat{Q}$ der Summe in [6] ist also die freie Energie der Lamelle gegeben durch

$$F = - kT \ln \hat{Q} = \hat{E} + (q-1)\,kT \ln \hat{g}''$$
$$- kT \ln \hat{g}'. \qquad [13]$$

Hieraus folgen durch Differentiation nach den freien Variablen, d.h. nach den Komponenten von z in [7], und Nullsetzen der Ableitungen die *Gleichungen des quasi-chemischen Gleichgewichtes*

$$\left(\frac{\hat{x}_0}{\hat{x}_l} \right)^{q-1} \left(\frac{\hat{x}''_{ll}}{\hat{x}_{00}} \right)^{q/2}$$
$$= (1-\Phi_{ll})^{q/2}\,\gamma_l\,e^{-l(\Delta u/kT)}\,K''_1(l; 0); \quad l>0, \qquad [14a]$$

$$\frac{\hat{x}_{l0}^2}{\hat{x}_{00}\,\hat{x}''_{ll}} = \frac{4}{1-\Phi_{ll}}\,K''_2(l; 0); \quad l>0, \qquad [15a]$$

$$\frac{\hat{x}_{lm}''^2}{\hat{x}''_{ll}\,\hat{x}''_{mm}} = \frac{4\,(1-\Phi_{lm})^2}{(1-\Phi_{ll})\,(1-\Phi_{mm})}\,K''_3(l; m);$$
$$l>m>0, \qquad [16a]$$

$$\frac{\hat{x}'_{lm}}{\hat{x}''_{lm}} = \frac{\Phi_{lm}}{1-\Phi_{lm}}\,K''_4(l; m); \quad l \geqq m > 0, \qquad [17a]$$

mit den Gleichgewichtskonstanten

$$\ln K''_1(l; 0) \equiv - \frac{q}{2kT}\,(w''_{ll} - w_{00}), \qquad [14b]$$

$$\ln K''_2(l; 0) \equiv - \frac{1}{kT}\,[2\,w_{l0} - (w''_{ll} + w_{00})], \qquad [15b]$$

$$\ln K''_3(l; m) \equiv - \frac{1}{kT}\,[2\,w''_{lm} - (w''_{mm} + w''_{ll})], \qquad [16b]$$

$$\ln K''_4(l; m) \equiv - \frac{1}{kT}\,(w'_{lm} - w''_{lm}). \qquad [17b]$$

[14—17] sind vergleichbar mit den aus der Theorie der chemischen Reaktionen bekannten Massenwirkungsgesetzen.

Als freie Gleichgewichtsenergie pro Kette folgt damit aus [8] und [11—13] die einfache Beziehung

$$\frac{F}{N} = u_0 + \frac{q}{2}\,w_{00} + kT \ln \frac{\hat{x}_{00}^{q/2}}{\hat{x}_0^{q-1}}. \qquad [18]$$

Die Gln. [14—17] lassen eine ausgesprochen breite Verhaltensmannigfaltigkeit zu. Fassen wir die Lamelle als eine feste Substitutionslösung von (l)-Ketten auf, so können wir, wie in der Theorie der Lösungen, unterscheiden zwischen dem idealen, perfekten und nicht-idealen Fall und zwischen Athermie, Solvatation oder Aggregation in bezug auf eine oder mehrere der insgesamt $L+1$ Komponenten. Wie in der Theorie der Legierungen können sich Überstrukturen bilden oder Ausscheidungen auftreten.

In den folgenden Abschnitten soll vor allem der Fall der Aggregation behandelt werden. Dabei sind die energetisch günstigen Paare $(l; l)'$ in den Mittelpunkt zu stellen. Wir formen daher die Gln. [14—17] noch um zu

$$\left(\frac{\hat{x}_0}{\hat{x}_l} \right)^{q-1} \left(\frac{\hat{x}'_{ll}}{\hat{x}_{00}} \right)^{q/2}$$
$$= \Phi_{ll}^{q/2}\,\gamma_l\,e^{-l(\Delta u/kT)}\,K'_1(l;0); \quad l>0, \qquad [19a]$$

$$\frac{\mathring{x}_{l0}^2}{\mathring{x}_{00}\,\mathring{x}_{ll}'} = \frac{4}{\Phi_{ll}}\,K_2'(l;0); \quad l>0, \qquad [20\,\mathrm{a}]$$

$$\frac{x_{lm}'^2}{\mathring{x}_{ll}'\,\mathring{x}_{mm}'} = \frac{4\,\Phi_{lm}^2}{\Phi_{ll}\,\Phi_{mm}}\,K_3'(l;m); \quad l>m>0, \qquad [21\,\mathrm{a}]$$

$$\frac{\mathring{x}_{lm}''}{\mathring{x}_{lm}'} = \frac{1-\Phi_{lm}}{\Phi_{lm}}\,K_4'(l;m); \quad l\geqq m>0 \quad [22\,\mathrm{a}]$$

mit den Gleichgewichtskonstanten

$$\ln K_1'(l;0) \equiv -\frac{q}{2\,k\,T}\,(w_{ll}' - w_{00}), \qquad [19\,\mathrm{b}]$$

$$\ln K_2'(l;0) \equiv -\frac{1}{k\,T}\,[2\,w_{l0} - (w_{ll}' + w_{00})],\ [20\,\mathrm{b}]$$

$$\ln K_3'(l;m) \equiv -\frac{1}{k\,T}\,[2\,w_{lm}' - (w_{ll}' + w_{mm}')], \qquad [21\,\mathrm{b}]$$

$$\ln K_4'(l;m) \equiv -\frac{1}{k\,T}\,(w_{lm}'' - w_{lm}'). \qquad [22\,\mathrm{b}]$$

Im einfachsten Fall verschwindender zwischenmolekularer Wechselwirkungskräfte (ideales System; $K_\nu = 1$, $\Phi_{lm} = 1$) ergibt sich zunächst aus [22]

$$\mathring{x}_{lm}'' = 0,$$

d.h. nach [4]

$$\mathring{x}_{lm}' = \mathring{x}_{lm}; \quad l \geqq m > 0.$$

Aus [20] und [21] folgt dann

$$\mathring{x}_{lm}^2 = 4\,\mathring{x}_{ll}\,\mathring{x}_{mm}; \quad l>m\geqq 0$$

und aus [19]

$$\frac{\mathring{x}_l}{\mathring{x}_0} = \gamma_l\,e^{-l(\varDelta u/kT)}\left(\frac{\mathring{x}_{00}\,\mathring{x}_l^2}{\mathring{x}_0^2\,\mathring{x}_{ll}}\right)^{q/2}; \quad l>0. \qquad [23]$$

Für $L=1$ sind damit auf Grund der Nebenbedingungen [1–3] die einzelnen Komponenten notwendig statistisch verteilt. Für $L>1$ ist die statistische Verteilung

$$\left.\begin{aligned}\mathring{x}_{ll} &= \mathring{x}_l^2; & l&\geqq 0 \\ \mathring{x}_{lm} &= 2\,\mathring{x}_l\,\mathring{x}_m; & l&>m\geqq 0\end{aligned}\right\} \qquad [24]$$

eine mögliche Lösung.

Ist

$$\gamma_l = \binom{L}{l},$$

so führt die statistische Verteilung auf ein *Schottky*-System (ideales 2-Niveau-System) [vgl. (1)]. Ist

$$\gamma_l = 2^l \binom{Z-2\,l}{l}$$

(Z: Anzahl der Kettenglieder; vgl. [56] in Abschnitt 6), so ist das Verhalten der idealen Lamelle dem Verhalten eines *Schottky*-Systems qualitativ ähnlich. Für $L=1$ stimmt die ideale Lamelle in jedem Falle exakt mit einem *Schottky*-System überein.

Für $T\to\infty$ folgt aus [23], da das Ergebnis nicht von der Koordinationszahl q abhängen kann,

$$\frac{\mathring{x}_{00}\,\mathring{x}_l^2}{\mathring{x}_0^2\,\mathring{x}_{ll}} = 1\,.$$

Zusammen mit den Nebenbedingungen [1–3] erhalten wir daraus auch für $L>1$ die statistische Verteilung.

Ferner folgt

$$\mathring{x}_l = \gamma_l\,\mathring{x}_0, \qquad [25]$$

d.h. nach [1]

$$\mathring{x}_0 = \frac{1}{1+\sum_l' \gamma_l} \qquad [26]$$

und

$$\mathring{x}_l = \frac{\gamma_l}{1+\sum_l' \gamma_l}\,.$$

[25] ist ein Äquivalent für den sog. Gleichverteilungssatz. Eine Gleichverteilung unter den (l)-Ketten ist allerdings nur dann gegeben, wenn für alle l $\gamma_l = 1$, d.h. $\sum_l' \gamma_l = L$ gilt. Ist $\sum_l' \gamma_l = 1$, so besteht eine Gleichverteilung zwischen den defektfreien und defekten Ketten.

3. Laterale Nah- und Fernordnung

Als Maß für die Ordnung eines Gitters wählt man gewöhnlich einen Parameter σ, der bei vollständiger Ordnung auf 1 normiert ist und im Falle vollständiger Unordnung verschwindet. Für die in (1) als Parameter der lateralen Nah- und Fernordnung verwendeten Variablen $\mathring{x}_{00}$ und $\mathring{x}_0$ trifft das nicht exakt zu, da nach [26]

$$\lim_{T\to\infty} \mathring{x}_0 = \frac{1}{1+\sum_l' \gamma_l}$$

und mit [24]

$$\lim_{T\to\infty} \mathring{x}_{00} = \frac{1}{(1+\sum_l' \gamma_l)^2}$$

gilt. Als Maß σ_n der lateralen Nahordnung wird man daher besser

$$\sigma_n \equiv \frac{1}{(1 + \sum_l{}' \gamma_l)^2 - 1}[(1 + \sum_l{}' \gamma_l)^2 \hat{x}_{00} - 1] \quad [27]$$

und als Maß σ_f der lateralen Fernordnung

$$\sigma_f \equiv \frac{1}{\sum_l{}' \gamma_l}[(1 + \sum_l{}' \gamma_l) \hat{x}_0 - 1] \quad [28]$$

definieren.

Ist $\sum_l{}' \gamma_l = 1$, so erhält man die aus der Theorie der binären Legierungen bekannten Maßzahlen

$$\sigma_n = \tfrac{1}{3}(4 \hat{x}_{00} - 1),$$
$$\sigma_f = 2 \hat{x}_0 - 1.$$

Für uns ist das jedoch ein weitgehend hypothetischer Fall, der nur für $L=1$, $\gamma_1 = 1$ Gültigkeit haben kann und z.B. auf einzelne Segmentschichten im Sinne von *Blasenbrey* und *Pechhold* anwendbar wäre.

Im allgemeinen ist

$$\sum_l{}' \gamma_l \gg 1$$

anzunehmen, so daß aus [27—28] — wie in (1) vorausgesetzt —

$$\sigma_n \approx \hat{x}_{00} \quad \text{und} \quad \sigma_f \approx \hat{x}_0$$

folgt.

Mit Hilfe von [27] und [28] läßt sich die freie Energie [18] als Funktion des Nah- und Fernordnungsparameters ausdrücken. Es gilt

$$\frac{F}{N} = u_0 + \frac{q}{2} w_{00}$$
$$+ kT \ln \frac{\{[(1 + \sum_l{}' \gamma_l)^2 - 1]\sigma_n + 1\}^{q/2}}{(1 + \sum_l{}' \gamma_l)[(\sum_l{}' \gamma_l)\sigma_f + 1]^{q-1}}.$$

Zu bemerken ist, daß der durch [27] und [28] definierte Grad der lateralen Ordnung sich allein auf das Verhältnis *defekt : nicht defekt* bezieht. Da wir unter den defekten Ketten zwischen Ketten mit verschiedener Anzahl von Defekten unterscheiden, wäre im Prinzip auch der Ordnungsgrad noch weiter zu differenzieren. Dem entspricht, daß das Verhältnis

$$\frac{\hat{x}_{00}{}^{q/2}}{\hat{x}_0{}^{q-1}} = \left(\frac{\hat{x}_{00}}{\hat{x}_0^2}\right)^{q/2} \hat{x}_0 \approx \left(\frac{\sigma_n}{\sigma_f^2}\right)^{q/2} \sigma_f$$

durchaus nicht mit σ_n, $\sigma_f \to 0$ verschwinden muß. Hierauf werden wir in Abschnitt 6 zurückkommen.

4. Lamelle aus Ketten mit maximal einem Konformationsfehler ($L=1$)

Können die Ketten maximal nur einen Konformationsfehler aufnehmen, sind also nur Paare

$$(0;0)\ (1;0)\ (1;1)'\ (1;1)''$$

möglich, und beträgt die Koordinationszahl

$$q = 6,$$

so reduzieren sich die Gleichungen des quasichemischen Gleichgewichts [19—22] mit den Nebenbedingungen [1—4] auf

$$\left(\frac{\hat{x}_0}{\hat{x}_1}\right)^5 \left(\frac{\hat{x}_{11}}{\hat{x}_{00}}\right)^3 = B_1(1 + D_{11})^3, \quad [29a]$$

$$\hat{x}_{10}^2 = C_{10}\hat{x}_{11}\hat{x}_{00}, \quad [30a]$$

$$\hat{x}_{11}'' = D_{11}\hat{x}_{11}' \quad [31a]$$

mit den Konstanten

$$\ln B_1 \equiv \ln(\Phi_{11}^3 \gamma_1) - \frac{1}{kT}[3(w_{11}' - w_{00}) + \Delta u], \quad [29b]$$

$$\ln C_{10} \equiv \ln\left[\frac{4}{\Phi_{11}(1 + D_{11})}\right]$$
$$- \frac{1}{kT}[2w_{10} - (w_{11}' + w_{00})], \quad [30b]$$

$$\ln D_{11} \equiv \ln\left(\frac{1 - \Phi_{11}}{\Phi_{11}}\right) - \frac{1}{kT}(w_{11}'' - w_{11}') \quad [31b]$$

und den Nebenbedingungen

$$\hat{x}_0 = 1 - \hat{x}_1,$$
$$\hat{x}_{00} = 1 - \hat{x}_1 - \tfrac{1}{2}\hat{x}_{10},$$
$$\hat{x}_{11} = \hat{x}_{11}' + \hat{x}_{11}'' = \hat{x}_1 - \tfrac{1}{2}\hat{x}_{10}.$$

Unter Verwendung der Transformation

$$\hat{x}_1 \equiv \tfrac{1}{2}(1 + y_1)$$

und der Abkürzung

$$C_{10}'(T) \equiv \frac{C_{10}}{4 - C_{10}} \quad [32]$$

erhalten wir daraus zur *Bestimmung der Gleichgewichtskonzentrationen* die Gleichungen

$$A_1 \equiv 5 \ln\left(\frac{1 - y_1}{1 + y_1}\right) - 3 \ln\left(\frac{1 - y_1 - \hat{x}_{10}}{1 + y_1 - \hat{x}_{10}}\right)$$
$$= \ln B_1(1 + D_{11})^3, \quad [33]$$

$$\hat{x}_{10} = [C_{10}'^2 + C_{10}'(1 - y_1^2)]^{1/2} - C_{10}', \quad [34]$$
$$\hat{x}_0 = \tfrac{1}{2}(1 - y_1),$$
$$\hat{x}_{00} = \tfrac{1}{2}(1 - y_1 - \hat{x}_{10}),$$

$$\hat{x}_{11} = \tfrac{1}{2}\,(1 + y_1 - \hat{x}_{10})\,,$$

$$\hat{x}'_{11} = \frac{1 + y_1 - \hat{x}_{10}}{2\,(1 + D_{11})}\,,$$

$$\hat{x}''_{11} = \frac{D_{11}\,(1 + y_1 - \hat{x}_{10})}{2\,(1 + D_{11})}\,.$$

Als einzige echte Variable bleibt

$$y_1 = 2\,\hat{x}_1 - 1\,,$$

deren Gleichgewichtswert durch [33] festgelegt ist und mit deren Hilfe alle anderen Gleich-

gewichtskonzentrationen ausgerechnet werden können.

Wie man unmittelbar sieht, besitzen [33] und [34] die Symmetrieeigenschaften

$$\hat{x}_{10}(y_1) = \hat{x}_{10}(-y_1)\,, \qquad\qquad [35]$$

$$-A_1(y_1) = A_1(-y_1)\,. \qquad\qquad [36]$$

Ferner besitzt $A_1(y_1)$ in dem physikalisch allein zulässigen Bereich

$$-1 \leqq y_1 \leqq +1$$

bei bestimmten Werten von $C'_{10}(T)$ die von Umwandlungserscheinungen her bekannte s-förmige Gestalt (vgl. Abb. 1). Die Lamelle ist also unter gewissen Voraussetzungen bei einer bestimmten Temperatur einer sprunghaften Änderung des Parameters y_1 und damit einer sprunghaften

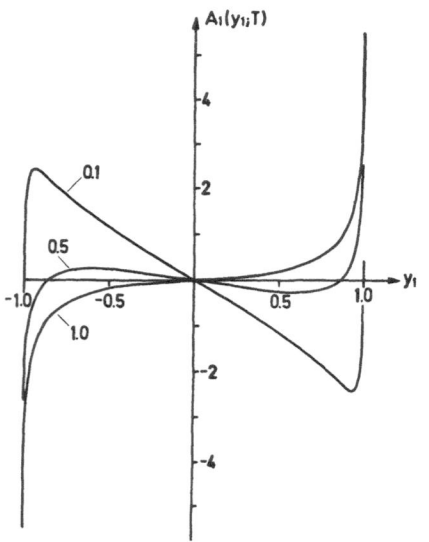

Abb. 1. A_1 als Funktion von y_1 nach [33] mit C'_{10} als Parameter (vgl. Abschnitt 7)

Änderung der Gleichgewichtskonzentrationen $\hat{x}_0$, $\hat{x}_1$, $\hat{x}_{00}$, $\hat{x}'_{11}$ und $\hat{x}''_{11}$ fähig. Nach [36] ist die Umwandlung durch den virtuellen Wert $y = 0$ gekennzeichnet. Wegen [35] kann daher $\hat{x}_{10}$ keinen Sprung erleiden.

Als *Bestimmungsgleichung für die Umwandlungstemperatur* T_{U1} erhält man mit $y_1 = 0$ aus [33]

$$\ln B_1(T_{U1})\,[1 + D_{11}(T_{U1})]^3 = 0 \qquad [37\,\mathrm{a}]$$

oder nach [29 b] und [31 b]

$$k\,T_{U1}\ln\gamma_1\left[\varPhi_{11} + (1-\varPhi_{11})\exp\left\{-\frac{w''_{11}-w'_{11}}{k\,T_{U1}}\right\}\right]^3 = 3\,(w'_{11}-w_{00}) + \varDelta u\,. \qquad [37\,\mathrm{b}]$$

Die Temperaturlage der Umwandlung hängt ab von den Energiedifferenzen $w''_{11}-w'_{11}$, $w'_{11}-w_{00}$, $\varDelta u$, dem Bruchteil $\varPhi_{11}$ der energetisch günstigen $(1;1)$-Konfigurationen und der Anzahl γ_1 der unterscheidbaren (1)-Isomeren. Sie ist hingegen von der Differenz $2w_{10}-(w'_{11}+w_{00})$ unabhängig.

Bei der aus [37] bestimmbaren Temperatur T_{U1} setzt jedoch nur dann eine Umwandlung ein, wenn zusätzlich die Bedingung

$$C'_{10}(T_{U1}) < \frac{4}{5} \qquad\qquad [38\,\mathrm{a}]$$

bzw. nach [32]

$$C_{10}(T_{U1}) < \frac{16}{9} \qquad\qquad [38\,\mathrm{b}]$$

oder nach [30 b] und [31 b]

$$2\,w_{10} - (w'_{11}+w_{00}) > -\,k\,T_{U1}\ln\frac{4}{9}\,\varPhi_{11}(1+D_{11})$$

$$= -\,k\,T_{U1}\ln\frac{4}{9}\left[\varPhi_{11} + (1-\varPhi_{11})\right.$$

$$\left.\times \exp\left\{-\frac{w''_{11}-w'_{11}}{k\,T_{U1}}\right\}\right] \qquad [38\,\mathrm{c}]$$

erfüllt ist. Die *Existenz einer Umwandlung* hängt also wesentlich auch von der Differenz

$$2\,w_{10} - (w'_{11} + w_{00})$$

ab. Da die rechte Seite der Ungleichung [38 c] auf Grund unserer Voraussetzung $w''_{11} > w'_{11}$ stets positiv ist, muß außerdem

$$2\,w_{10} - (w'_{11} + w_{00}) > 0$$

gelten. Eine Umwandlung setzt nur dann ein, wenn unter den defekten Ketten, die zu den energetisch günstigen $(1;1)$-Paaren führen, eine

Aggregationstendenz besteht und diese Aggregationstendenz den kritischen Wert

$$W_{\text{krit}} \equiv k\,T_{\text{U1}}\ln\left[\frac{9}{4\,\Phi_{11}(1+D_{11})}\right] > 0 \qquad [39]$$

übersteigt.

Bilden sich unterhalb und im Bereich der Umwandlungstemperatur unter den (1; 1)-Paaren praktisch nur die energetisch günstigen Paare aus, d. h. gilt

$$w_{11}'' - w_{11}' \gg k\,T_{\text{U1}} \quad \text{bzw.} \quad D_{11}(T_{\text{U1}}) \approx 0,$$

so tritt zur Bestimmung der Umwandlungstemperatur an die Stelle von [37] die Gleichung

$$T_{\text{U1}} = \frac{3\,(w_{11}' - w_{00}) + \Delta u}{k\ln(\Phi_{11}^3\,\gamma_1)}$$

und an die Stelle der Bedingung [38c] für die Existenz der Umwandlung die Ungleichung

$$2\,w_{10} - (w_{11}' + w_{00}) > k\,T_{\text{U1}}\ln\left(\frac{9}{4\,\Phi_{11}}\right).$$

Ist die Aggregationstendenz so stark bzw. der energetische Unterschied zwischen den ungünstigen und günstigen (1; 1)-Paaren so klein, daß

$$2\,w_{10} - (w_{11}' + w_{00}) \gg w_{11}'' - w_{11}' \qquad [40]$$

gilt, so folgt aus [38]

$$2\,w_{10} - (w_{11}' + w_{00}) > k\,T_{\text{U1}}\ln\left(\frac{9}{4}\right).$$

Die Existenz der Umwandlung ist in diesem Falle unabhängig von Φ_{11}.

Gilt hingegen

$$2\,w_{10} - (w_{11}' + w_{00}) \leq w_{11}'' - w_{11}', \qquad [41]$$

so muß, wenn eine Umwandlung in Erscheinung treten soll, der Bruchteil Φ_{11} um so größer sein, je schwächer die Aggregationstendenz ist. So tritt z. B. mit

$$2\,w_{10} - (w_{11}' + w_{00}) = 450\,k \approx 900\;\text{cal/Mol}$$
$$\leq w_{11}'' - w_{11}'$$

nur dann eine Umwandlung auf, wenn $\Phi_{11} > 0{,}4$ ist (vgl. Abb. 2). Für das Gleichheitszeichen in [41] folgt aus [38] als Bedingung für die Existenz einer Umwandlung

$$2\,w_{10} - (w_{11}' + w_{00}) > k\,T_{\text{U1}}\ln\left(\frac{5 + 4\,\Phi_{11}}{4\,\Phi_{11}}\right).$$

Unabhängig von [40] und [41] hängt die Lage der Umwandlungstemperatur stets von Φ_{11} ab.

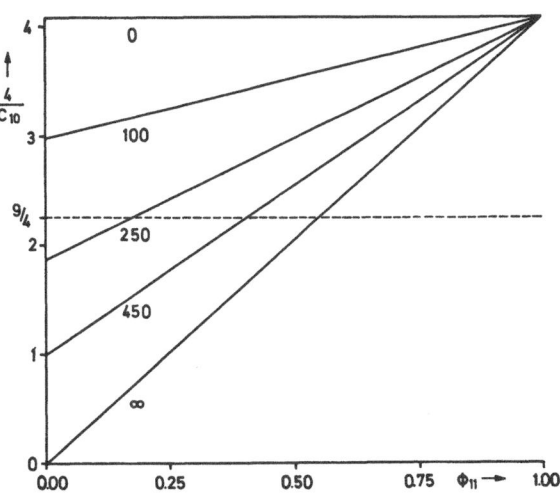

Abb. 2. $4/C_{10}$ als Funktion von Φ_{11} nach [30b], [31b] mit $2\,w_{10} - (w_{11}' + w_{00}) = 450\,k$ und $w_{11}'' - w_{11}'$ als Parameter. Eine Umwandlung tritt nach [38b] nur dann in Erscheinung, wenn $(4/C_{10}) > 9/4 = 2{,}25$ ist

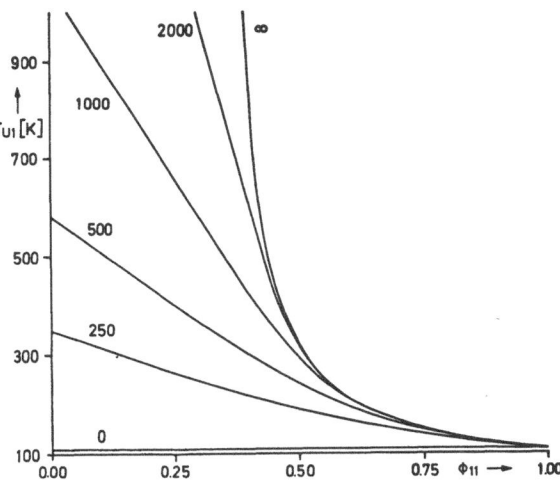

Abb. 3. Umwandlungstemperatur T_{U1} als Funktion von Φ_{11} nach [37b] mit $\gamma_1 = 24$, $3\,(w_{11}' - w_{00}) + \Delta u = 350\,k$[2]) und $w_{11}'' - w_{11}'$ als Parameter

In Abb. 3 ist als Beispiel $T_{\text{U1}} = T_{\text{U1}}(\Phi_{11})$ nach [37b] mit

$$3\,(w_{11}' - w_{00}) + \Delta u = 350\,k \approx 700\;\text{cal/Mol},$$

$\gamma_1 = 24$ und einigen Werten von $w_{11}'' - w_{11}'$ als Parameter dargestellt[2]). Bei größerer Energiedifferenz $w_{11}'' - w_{11}'$ fällt die Umwandlungstempe-

---

[2]) Der Wert $\gamma_1 = 24$ entspricht dem Ansatz [56] mit $Z = 14$; der Wert $3\,(w_{11}' - w_{00}) + \Delta u = 350\,k$ entspricht etwa der Umwandlungswärme, die von *Lagaly, Fitz* und *Weiss* bei n-Tetradecylammonium-n-Tetradecanol-Beidellit gefunden wird (vgl. Abschnitt 6).

ratur nur dann in den Bereich von 200 bis 400 K, wenn Φ_{11} nicht zu klein ist. Die Umwandlung zu einer idealen Blockstruktur (große Differenz $w''_{11} - w'_{11}$, kleines Φ_{11}) ist nur bei weit höheren Temperaturen möglich.

Es tritt hier die Frage auf, wie groß Φ_{11} überhaupt werden kann. Prinzipiell ist Φ_{11} eingeschränkt auf

$$0 \leqq \Phi_{11} \leqq 1 .$$

Nimmt man jedoch mit *Blasenbrey* und *Pechhold* (2, 4—5) an, daß die durch die Konformationsfehler erzeugte Störung längs der Kettenachsen schon nach wenigen CH_2-Gruppen verschwindet, so können nur solche Isomere zu energetisch begünstigten (1; 1)-Paaren führen, bei denen die Defekte nahe beieinander liegen. Der maximal mögliche Bruchteil der energetisch günstigen (1; 1)-Paare ist dann wesentlich kleiner als Eins. Sind z.B. unter den γ_1^2 Kombinationen zweier (1)-Ketten nur die γ_1 Paare aus identischen Ketten energetisch begünstigt, so gilt

$$\Phi_{11} \gamma_1^2 = \gamma_1 ,$$

also

$$\Phi_{11} = 1/\gamma_1 ,$$

d.h. mit $\gamma_1 = 24$ z.B. $\Phi_{11} \approx 0{,}04$. Alle anderen denkbaren Werte von Φ_{11} liegen in der gleichen Größenordnung.

Nimmt man dagegen an, daß die durch die Konformationsfehler hervorgerufenen Störungen längs der Kettenachsen sehr weitreichend sind[3]), so können auch (1)-Ketten mit weiter voneinander entfernt liegenden Defekten zu einer energetisch günstigen Paar-Kombination führen. In diesem Falle scheinen Werte bis in die Größenordnung $\Phi_{11} \approx 0{,}5$ möglich.

5. Der Charakter der Umwandlung für $L = 1$ und $q = 6$

In den Abb. 4 und 5 sind die Verhältnisse nach [18] und [33], [34] für den Fall

$$\left.\begin{aligned}
\gamma_1 &= 14 \\
\Phi_{11} &= 0{,}5 \\
3\,(w'_{11} - w_{00}) + \Delta u &= 350\,k \approx 700 \text{ cal/Mol} \\
2\,w_{10} - (w'_{11} + w_{00}) &= 450\,k \approx 900 \text{ cal/Mol} \\
w''_{11} - w'_{11} &= 500\,k \approx 1000 \text{ cal/Mol}
\end{aligned}\right\} \text{ [42]}$$

---

[3]) Die Resultate der *Bragg-Williams*-Näherung sprechen dafür (6). Experimentell sollten sich weitreichende Störungen längs der Kettenachsen durch eine gewisse Volumenvergrößerung bemerkbar machen.

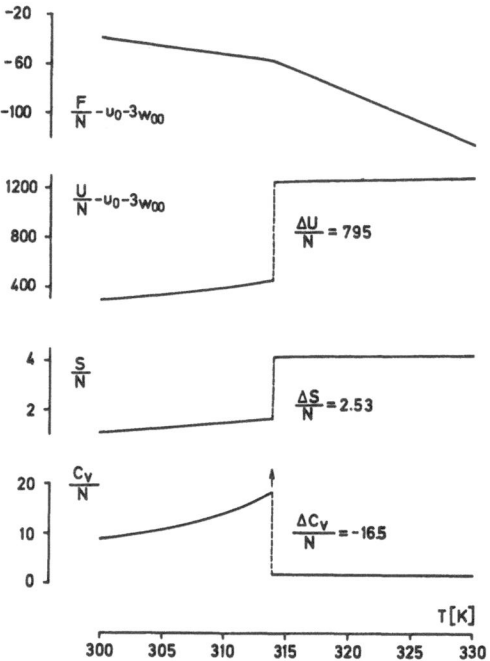

Abb. 4. Freie Energie F/N, innere Energie U/N, Entropie S/N und Wärmekapazität C_V/N pro Kette der Lamelle [42] als Funktion der Temperatur nach [18] und [33—34]. u_0: innere Energie einer defektfreien Kette; w_{00}: Wechselwirkungsenergie eines Paares defektfreier Ketten. Der Faktor 3 vor w_{00} rührt von der Koordinationszahl $q = 6$ her. Bei $T = 313{,}9$ K durchläuft die Lamelle eine Umwandlung erster Ordnung. Energieeinheit: cal/Mol

dargestellt. γ_1 wurde dem vereinfachten Ansatz $\gamma_l = \binom{Z}{l}$ [vgl. (1)] mit $Z = 14$ entnommen. Die Wechselwirkungsparameter wurden so gewählt, daß sich für alle Konzentrationen eine merkbare Änderung innerhalb der Zeichengenauigkeit ergab.

Das System durchläuft bei $T_{U1} = 313{,}9$ K eine diskontinuierliche Umwandlung. Die Temperaturabhängigkeit der freien Energie F, der Entropie $S = -\partial F/\partial T$ und der inneren Energie $U = F + TS$ verdeutlicht, daß es sich dabei um eine Umwandlung erster Ordnung im Sinne des *Ehrenfest*schen Schemas handelt. Die Umwandlung ist also mit einer latenten Wärme verbunden, und die Wärmekapazität

$$C_V = -T\,(\partial^2 F/\partial T^2)$$

wird bei T_{U1} unendlich groß. Der untere und obere Grenzwert der Wärmekapazität bleiben jedoch endlich. Die Umwandlung kann daher

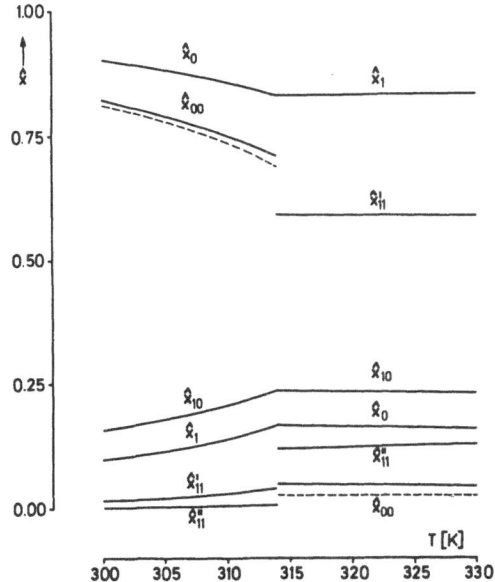

Abb. 5. *Ausgezogene Kurven:* Gleichgewichtskonzentrationen der verschiedenen Ketten und Kettenpaare der Lamelle [42] als Funktion der Temperatur nach [33—34] und den auf [34] folgenden Gleichungen. *Gestrichelte Kurven:* Gleichgewichtskonzentration $\hat{x}_{00}$ der Paare defektfreier Ketten bei statistischer Verteilung der Ketten ($\hat{x}_{00} = \hat{x}_0^2$)

durch die endliche Differenz

$$\Delta C_V \equiv \lim_{\delta \to 0}[C_V(T_{U1} + \delta) - C_V(T_{U1} - \delta)]$$

charakterisiert werden.

Der Anstieg der Wärmekapazität (das einphasige Vorschmelzen) ist vor der Umwandlung nicht sehr ausgeprägt und erstreckt sich über einen verhältnismäßig großen Temperaturbereich. Zum Teil ist das sicher eine Folge des Verhältnisses der Wechselwirkungsparameter [42] (vgl. z. B. mit Abb. 8), zum Teil sind dafür aber wahrscheinlich auch die Vernachlässigungen der quasi-chemischen Methode verantwortlich [vgl. *Münster* (7)]. Ein dritter Grund kann in der Bedingung $L = 1$ liegen. Zumindest fanden wir für $L \gg 1$ schon in der *Bragg-Williams*-Näherung einen weit schärfer ausgeprägten Vorschmelzeffekt und eine Umwandlung, die vom Typ der C_V-Kurve her gesehen, bis auf die Unendlichkeitsstelle bei T_U, mehr einer λ-Umwandlung glich (1), (6).

Interessant scheint die Feststellung, daß sich die Umwandlung mit steigender Aggregation aus einer *Schottky*-Anomalie entwickelt. Das soll u. a. im folgenden an Hand der Wärmekapazität der

Lamelle

$$\gamma_1 = 14$$
$$\Phi_{11} = 0,58$$
$$3(w'_{11} - w_{00}) + \Delta u = 350\,k \approx 700\,\text{cal/Mol}$$
$$w''_{11} - w'_{11} = 1000\,k \approx 2000\,\text{cal/Mol}$$ [43]

gezeigt werden. Die potentielle Umwandlungstemperatur dieser Lamelle liegt nach [37] bei $T_{U1} = 318,8$ K, der kritische Wert [39] der Aggregation bei

$$W_{\text{krit}} = 422{,}345\,k \approx 839\,\text{cal/Mol} .$$ [44]

Sind keine zwischenmolekularen Wechselwirkungskräfte vorhanden, so verhält sich die Lamelle wie ein *Schottky*-System. Insbesondere besitzt ihre Wärmekapazität ein Maximum (*Schottky*-Anomalie), dessen Temperaturlage T_A durch die Bedingung

$$e^{-\eta} = \frac{1}{\gamma_1}\frac{\eta - 2}{\eta + 2} ; \quad \eta \equiv \frac{\Delta u}{k\,T_A}$$

festgelegt ist. Mit $\gamma_1 = 14$ und $\Delta u = 600$ cal/Mol wird $T_A = 79{,}3$ K[4]).

Schalten wir das Wechselwirkungsfeld [43] ein, jedoch so, daß $2w_{10} - (w'_{11} + w_{00}) = 0$ gilt, so verschiebt sich das Maximum zu etwas höheren Temperaturen. Lassen wir zusätzlich eine schwache Aggregation zu, so wird das Maximum nochmals zu etwas höheren Temperaturen hin verschoben, bleibt aber bis

$$2w_{10} - (w'_{11} + w_{00}) = 300\,k$$

hin weit unterhalb der potentiellen Umwandlungstemperatur T_{U1} (in Abb. 6, oben, links außerhalb des dort erfaßten Temperaturbereiches).

Die Folge, die man nun erhält, wenn man die Aggregationstendenz weiter von $300\,k$ bis auf $900\,k$ steigert, ist in den Abb. 6 bis 9 dargestellt.

---

[4]) Den Wert $\Delta u = 600$ cal/Mol entnehmen wir der Anpassung der *Bragg-Williams*-Näherung an die Umwandlungstemperaturen der n-Paraffine (1). Für $\gamma_1 = 24$ (nach [56] mit $Z = 14$) wird $T_A = 71{,}7$ K. Für längere Ketten (größeres γ_1) verschiebt sich die *Schottky*-Anomalie zu noch tieferen Temperaturen, bis in den Bereich der Anomalien polymerer Gläser. Zu bemerken ist jedoch, daß die oben angegebene Bestimmungsgleichung mit beliebigem γ_1 nur für $L = 1$ gilt. Für $L > 1$ befindet sich die *Schottky*-Anomalie unter Umständen wieder bei höheren Temperaturen. Gilt speziell $\gamma_l = \binom{L}{l}$, so ist die Lage der *Schottky*-Anomalie unabhängig von L durch $e^{-\eta} = (\eta - 2)/(\eta + 2)$ bzw. durch $T_A = \Delta u/2{,}40\,k$ gegeben [vgl. (1)]. Mit $\Delta u = 600$ cal/Mol folgt daraus z. B. $T_A = 125{,}8$ K.

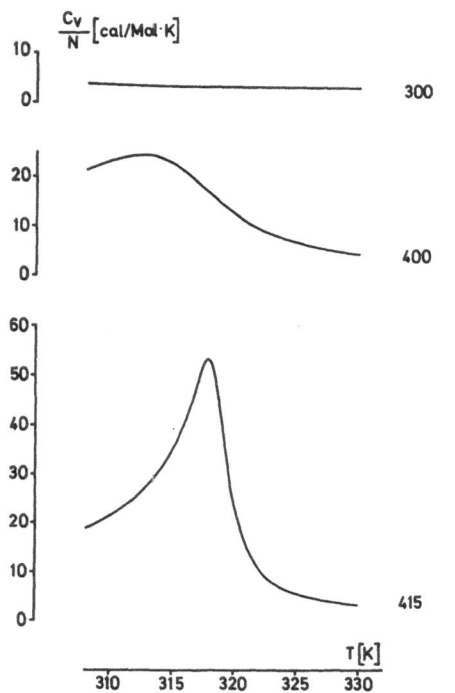

Abb. 6. Wärmekapazität pro Kette der Lamelle [43] als Funktion der Temperatur nach [18] und [33—34] mit $[2w_{10} - (w'_{11} + w_{00})]/k$ als Parameter. Das Maximum der beiden unteren Kurven entwickelt sich aus einer *Schottky*-Anomalie, die im Falle $2w_{10} - (w'_{11} + w_{00}) = 300\,k$ links außerhalb des angegebenen Temperaturbereiches liegt

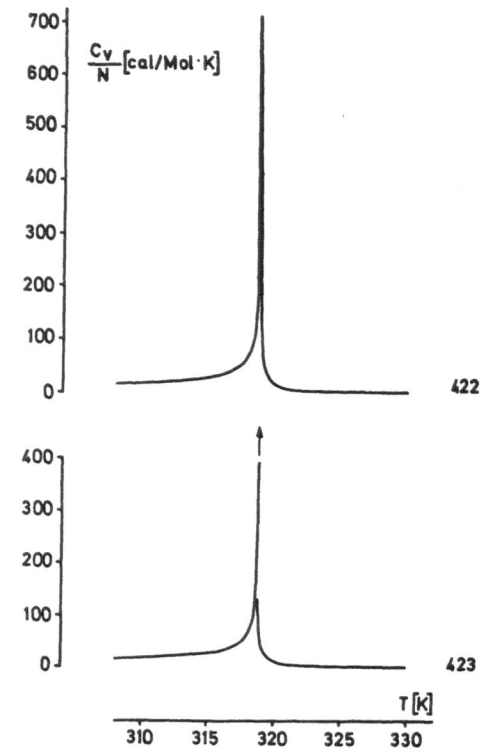

Abb. 7. Wärmekapazität pro Kette der Lamelle [43] als Funktion der Temperatur nach [18] und [33—34] kurz unterhalb und kurz oberhalb des kritischen Aggregationswertes [44]. Die obere Kurve ist eine kontinuierliche Funktion, deren Maximum mit einer *Schottky*-Anomalie vergleichbar ist. Die untere Kurve enthält bei $T = 318{,}8$ K einen diskontinuierlichen Sprung $\Delta C_V = 258$ cal/Mol · K. Da die zugehörige innere Energie bei $T = 318{,}8$ K ebenfalls einen Sprung erleidet, tritt dort zusätzlich eine latente Wärme auf (Pfeil). Experimentell dürften die beiden Kurven wohl kaum zu unterscheiden sein

Die Steigerung der Aggregation bewirkt zunächst eine weitere Verschiebung des Maximums nach T_{U1} hin, wobei seine Flanken immer steiler und seine Höhe immer größer werden (Abb. 6, Mitte und unten). Die Ursache für diese Entwicklung ist darin zu sehen, daß mit steigender Differenz $2w_{10} - (w'_{11} + w_{00})$ die Bildung von (1; 0)-Paaren, die notwendig am Beginn jeder kontinuierlich verlaufenden Defekterzeugung steht, energetisch immer ungünstiger wird, daß aber, solange die Aggregationstendenz unterhalb der kritischen Aggregation bleibt, sich doch noch bei T_{U1} die maximal mögliche Anzahl von (1; 0)-Paaren ausbilden kann.

Wird dann der kritische Wert [44] überschritten, so ist die Bildung einer genügenden Anzahl von (1; 0)-Paaren derart erschwert, daß die dem Gleichgewicht entsprechende Anzahl von defekten Ketten nur noch in einem diskontinuierlichen Sprung unter Bildung von mehr oder minder ausgedehnten (1; 1)-Aggregaten mit einer verminderten Anzahl von (1; 0)-Paaren an ihrer Oberfläche entstehen kann. Die *Schottky*-

Anomalie geht damit in eine Umwandlung erster Ordnung über (Abb. 7).

Übersteigt die Aggregationstendenz den kritischen Wert nur wenig, so werden die (1; 0)-Paare jedoch keineswegs vollständig unterdrückt. Wie Abb. 5 zeigt, kann die Zahl der (1; 0)-Paare immer noch 25% betragen. Erst weitere Steigerung der Aggregation verringert die Zahl der (1; 0)-Paare und damit das Vorschmelzen wesentlich. Hierdurch nehmen die Wärmeaufnahmefähigkeit der Lamelle vor der Umwandlung und die Sprunghöhe ΔC_V bei der Umwandlung immer mehr ab, bis die Wärmekapazität, wie bei einer Umwandlung dritter Ordnung, nur noch eine Knickstelle aufweist (Abb. 8 und 9, oben). Danach wächst ΔC_V mit der Aggregation wieder an, nun aber mit umgekehrtem Vorzeichen, so daß die Wärmekapazität vor der Umwandlung

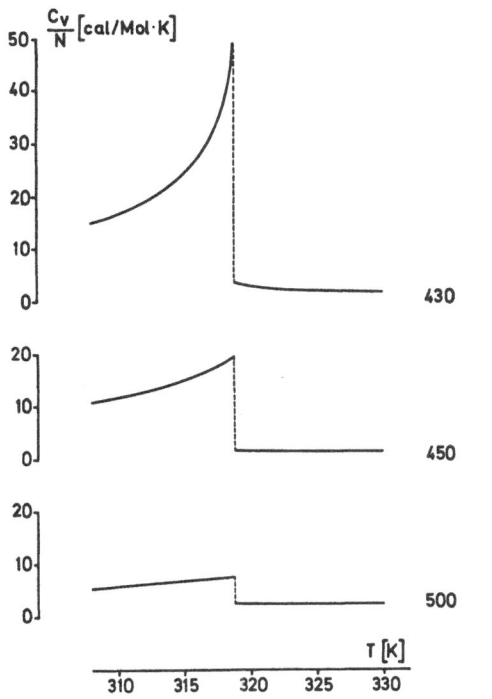

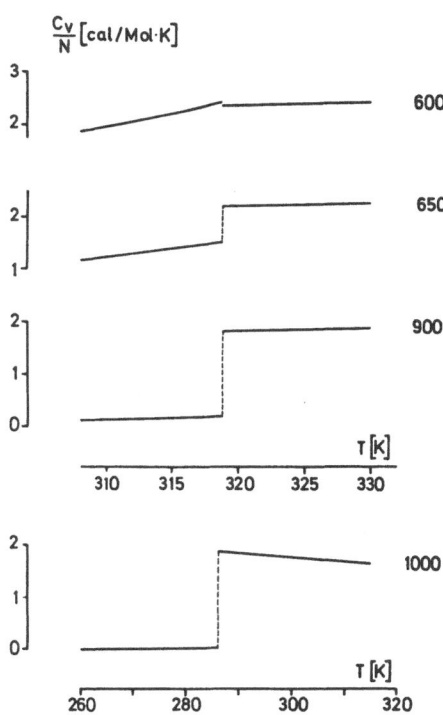

Abb. 8. Wärmekapazität der Lamelle [43] als Funktion der Temperatur nach [18] und [33—34] bei mittlerer Aggregationstendenz. Parameter ist $[2w_{10} - (w'_{11} + w_{00})]/k$

Abb. 9. Wärmekapazität pro Kette der Lamelle [43] als Funktion der Temperatur nach [18] und [33—34] bei starker Aggregationstendenz. Parameter ist $[2w_{10} - (w'_{11} + w_{00})]/k$. In der untersten Abbildung wurde $w''_{11} - w'_{11} = 1000\,k$ durch $w''_{11} - w'_{11} = 400\,k$ ersetzt. Der Umwandlungspunkt wird hierdurch nach $T = 286,3$ K verschoben

kleiner ist als nach der Umwandlung (Abb. 9, Mitte). Bei sehr starker Aggregation wird schließlich die Bildung der (1; 0)-Paare fast vollständig unterdrückt. Die Umwandlung verläuft dann praktisch ohne Vorschmelzen in der reinen Form

$$0;0) \xrightarrow{T_{U1}} (1;1).$$

Typische Werte am Umwandlungspunkt der Lamelle [43] bei starker Aggregation sind in Tabelle 1 zusammengestellt. Die Zahl der (1; 0)-Paare beträgt hier nur noch 0,08%. Das Energie-

niveau der (1; 0)-Paare liegt bereits derart hoch, daß bei der Umwandlung weniger defektfreie Ketten bestehen bleiben, als dem Grenzwert [26]

$$\lim_{T \to \infty} \hat{x}_0 = \frac{1}{1 + \gamma_1} = 0,066\,667 \qquad [45]$$

entsprechen. Die Ordnungsparameter [27] und [28] nehmen daher nach der Umwandlung negative Werte an. Sobald das Energieäquivalent der Temperatur die Aggregation überspielt, müssen sich notwendig einige Defekte wieder zurückbilden.

Gilt, wie im Beispiel [43], $w''_{11} - w'_{11} \gg k\,T_{U1}$, so entstehen am Umwandlungspunkt vor allem (1; 1)'-Paare (vgl. Tabelle 1). Ist jedoch $w''_{11} - w'_{11}$ nicht zu weit von $k\,T_{U1}$ entfernt, so hat das System bei weiterer Temperaturerhöhung die Möglichkeit, durch Bildung von (1; 1)''-Paaren noch weiter Energie zu speichern. Bei starker Aggregation verläuft die Wärmekapazität daher stufenförmig, wie z.B. in Abb. 9, Mitte.

Gilt dagegen $w''_{11} - w'_{11} \lesssim k\,T_{U1}$, entsteht schon am Umwandlungspunkt eine große Zahl von

Tabelle 1. Charakteristische Werte der Lamelle [43] am Umwandlungspunkt $T_{U1} = 318,8$ K bei starker Aggregation. $2w_{10} - (w'_{11} + w_{00}) = 1000\,k \approx 2000$ cal/Mol; Energieeinheit: cal/Mol

	$T_{U1} - \delta$	$T_{U1} + \delta$
$\hat{x}_0$	0,999 606	0,000 394
$\hat{x}_1$	0,000 394	0,999 606
$\hat{x}_{00}$	0,999 214	0,000 002
$\hat{x}_{10}$	0,000 784	0,000 784
$\hat{x}'_{11}$	0,000 002	0,968 745
$\hat{x}''_{11}$	0,000 000	0,030 469
$(F/N) - u_0 - 3w_{00}$	− 0,245 916	− 0,245 916
$(U/N) - u_0 - 3w_{00}$	2,610 82	878,706
S/N	0,008 96	2,756 84

(1; 1)''-Paaren. Oberhalb T_{U1} besitzt das System dann eine geringere Energieaufnahmefähigkeit. Die Wärmekapazität fällt nach der Umwandlung mit steigender Temperatur wieder ab, wie z. B. in Abb. 9, unten. Es scheint denkbar, daß auf diese Weise bei starker Aggregation auch eine „spiegelbildliche λ-Umwandlung" zustande kommen kann, wie sie von *Lagaly*, *Fitz* und *Weiss* (8) und (9) bei den Umwandlungen von n-Alkanolkomplexen der n-Alkylammonium-Beidellite beobachtet wird[5]).

In Abb. 10 ist schließlich für die Lamelle [43] die Konzentration $\hat{x}_0$ der defektfreien Ketten als

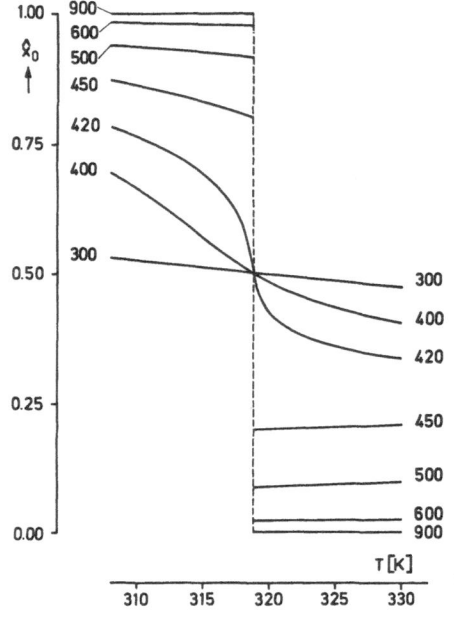

Abb. 10. Gleichgewichtskonzentration $\hat{x}_0$ der defektfreien Ketten in der Lamelle [43] als Funktion der Temperatur nach [33—34] für verschiedene Werte der Aggregation (vgl. Abb. 6—9). Oberhalb des kritischen Wertes $[2w_{10} - (w'_{11} + w_{00})] = 422{,}345\,k$ verläuft $\hat{x}_0(T)$ in einer s-förmigen Kurve, deren physikalisch irrelevanter Teil nicht eingezeichnet wurde

---

[5]) Die Berechnung der Wärmekapazität bzw. der zweiten Ableitung der freien Energie nach der Temperatur erweist sich bei starker Aggregationstendenz als äußerst schwierig, da dann y_1 stets nahe bei ± 1 liegt und infolgedessen an den Rechner die Forderung eines außerordentlich hohen Genauigkeitsgrades zu stellen ist. Aus diesem Grunde konnten für den Fall hoher Aggregation nur wenige Beispiele gesammelt werden, die alle vom Typ Abb. 9 sind. — Für die Bewältigung des rechnerischen Problems möchte ich an dieser Stelle den Herren Dr. *F. Heinrich* und *H. Schneider* (Ludwigshafen am Rhein) danken. Herrn *H. Schneider* danke ich ferner für die Besorgung der numerischen Rechnungen zu Abb. 6 bis 10 und Tabelle 1.

Funktion der Temperatur mit $2w_{10} - (w'_{11} + w_{00})$ als Parameter dargestellt. Es fällt auf, daß auch bei mittlerer Aggregation $\hat{x}_0$ nach der Umwandlung wieder leicht ansteigt. Das kann nicht mit der „Gleichverteilung" [25] und [26] erklärt werden, da die Konzentrationen $\hat{x}_0$ in diesen Fällen größer sind, als der Grenzwert [45] vorschreibt. Die Ursache scheint vielmehr in einem gleich nach der Umwandlung einsetzenden, durch die Bildung von (1; 1)''- und (1; 0)-Paaren bedingten Abbau der (1; 1)'-Aggregate zu liegen. Jedenfalls steigen mit $\hat{x}_0$ auch $\hat{x}_{00}$, $\hat{x}_{10}$ und $\hat{x}''_{11}$ schwach an, während $\hat{x}_1$ und $\hat{x}'_{11}$ leicht zurückgehen[6]).

6. Zur Theorie des Lagaly-Fitz-Weiss-Effektes

Im ersten Teil (1) und in (6) wurde gezeigt, daß sich die Temperatur- und Kettenlängenabhängigkeit der Umwandlungserscheinungen in n-Paraffinen schon mit der einfacheren *Bragg-Williams*-Näherung befriedigend erklären lassen. Die quasi-chemische Näherung sollte darüber hinaus verbesserte Aussagen über die Konzentrationen der einzelnen Komponenten und zusätzliche Aussagen über die Nachbarschaftsverhältnisse der Komponenten liefern. Ferner sollte die quasi-chemische Näherung aber auch eine Erklärung der ganz anders gearteten Umwandlungserscheinungen ermöglichen, die in bimolekularen Filmen aus n-Alkyl-Ketten zwischen glimmerartigen Schichtsilikaten beobachtet und von *Lagaly*, *Fitz* und *Weiss* (8) und (9) auf die Bildung von Kinkblockstrukturen zurückgeführt werden. Dem letzten Problem wollen wir uns abschließend zuwenden.

Der Einfachheit halber übertragen wir die Ergebnisse von *Lagaly*, *Fitz* und *Weiss* auf eine einzelne Lamelle aus n-Alkyl-Ketten. Unter dem Einfluß gewisser Oberflächenkräfte sollte eine solche Lamelle nach den Vorstellungen von *Lagaly*, *Fitz* und *Weiss* im Idealfalle einer endlichen Umwandlungsfolge

$$(0; 0) \xrightarrow{T_{U1}} (1; 1) \xrightarrow{T_{U2}} (2; 2) \xrightarrow{T_{U3}} (3; 3) \longrightarrow \cdots \quad [46]$$

fähig sein. Die Frage ist, ob eine solche Folge auch eine mögliche Lösung der Theorie ist. Von

---

[6]) Zusätzlich hätten wir uns hier der Stabilität der Zustände zu versichern. Wir verzichten zunächst darauf, weil die Wärmekapazität positiv ist und ähnliche Zustandsänderungen (mit schwach ansteigendem $\hat{x}_0$ oberhalb der Umwandlungstemperatur) sich in der *Bragg-Williams*-Näherung als stabil erweisen. Eine korrekte Stabilitätsbetrachtung soll folgen.

vornherein ist klar, daß das nur dann zutreffen kann, wenn die Gleichgewichtsgleichungen eine ganz spezielle Struktur aufweisen.

In der von *Blasenbrey* und *Pechhold* (2) aufgestellten Theorie ist die Folge [46] keine Lösung. Speziell im Segmentmodell kann die Folge nicht zustande kommen, weil dort erstens in den als völlig unabhängig voneinander betrachteten Segmentschichten eine Information darüber fehlt, wieviele Kinken in einer Kette bereits vorhanden sind, und zweitens die Segmente höchstens eine Kinke enthalten und folglich die einzelnen Segmentschichten nur eine einzige Umwandlung durchlaufen können. Da die Segmentschichten als identisch angesehen werden, ist auch die ganze Lamelle nur einer einzigen Umwandlung fähig.

In bezug auf die in Abschnitt 2 aufgestellten Gleichgewichtsgleichungen ist zunächst festzustellen, daß die freie Energie [18] mit verschwindenden Konzentrationen $\hat{x}_0$ und $\hat{x}_{00}$ keineswegs schon ihren kleinsten Wert erreichen muß. Vielmehr gilt nach [19] für alle $l > 0$

$$\frac{\hat{x}_{00}{}^{q/2}}{\hat{x}_0{}^{q-1}} = \frac{1}{B_l} \cdot \frac{\hat{x}_{ll}'{}^{q/2}}{\hat{x}_l'{}^{q-1}}$$

mit

$$\ln B_l \equiv \ln\left(\Phi_{ll}{}^{q/2} \gamma_l\right) - \frac{1}{kT}\left[\frac{q}{2}\left(w_{ll}' - w_{00}\right) + l\,\Delta u\right], \qquad [47]$$

also an Stelle von [18] auch

$$\frac{F}{N} = u_0 + \frac{q}{2}\,w_{ll}' + l\,\Delta u - kT\ln\left(\Phi_{ll}{}^{q/2}\gamma_l\right) + kT\ln\left[\frac{\hat{x}_{ll}'{}^{q/2}}{\hat{x}_l'{}^{q-1}}\right]. \qquad [48]$$

Die freie Energie kann als Funktion jedes beliebigen Variablenpaares $\hat{x}_l$, $\hat{x}_{ll}'$ ($l \geqq 0$) ausgedrückt und daher für jedes beliebige Intervall $T_{U,l-1} \leqq T \leqq T_{Ul}$ der Folge [46] angegeben werden.

Außerdem ist festzustellen, daß, ohne einen Widerspruch zu erzeugen, angenommen werden darf, daß nur Paare

$$(l-1; l-1)\,(l-1; l)\,(l; l)$$

anwesend sind. Die Gleichungen des quasi-chemischen Gleichgewichtes [19—22] reduzieren sich dann auf

$$\hat{x}_{l-1}{}^{q-1}\hat{x}_{ll}'{}^{q/2} = \frac{B_l}{B_{l-1}}\,\hat{x}_l{}^{q-1}\hat{x}_{l,l-1}'{}^{\frac{q}{2}-1},$$

$$\hat{x}_{l,l-1}'^2 = \frac{4\,\Phi_{l,l-1}^2}{\Phi_{ll}\Phi_{l-1,l-1}}\,K_3'(l;\,l-1)\,\hat{x}_{ll}'\,\hat{x}_{l-1,l-1}',$$

$$\hat{x}_{l-1,l-1}'' = \frac{1-\Phi_{l-1,l-1}}{\Phi_{l-1,l-1}} \times K_4'(l-1;\,l-1)\,\hat{x}_{l-1,l-1}',$$

$$\hat{x}_{l,l-1}'' = \frac{1-\Phi_{l,l-1}}{\Phi_{l,l-1}}\,K_4'(l;\,l-1)\,\hat{x}_{l,l-1}',$$

$$\hat{x}_{ll}'' = \frac{1-\Phi_{ll}}{\Phi_{ll}}\,K_4'(l;\,l)\,\hat{x}_{ll}',$$

die Nebenbedingungen [1—4] auf

$$\hat{x}_{l-1} + \hat{x}_l = 1,$$

$$\hat{x}_{l-1,l-1} + \tfrac{1}{2}\,\hat{x}_{l,l-1} = \hat{x}_{l-1},$$

$$\hat{x}_{ll} + \tfrac{1}{2}\,\hat{x}_{l,l-1} = \hat{x}_l,$$

$$\hat{x}_{l-1,l-1}' + \hat{x}_{l-1,l-1}'' = \hat{x}_{l-1,l-1},$$

$$\hat{x}_{l,l-1}' + \hat{x}_{l,l-1}'' = \hat{x}_{l,l-1},$$

$$\hat{x}_{ll}' + \hat{x}_{ll}'' = \hat{x}_{ll}.$$

Mit den Abkürzungen

$$\ln D_{mn} \equiv \ln\left[\frac{1-\Phi_{mn}}{\Phi_{mn}}\right] - \frac{1}{kT}\left(w_{mn}'' - w_{mn}'\right)$$
$$(m = l,\,l-1;\; n = l,\,l-1),$$

$$\ln C_{l,l-1} \equiv$$
$$\ln\left[\frac{4\,\Phi_{l,l-1}^2(1+D_{l,l-1})^2}{\Phi_{ll}\Phi_{l-1,l-1}(1+D_{ll})(1+D_{l-1,l-1})}\right]$$
$$- \frac{1}{kT}\left[2\,w_{l,l-1}' - (w_{ll}' + w_{l-1,l-1}')\right],$$

$$C_{l,l-1}' \equiv \frac{C_{l,l-1}}{4 - C_{l,l-1}}$$

und der Transformation

$$\hat{x}_l \equiv \tfrac{1}{2}\,(1 + y_l)$$

folgen daraus zur Bestimmung der Gleichgewichtskonzentrationen die Gleichungen

$$A_l \equiv$$
$$(q-1)\ln\left[\frac{1-y_l}{1+y_l}\right] - \frac{q}{2}\ln\left[\frac{1-y_l-\hat{x}_{l,l-1}}{1+y_l-\hat{x}_{l,l-1}}\right]$$
$$= \ln\left[\frac{B_l(1+D_{ll})^{q/2}}{B_{l-1}(1+D_{l-1,l-1})^{q/2}}\right], \qquad [49]$$

$$\hat{x}_{l,l-1} = \left[C_{l,l-1}'^2 + C_{l,l-1}'(1-y_l^2)\right]^{1/2} - C_{l,l-1}', \qquad [50]$$

$$\hat{x}_{l-1} = \tfrac{1}{2}\,(1 - y_l),$$

$$\hat{x}_{l-1,l-1} = \tfrac{1}{2}\,(1 - y_l - \hat{x}_{l,l-1}),$$

$$\hat{x}_{ll} = \tfrac{1}{2}\,(1 + y_l - \hat{x}_{l,l-1}),$$

$$\hat{x}_{l-1,l-1}' = \frac{1 - y_l - \hat{x}_{l,l-1}}{2(1 + D_{l-1,l-1})},$$

$$\hat{x}''_{l-1,l-1} = \frac{D_{l-1,l-1}(1 - y_l - \hat{x}_{l,l-1})}{2(1 + D_{l-1,l-1})},$$

$$\hat{x}'_{ll} = \frac{1 + y_l - \hat{x}_{l,l-1}}{2(1 + D_{ll})},$$

$$x''_{ll} = \frac{D_{ll}(1 + y_l - \hat{x}_{l,l-1})}{2(1 + D_{ll})}.$$

Für $q = 6$ sind die Gln. [49] und [50] formal identisch mit den Gln. [33] und [34], die wir für das System $(0;0)$ $(0;1)$ $(1;1)$ gefunden haben. Mit $\Phi_{00} = \Phi_{10} = \gamma_0 = 1$ und $l = 1$ gehen sie direkt in diese über. Dementsprechend sind auch die Lösungen formal identisch.

Überschreitet die Differenz $2w'_{l,l-1} - (w''_{ll} + w'_{l-1,l-1})$ einen bestimmten kritischen Wert W_{krit}, der für $q = 6$ wieder aus

$$C'_{l,l-1}(T_{Ul}) < \frac{4}{5}$$

entnommen werden kann, so besitzt das System bei T_{Ul} eine Umwandlung erster Ordnung[7]). Die Umwandlungstemperatur ist wegen der Antisymmetrie von A_l gegeben durch $y_l = 0$ bzw.

$$\ln \frac{B_l}{B_{l-1}} \left[\frac{1 + D_{ll}}{1 + D_{l-1,l-1}} \right]^{q/2} = 0.$$

Gilt $w''_{ll} - w'_{ll} \gg k T_{Ul}$ und

$$w''_{l-1,l-1} - w'_{l-1,l-1} \gg k T_{Ul},$$

d.h. $D_{ll} \approx D_{l-1,l-1} \approx 0$, so folgt

$$\ln \frac{B_l(T_{Ul})}{B_{l-1}(T_{Ul})} = 0$$

oder nach [47]

$$T_{Ul} = \frac{\frac{q}{2}(w'_{ll} - w'_{l-1,l-1}) + \Delta u}{k[\ln(\Phi_{ll}{}^{q/2}\gamma_l) - \ln(\Phi_{l-1,l-1}^{q/2}\gamma_{l-1})]}.$$

[51]

Bei genügend starker Aggregation zwischen den (l)- bzw. $(l-1)$-Ketten erhalten wir wiederum einen scharfen Übergang

$$(l-1; l-1) \xrightarrow{T_{Ul}} (l; l),$$

praktisch ohne Bildung von $(l-1; l)$-Paaren. Die Gleichungen des quasi-chemischen Gleich-

---

gewichtes [19—22] enthalten also unter der Bedingung

$$2w'_{l,l-1} - (w''_{ll} + w'_{l-1,l-1}) \gg W_{krit}(l; l-1)$$

Lösungen der Form

$$(0; 0) \xrightarrow{T_{U1}} (1; 1),$$
$$(1; 1) \xrightarrow{T_{U2}} (2; 2),$$
$$(2; 2) \xrightarrow{T_{U3}} (3; 3),$$
$$\dots \text{usw.} \dots$$

Gilt $T_{U,l-1} < T_{Ul}$ $(l > 1)$ und sind $T_{U,l-1}$ und $T_{U,l}$ nicht zu weit voneinander entfernt, so lassen sich diese Lösungen widerspruchsfrei zu der Folge [46] zusammensetzen. Die Möglichkeit, mehrere Umwandlungen in ein und derselben Lamelle zu finden, resultiert aus der Eigenschaft der Lamelle, sich wie ein Viel-Komponenten-System zu verhalten, dessen Gleichungen des quasi-chemischen Gleichgewichtes bei sehr starker Aggregation der einzelnen Komponenten entkoppeln, derart, daß jede Komponente für sich eine Umwandlung durchläuft.

Bei der idealen Folge [46] müssen im übrigen im Intervall $T_{Ul} < T < T_{U,l+1}$ alle Variablen bis auf $\hat{x}_l$ und $\hat{x}_{ll}$ exakt verschwinden. Ferner muß $\hat{x}_l = \hat{x}_{ll} = 1$ sein. Nehmen wir außerdem an, daß sich am Umwandlungspunkt T_{Ul} nur die energetisch günstigen $(l; l)'$-Paare bilden, setzen wir also generell für alle l

$$w''_{ll} - w'_{ll} \gg k T \quad \text{bzw.} \quad D_{ll} \approx 0$$

voraus, so wird $x'_{ll} = \hat{x}_{ll} = 1$ und $\hat{x}''_{ll} = 0$. Für die freie Energie der Lamelle im Intervall $T_{Ul} \leqq T \leqq T_{U,l+1}$ folgt damit aus [48]

$$\frac{F}{N} = u_0 + \frac{q}{2} w'_{ll} + l \Delta u - k T \ln(\Phi_{ll}{}^{q/2} \gamma_l) \quad [52]$$

und hieraus für den Sprung der inneren Energie bzw. der Entropie am Umwandlungspunkt T_{Ul}

$$\Delta_l \left(\frac{U}{N} \right) = \frac{q}{2}(w'_{ll} - w'_{l-1,l-1}) + \Delta u, \quad [53]$$

$$\Delta_l \left(\frac{S}{N} \right) = k[\ln(\Phi_{ll}{}^{q/2} \gamma_l) - \ln(\Phi_{l-1,l-1}^{q/2} \gamma_{l-1})].$$

[54]

Ein Vergleich mit [51] liefert

$$T_{Ul} = \Delta_l \left(\frac{U}{N} \right) \Big/ \Delta_l \left(\frac{S}{N} \right). \quad [55]$$

Die Wärmekapazität der idealen Folge [46] wird proportional zu einer Summe von *Dirac*schen Delta-Funktionen:

$$C_V \sim \sum_l{}' \delta(T - T_{Ul}).$$

Um zu einem abschätzenden numerischen Vergleich mit den experimentellen Ergebnissen von *Lagaly*, *Fitz* und *Weiss* zu kommen, vernachlässigen wir Druck- und Volumeneffekte sowie vor allem die Unterschiede, die zwischen den von *Lagaly*, *Fitz* und *Weiss* untersuchten bimolekularen Filmen und unserer Einzellamelle bestehen.

Identifizieren wir $T_{\mathrm{U}l}$ und $\Delta_l(U/N)$ mit den von *Lagaly*, *Fitz* und *Weiss* an *n*-Tetradecylammonium-*n*-Tetradecanol-Beidellit gefundenen Umwandlungstemperaturen und Umwandlungswärmen, so gelangen wir über [52—55] zu dem in Abb. 11 dargestellten Bild. Es zeigt sich, daß die *Lagaly-Fitz-Weiss*schen Umwandlungsgrößen ΔS und ΔH [Tabelle 1 in (8)] nicht der Relation [55] genügen. Das mag in erster Linie daran liegen, daß ΔS und ΔH aus einer Integration über den Temperaturbereich $T_{\mathrm{U}l} \leqq T \leqq T_{\mathrm{U},\,l+1}$ gewonnen wurden und somit streng nicht mit $\Delta_l(S/N)$ und $\Delta_l(U/N)$ vergleichbar sind.

Über die Größenordnung der molekularen Parameter w'_{ll} und Φ_{ll} erfahren wir aus der Identifikation nur dann etwas, wenn wir zusätzlich die Parameter q, Δu und γ_l festlegen. Zu diesem Zweck nehmen wir $q = 6$, $\Delta u = 600$ cal/Mol [vgl. Anm. [4]] und

$$\gamma_l = 2^l \binom{Z-2l}{l} \qquad [56]$$

mit $Z = 14$ an. [56] entspringt der Voraussetzung, daß jeder Defekt drei Kettenglieder umfaßt (vgl. Abb. 12) und an jedem Platz zwei verschiedene Lagen einnehmen kann. Für $w'_{ll} - w_{00}$ und Φ_{ll}

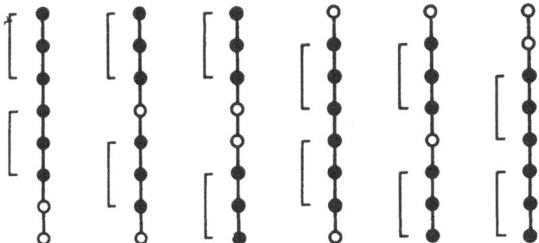

Abb. 12. Zur Erläuterung des Ansatzes [56]: Nehmen wir an, daß ein Konformationsfehler drei Kettenglieder umfaßt, so kann eine Kette mit acht Gliedern und zwei Konformationsfehlern sechs Isomere bilden. Allgemein kann eine Kette mit Z Gliedern und l Konformationsfehlern in $\binom{Z-2l}{l}$ isomeren Formen vorkommen

Tabellen 2 und 3. Gibt man die in den ersten Spalten angegebenen Werte vor, so erhält man aus [53—56] mit $q = 6$, $\Delta u = 600$ cal/Mol und $Z = 14$ die Werte der letzten Spalten. Die vorgegebenen Werte stimmen mit den von *Lagaly*, *Fitz* und *Weiss* an *n*-Tetradecylammonium-*n*-Tetradecanol-Beidellit gefundenen Werten überein. Energieeinheit: cal/Mol

Tabelle 2

l	vorgegeben		berechnet		
	$T_{\mathrm{U}l}$	$\Delta_l(U/N)$	$\Delta_l(S/N)$	$w'_{ll} - w_{00}$	Φ_{ll}
1	323	730	2,260	43,33	0,51
2	333	720	2,162	83,33	0,37
3	345	680	1,971	110,00	0,38

Tabelle 3

l	vorgegeben		berechnet		
	$T_{\mathrm{U}l}$	$\Delta_l(S/N)$	$\Delta_l(U/N)$	$w'_{ll} - w_{00}$	Φ_{ll}
1	323	2,3	742,9	47,63	0,51
2	333	2,3	765,9	102,93	0,38
3	345	2,3	793,5	167,43	0,41

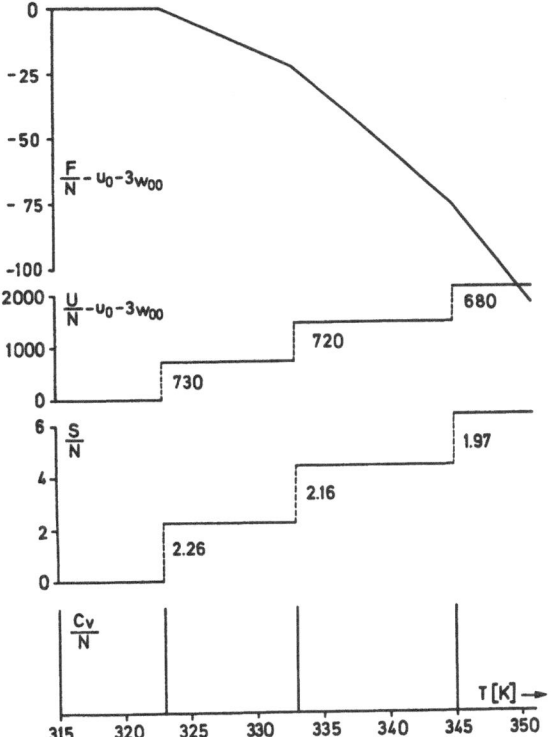

Abb. 11. Freie Energie F/N, innere Energie U/N, Entropie S/N und Wärmekapazität C_V/N pro Kette einer Lamelle, welche die ideale Umwandlungsfolge [46] durchläuft, als Funktion der Temperatur nach [52—55]. Vorausgesetzt wurde, daß sich an den Umwandlungspunkten nur die energetisch günstigen $(l; l)$-Paare bilden. Vorgegeben wurden $T_{\mathrm{U}1} = 323$ K, $T_{\mathrm{U}2} = 333$ K, $T_{\mathrm{U}3} = 345$ K und $\Delta_1(U/N) = 730$, $\Delta_2(U/N) = 720$, $\Delta_3(U/N) = 680$. Energieeinheit: cal/Mol

erhalten wir damit die in Tabelle 2 angegebenen Werte. Sowohl die Größenordnung als auch der mit l steigende Trend von $w'_{ll} - w_{00}$ und der mit l fallende Trend von Φ_{ll} scheinen unmittelbar plausibel.

Identifizieren wir T_{Ul} und $\Delta_l(S/N)$ mit den *Lagaly-Fitz-Weiss*schen Werten, so kommen wir zu den Daten in Tabelle 3. $w'_{ll} - w_{00}$ und Φ_{ll} zeigen die gleiche Größenordnung und den gleichen Trend wie in Tabelle 2.

Ein konstanter, von l unabhängiger Entropiesprung, wie er von *Lagaly*, *Fitz* und *Weiss* gefunden wird (vgl. Tabelle 3), ist allerdings nach [54] ohne weiteres nicht zu erwarten, da er die Proportionalität

$$\frac{\Phi_{ll}}{\Phi_{l-1,\,l-1}} \sim \left[\frac{\gamma_{l-1}}{\gamma_l}\right]^{2/q}$$

erfordern würde, die sicher nur unter ganz bestimmten, wenig wahrscheinlichen Bedingungen besteht. Die Annahme einer angenäherten Konstanz des Entropiesprunges verdeutlicht jedoch eine der möglichen Ursachen für den Abbruch der Folge [46]:

Setzen wir $\Delta_l(S/N) \approx \Delta(S/N) = $ const. voraus, so folgt aus [54]

$$\Phi_{ll} = \Phi_{l-1,\,l-1}\left[\frac{\gamma_{l-1}}{\gamma_l}\right]^{2/q} \exp\left\{\frac{2\,\Delta(S/N)}{qk}\right\}. \qquad [57]$$

Die daraus mit $\Delta(S/N) = 2{,}3$ cal/Mol $\cdot$ K, $q = 6$, $\Phi_{00} = \gamma_0 = 1$ und [56] für $Z = 10$ bis 22 resultierende Funktion Φ_{ll} von l ist in Abb. 13 dargestellt. Da γ_l als Funktion von l ein Maximum besitzt, müßte Φ_{ll} ein Minimum aufweisen. Andererseits ist aber anzunehmen, daß Φ_{ll} tatsächlich eine mit steigendem l monoton abfallende Funktion ist. Die mit l steigenden Φ_{ll}-Werte (in Abb. 13 die ausgefüllten Kreise) haben daher nur hypothetischen Charakter. Unter Voraussetzung eines konstanten Entropiesprunges können die Lamellen $Z = 10$ bis 15 höchstens zwei, die Lamellen $Z = 16$ bis 21 höchstens drei und die Lamelle $Z = 22$ höchstens vier Umwandlungen durchlaufen (vgl. die leeren Kreise in Abb. 13). Qualitativ stimmt diese Aussage mit den Beobachtungen von *Lagaly*, *Fitz* und *Weiss* (9) überein.

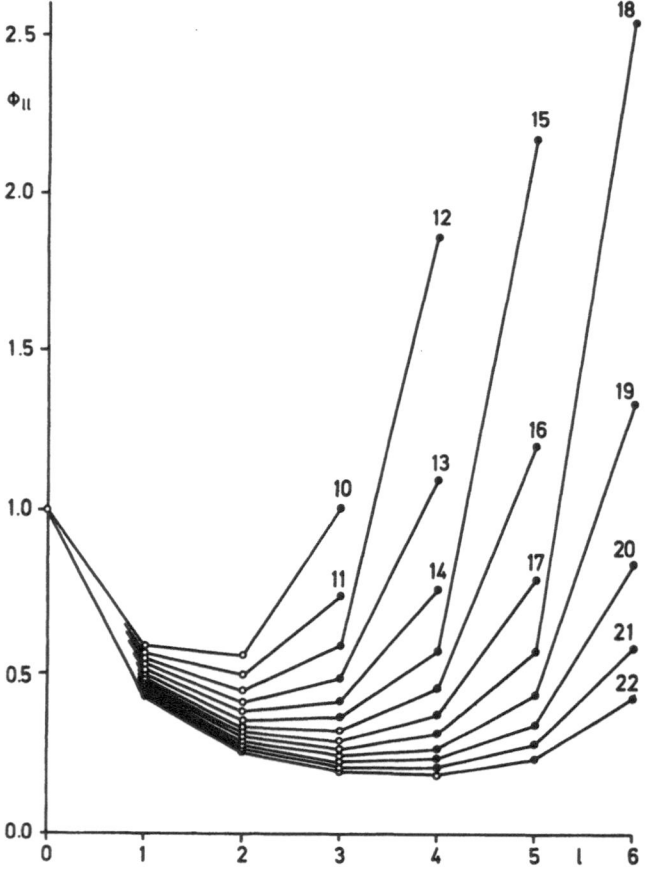

Abb. 13. Φ_{ll} als Funktion von l nach [56–57] mit $q = 6$, $\Phi_{00} = \gamma_0 = 1$, $\Delta(S/N) = 2{,}3$ cal/Mol $\cdot$ K und der Kettenlänge Z als Parameter. Werte $\Phi_{ll} > 1$ können nicht realisiert werden. Ebenso können aber auch die durch die ausgefüllten Kreise gekennzeichneten Werte nicht realisiert werden, wenn Φ_{ll} eine mit l monoton abfallende Funktion ist. Die Folge [46] bricht dann notwendig schon bei $l < L$ ab

Zu einer ähnlichen Folgerung gelangt man, wenn man die allgemeinere Abbruch-Bedingung

$$T_{Ul} = T_{U,l+1}$$

bzw. nach [55] und [54]

$$\left[\frac{\Phi_{ll}^{q/2} \gamma_l}{\Phi_{l-1,l-1}^{q/2} \gamma_{l-1}}\right]^{k \Delta_{l+1}(U/N)} = \left[\frac{\Phi_{l+1,l+1}^{q/2} \gamma_{l+1}}{\Phi_{ll}^{q/2} \gamma_l}\right]^{k \Delta_l(U/N)}$$

heranzieht und $\Delta_l(U/N) = $ const. voraussetzt. Die Folge [46] bricht dann ab, wenn

$$\frac{\Phi_{ll}^2}{\Phi_{l-1,l-1} \Phi_{l+1,l+1}} = \left[\frac{\gamma_{l-1} \gamma_{l+1}}{\gamma_l^2}\right]^{2/q}$$

gilt.

Als eine mögliche Ursache des Abbruchs der Folge [46], bevor $l = L$ erreicht wird, erscheint danach ganz allgemein, daß die Zahl γ_l der Isomeren, die eine mit l Defekten behaftete Kette bilden kann, zwischen $l = 0$ und $l = L$ notwendig ein Maximum aufweist, während der Bruchteil Φ_{ll} der energetisch günstigen $(l; l)$-Paare mit steigender Zahl l der Defekte pro Kette monoton abnimmt.

Prinzipiell läßt sich also der *Lagaly-Fitz-Weiss*-Effekt auf der Basis der Gln. [19—22] verstehen, wenn man zwischen den Ketten mit gleicher Anzahl von Konformationsfehlern, die zu einer energetisch günstigen Paarbildung führen, eine sehr starke Aggregation annimmt. Es bleibt die Frage, woher diese starke Aggregationstendenz rührt, da vergleichbare *n*-Paraffin-Lamellen den Effekt nicht zeigen. Es liegt auf der Hand, die Ursache für die starke Aggregation direkt oder indirekt (z.B. über die Gitterstruktur) mit der Oberflächenladung der Filme in Verbindung zu bringen.

7. Anhang

Nach [34] gilt

$$\lim_{y_1 \to +1} \hat{x}_{10}(y_1) = 0.$$

Einsetzen dieses Grenzwertes in [33] scheint für A_1 den Grenzwert

$$\lim_{y_1 \to +1} A_1(y_1) = -\infty \qquad [?]$$

zu liefern. Das ist jedoch falsch.

Für $y_1 \to +1$ ist vielmehr $\hat{x}_{10}$ in der folgenden Weise zu entwickeln:

$$\hat{x}_{10} = [C_1'^2 + C_1'(1 - y_1^2)]^{1/2} - C_1'$$
$$= C_1'\left[1 + \frac{1 - y_1^2}{C_1'}\right]^{1/2} - C_1'$$

$$\approx C_1'\left[1 + \frac{1 - y_1^2}{2C_1'}\right] - C_1'$$
$$= \tfrac{1}{2}(1 - y_1)(1 + y_1).$$

Für $y_1 \to +1$ gilt daher

$$1 - y_1 - \hat{x}_{10} \approx \tfrac{1}{2}(1 - y)^2,$$
$$1 + y_1 - \hat{x}_{10} \approx \tfrac{1}{2}(1 + y)^2.$$

Setzen wir diese Werte in [33] ein, so folgt

$$\lim_{y_1 \to +1} A_1 = \lim_{y_1 \to +1}\left[5 \ln\left(\frac{1 - y}{1 + y}\right) - 3 \ln\left(\frac{1 - y}{1 + y}\right)^2\right]$$
$$= \lim_{y_1 \to +1}\left[\ln\left(\frac{1 + y}{1 - y}\right)\right] = +\infty,$$

wie in Abschnitt 4 und speziell in Abb. 1 behauptet.

Zusammenfassung

Betrachtet werden Lamellen aus *n*-Alkyl-Ketten, die mit Konformationsfehlern behaftet sein können. Die Thermodynamik dieser Lamellen wird in der quasi-chemischen Näherung von *Fowler* und *Guggenheim* behandelt. Die Gleichungen des quasi-chemischen Gleichgewichtes lassen eine ausgesprochen breite Verhaltensmannigfaltigkeit zu, vergleichbar mit der fester Lösungen und Legierungen.

Im Falle stärkerer Aggregation unter den defekten Ketten sind die Lamellen einer Umwandlung erster Ordnung fähig. Es wird gezeigt, daß sich die Umwandlung mit zunehmender Aggregation aus einer *Schottky*-Anomalie entwickelt.

Bei sehr starker Aggregation unter den Ketten mit gleicher Anzahl von Defekten können die Gleichungen des quasi-chemischen Gleichgewichtes entkoppeln, derart, daß jede Komponente des Systems für sich eine Umwandlung durchläuft. Das entspricht den Beobachtungen von *Lagaly*, *Fitz* und *Weiss* an Alkanol-komplexen des *n*-Alkylammonium-Beidellit.

Summary

There are considered lamellae out of *n*-alkyl-chains which may be affected with conformational defects. The thermodynamics of these lamellae are treated in the quasi-chemical approximation of *Fowler* and *Guggenheim*. The equations of the quasi-chemical equilibrium allow a broad variety of behaviour, comparable to that of solid solutions and alloys.

In the case of a stronger aggregation among the defect chains the lamellae are capable of a first order transition. It is shown that with growing aggregation the transition develops from a *Schottky* anomaly.

In the case of a very strong aggregation among the chains with the same number of defects the equations of the quasi-chemical equilibrium can decouple in such a manner that each component of the system passes through a transition of its own. That corresponds to the observations of *Lagaly*, *Fitz* and *Weiss* as to alkanol complexes of *n*-alkyl-ammonium beidellite.

Literatur

1) *Baur, H.*, Colloid Polymer Sci. **252**, 899 (1974).
2) *Blasenbrey, S.* und *W. Pechhold*, Rheol. Acta **6**, 174 (1967).
3) *Fowler, R. H.* und *E. A. Guggenheim*, Statistical Thermodynamics, Reprint (Cambridge, 1960).
4) *Pechhold, W.* und *S. Blasenbrey*, Kolloid-Z. u. Z. Polymere **216/217**, 235 (1967).
5) *Blasenbrey, S.* und *W. Pechhold*, Ber. Bunsenges. **74**, 784 (1970).
6) *Baur, H.*, Colloid & Polymer Sci. **252**, 641 (1974).
7) *Münster, A.*, Statistische Thermodynamik, S. 566. (Berlin-Göttingen-Heidelberg, 1956).
8) *Lagaly, G., S. Fitz* und *A. Weiss*, Progr. Colloid & Polymer Sci. **57**, 54 (1975).
9) *Lagaly, G., S. Fitz* und *A. Weiss*, Clays Clay Min. **23**, 45 (1975).

Anschrift des Verfassers:

Dr. *H. Baur*
WHM/G 201, BASF Aktiengesellschaft
D-6700 Ludwigshafen am Rhein

Progr. Colloid & Polymer Sci. **58**, 19–29 (1975)

Institut Laue-Langevin, Grenoble, France

Smooth frequency distribution derived from complex phases
A new technique for phonons in polymer chains

Christhard Schmid

With 1 figure and 2 tables

(Received May 2, 1974)

1. Introduction

The thermal motion of the various nuclei in a polymer is usually treated in (quasi) harmonic approximation, i.e. as a superposition of plane waves (phonons) characterized by their wave vector $\boldsymbol{\varphi}$ (which is the phase difference between two adjacent chemical repeat units), frequency $\omega_\beta(\boldsymbol{\varphi})$ and polarization vector. The subscript β refers to the β-th branch of normal modes. If the repeat unit contains p nuclei, then there are $\gamma = 3p$ branches in the *Brillouin* zone. The frequencies $\omega_\beta^2(\boldsymbol{\varphi})$ are the $3p$ eigenvalues of the reduced dynamical matrix $\boldsymbol{D}(\boldsymbol{\varphi})$ which embraces in a $3p \times 3p$ array the whole of the vibrational properties (kinetic and potential coupling among coordinates, modulated by the phase factor $\boldsymbol{\varphi}$) of the polymer.

If the reduced dynamical matrix is given for a particular polymer then there is no difficulty in deriving the $3p$ dispersion relations. In some cases (e.g. for an isolated polyethylene skeleton (1, 2, 3)) it is possible to factorize the characteristic equation of $\boldsymbol{D}(\boldsymbol{\varphi})$ and to obtain the dispersion relations in closed form. In general the dispersion relations are obtained by numerical diagonalization of the reduced dynamical matrix.

Whereas the diagonalization of this (small) $3p \times 3p$ matrix is almost trivial for a modern computer, the derivation of a frequency spectrum from the dispersion relations remains a formidable problem. The frequency distribution (density of states) $g(\omega)$ is defined as the fraction of normal modes with frequencies in the interval between ω and $\omega + d\omega$ in the limit $d\omega \to 0$. An obvious formal expression for this is (4)

$$g(\omega^2) \equiv \frac{1}{2\omega} g(\omega) = \frac{1}{\gamma N} \sum_{\beta=1}^{\gamma} \sum_{\varphi} \delta(\omega_\beta^2(\boldsymbol{\varphi}) - \omega^2) \quad [1]$$

where N is the number of unit cells. Assuming periodic boundary conditions, the φ-sum is over N equidistant points in the first *Brillouin* zone. For $N \to \infty$, the density of states $g(\omega^2)$ is a smooth curve with a finite number of singularities (5).

Because of the fundamental importance of the density of states (all thermodynamic properties are derived from it) and related quantities, a large number of exact analytical and approximate numerical spectrum calculations have been made in lattice dynamics (an excellent review is given by *Maradudin, Montroll, Weiss* and *Ipatova* (6)).

The currently used *approximate numerical methods* are more or less sophisticated versions of a graphical procedure, known as root sampling methods. The simplest version consists in solving the secular determinant at a large number of equally spaced points in the *Brillouin* zone and plotting a histogram of the frequencies obtained. In graphical form, this has first been applied to polyethylene by *Wunderlich* (7). In a modified version due to *Gilat* and *Dolling* (8) the normal mode frequencies are calculated only at a coarse mesh of points and interpolated at a large number of fine mesh points. This version, which requires less computing time, has been widely applied to polymer chains and crystals (e.g. (9), (10)). For the sake of completeness we note that the version of *Gilat* and *Raubenheimer* (11) has not yet been applied to polymers, although it is the most efficient one, since it reduces the statistical noise drastically.

Whereas the root sampling methods are rather general, there are necessary (but not sufficient) conditions for the application of the known *exact analytical methods* of solid state theory (6):

2*

1. The lattice is one- (or at most two-) dimensional;

2. The dispersion relation is given in closed form;

3. The number n_0 of directly interacting neighbours is 1 (in a few exceptional cases $n_0 = 2$).

Fortunately, the most striking condition, number 1, is fulfilled. For frequencies which are not too low, a chain-polymer crystal behaves in a very good approximation like a one-dimensional crystal (vibrating in 3 dimensions, but having a one-dimensional *Brillouin* zone). Most papers about phonons in polymers do not take into account interchain interactions at all. The second condition is obviously in general not fulfilled, and the third condition is never fulfilled, since n_0 is at least 3 due to torsion (the torsional coordinate extends over 4 atoms). (Note that long range *Coulomb* forces are not considered in this paper.)

Very recently *Schmid* and *Hölzl* (12, 13) calculated exactly in closed form the vibrational *Green* function (the trace of which contains the density of states as imaginary part) of a polyethylene skeleton with simple valence force field. This model seems to be at the limit where a purely analytical calculation is possible. For if we include other degrees of freedom or off-diagonal terms in the valence force field, we either obtain a numerical dispersion relation or at least a characteristic equation of higher order than cubic. Thus we cannot hope to find a purely analytic method for a general polymer.

However, besides the very general approximate numerical methods and the very specialized exact analytic methods, one can imagine a third category of *exact semianalytical methods* which are less general than the numerical and less specialized than the analytical methods. In connection with the above mentioned work, *Schmid* and *Hölzl* developed a semianalytic method (14) for the exact computation of the *Green* function for polymers having a block-diagonalized dynamical matrix with one- and two-dimensional blocks so that the dispersion relation can be written in closed form. There is no restriction for the range n_0 of interaction except that it should be finite. The method has all the advantages of exact analytical treatments except that a polynomial equation of maximal order $2n_0$ is to be solved by computer.

It has been argued by *Zerbi* (15) that those methods are to be preferred which apply to real polymers with all their degrees of freedom, and which are sufficiently general so that no tedious work has to be done when changing from one polymer to the other. Obviously the root sampling method has this advantage.

In this paper a generalization of the above mentioned previous work is presented, such that the postulates of *Zerbi* are fulfilled. With this new method we do not want to compete with the root sampling method when real three-dimensional crystals or long range forces are involved. But in the case of approximately one-dimensional crystals with short range forces such as polymer chains, the new method has decisive advantages over any kind of approximate numerical method, since it is exact.

The characteristics of the new method compared to root sampling methods is summarized in Table 1 which will be explained in terms of an example. We consider the in-plane motion of a polyethylene skeleton with a simple valence force field for stretching and bending (*Kirk-*

Table 1. Comparison between root sampling methods and the method of this paper

Method	Root Sampling	Complex Phases (this paper)
Range of interaction (n_0)	Arbitrary	Finite
Dimension	Arbitrary	One-dimensional *Brillouin*-zone
Degrees of freedom (γ)	Arbitrary	Arbitrary
Input	Fine mesh of real phases φ	Frequency ω (one or several)
Basic computational problem	Phase dependent secular equation = algebraic equation of order γ in ω^2	Frequency dependent secular equation = algebraic equation of order $\leq n_0 \gamma$ in φ
Number of basic cycles	Number of mesh-points	1 for each frequency needed
Intermediate result	The whole dispersion relation at the mesh-points	Only those points of the dispersion relation (plus additional complex phases) which have frequency ω
Resulting density of states	Approximate	Exact
Refinement	Loss of earlier results	Earlier results remain valid

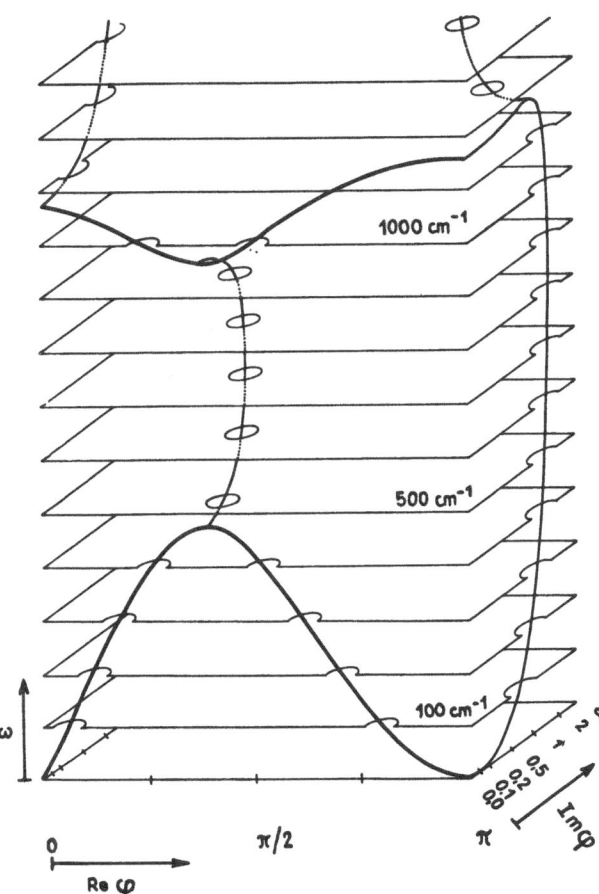

Fig. 1. Complex phase-frequency diagram for skeletal in-plane vibrations of poly-ethylene

wood's model (1), $n_0 = \gamma = 2$, force constants of *Zbinden* (16)) where the exact *Green* function is known (12). In Fig. 1 the solutions of the secular equation $\omega^2 - \omega_\beta^2(\varphi) = 0$ are plotted. This diagram is in principle identical with Fig. 5 of Ref. 12, but it is scaled in such a way that the normal dispersion relation appears at the front of the diagram (real phases φ, corresponding to undamped phonons). The imaginary part of the phase (which is plotted in a logarithmic scale proportional to $1 - \exp(-Im\ \varphi)$ corresponds to the damping. (The magnitude of the damping in the gap between 480 and 965 cm^{-1} is important for the position of localized modes due to imperfections (17).) The number of complex phases per energy (the number of holes in a sheet $\omega = $ const.) is in principle equal to $n_0 \cdot \gamma$ which is 4 in this model. However, one of these complex phases is independent of the frequency ($Re\ \varphi = 180°$, $Im\ \varphi = \infty$) and has not been drawn. Furthermore, in the gap and for very high frequencies we have holes appearing not at the edge, but inside the sheet. These

holes correspond to two complex conjugate phases. Thus, the number of complex phases in the diagram is 3 for all frequencies.

If the frequency distribution at a certain frequency ω is to be calculated, the knowledge of the dispersion relation at this frequency (the information on the front edge of the sheet $\omega = $ count) is apparently not sufficient, because the neighbourhood of these points plays an essential role. The root sampling method requires in principle all the information on the *vertical* plane $Im\ \varphi = 0$ (the front side of Fig. 1). In contrast to this, the method presented in this paper requires only the few complex phases $\varphi = \varphi(\omega)$ which is the information on the *horizontal* sheet of energy ω. The frequency distribution is shown to be the real part of a rational function of these complex phases (eq. [14]).

In the next section we give a brief but comprehensive description of the method from the point of view of the user. The statements of this section are elaborated and proved in

Sections 3—5. In Section 6 we treat the example of the polyatomic linear chain with nearest neighbour interaction which can be solved exactly (since the inverse characteristic equation is linear).

2. Short Formal Description of the Method

(a) Input: The Dynamical Matrix

Consider an ideal polymer chain with N chemical repeat units each consisting of p atoms. Let R_i^n describe the i-th displacement coordinate belonging to the n-th chemical unit. The coordinates R_i^n may be of any kind (internal or external), but without redundant ones (for more details see e.g. *Zerbi* (18)). The motion of the system as a whole about its center of mass plays no role if we consider very long ($N \to \infty$) polymers. Thus we have $3pN$ degrees of freedom. In harmonic approximation, the potential and kinetic energy are usually written as

$$V = \text{const} + \tfrac{1}{2} \sum_{\substack{n,n' \\ i,k}} F_{i,k}^{nn'} R_i^n R_k^{n'} , \qquad [2]$$

$$T = \tfrac{1}{2} \sum_{\substack{n,n' \\ i,k}} (G^{-1})_{ik}^{nn'} \dot{R}_i^n \dot{R}_k^{n'} , \qquad [3]$$

where F and G are symmetrical matrices which, due to the periodicity of the chain, depend only on $s = |n - n'| \leq n_0$. n_0 is the number of nearest directly interacting neighbouring cells. (Very often n_0 is assumed to be 3 due to torsional forces.) The solution of the *Lagrange* equations of motion

$$\sum_{n',k} \{(G^{-1})_{ik}^{nn'} \ddot{R}_k^{n'} + F_{ik}^{nn'} R_k^{n'}\} = 0 \qquad [4]$$

has the form of a plane wave

$$R_i^n = u_i\, e^{-i(\omega t + n\varphi)} . \qquad [5]$$

φ is the phase shift between two adjacent equivalent displacement coordinates, ω is the circular frequency.

Substitution of eq. [5] into eq. [4] yields a set of $\gamma = 3p$ linear equations for the amplitudes u_i with the solubility condition

$$S(\varphi, \omega^2) = 0 ,$$

where $S(\varphi, \omega^2)$ is the reduced secular determinant

$$S(\varphi, \omega^2) \equiv \prod_{\beta=1} \{\omega_\beta^2(\varphi) - \omega^2\}$$

$$= \det \left| \sum_{l=-n_0}^{n_0} D_{ij}^{l0}\, e^{-il\varphi} - \omega^2 \delta_{ij} \right| \qquad [6]$$

and D is the dynamical matrix whose elements are given by

$$D_{ij}^{lm} = \sum_n \sum_k G_{ik}^{ln} F_{kj}^{nm} . \qquad [7]$$

The dynamical matrix depends only on the difference $l - m$ and vanishes if $|l - m| > n_0$. For the following we will assume that it is symmetrical

$$D_{ij}^{lm} = D_{ji}^{ml} . \qquad [8]$$

This is true if the coordinates R_i^n are chosen to be mass adjusted cartesian ones since in that case G is the unity matrix. If for any reason other coordinates are to be preferred, the matrix equation $D = GF$ is to be replaced by the equation $D = \sqrt{G} F \sqrt{G}$.

These remarks are sufficient as a basis for the following brief, but complete, outline of the method from the practical point of view. We begin the theoretical discussion in the next section in order to prove the statements to be made.

(b) Derivatives of the Invariants of the Dynamical Matrix

Once the dynamical matrix is given in analytical or numerical form, $2n_0$ matrices $\boldsymbol{\alpha}^n$ ($n = 1, \ldots, 2n_0$) have to be derived with elements given by

$$\alpha_{ij}^n = \xi_0^n D_{ij}^{00} + \sum_{m=1}^{n_0} \xi_m^n (D_{ij}^{m0} + (-1)^n D_{ji}^{m0})$$
$$i, j = 1, \ldots, 3p . \qquad [9]$$

The coefficients ξ_m^n depend on the range of interaction, n_0, and are listed in Table 2 for $n_0 = 1, 2, 3$. The matrices $\boldsymbol{\alpha}^n$ are related to the *Fourier* transform of the dynamical matrix and are the input of the following algorithm

$$\boldsymbol{\beta}_{l-1}^k = \boldsymbol{\gamma}_{l-1}^k \qquad k = 0, 1, \ldots, 2(l-1)n_0,$$

$$\boldsymbol{\gamma}_l^k = \sum_{n=0}^{\min(k, 2n_0)} \boldsymbol{\beta}_{l-1}^{k-n} \boldsymbol{\alpha}^n \quad k = 0, 1, \ldots, 2ln_0, \qquad [10]$$

$$\eta_l^k = \frac{1}{l} \operatorname{tr} \boldsymbol{\gamma}_l^k \qquad k = 0, 1, \ldots, 2ln_0.$$

This algorithm runs from $l = 1$ to $l = \gamma$ with the initial condition $\eta_0^n = -\delta_{n0}$, $\boldsymbol{\gamma}_0^k = 0$. The output are the coefficients η_l^k which are related to the derivatives of the invariants of the dynamical matrix.

Table 2. The coefficients ξ_m^n of equation [19]

n_0	m	ξ_m^0	ξ_m^1	ξ_m^2	ξ_m^3	ξ_m^4	ξ_m^5	ξ_m^6
1	0	1	0	−1	0	0	0	0
	1	1	2	1	0	0	0	0
2	0	1	0	−2	0	1	0	0
	1	1	2	0	−2	−1	0	0
	2	1	4	6	4	1	0	0
3	0	1	0	−3	0	3	0	−1
	1	1	2	−1	−4	−1	2	1
	2	1	4	5	0	−5	−4	−1
	3	1	6	15	20	15	6	1

(c) Simplification of the Frequency-Dependent Secular Equation

Before entering into the frequency dependent part it is useful to transform the coefficients η_l^k into the form [11]

$$a_{lp} = (-1)^p \sum_{k=0}^{l n_0 - p} \binom{k+p}{k} \eta_l^{2(k+p)} \quad \begin{array}{l} l = 0, 1, \ldots, \gamma \\ p = 0, 1, \ldots, l n_0 \end{array}$$

and to look for vanishing coefficients. We define indices p_l ($l = 0, 1, \ldots, \gamma$) such that $a_{lp} = 0$ for $p \leq p_l$. Furthermore we determine an integer m_0 defined as the minimum of the expression $p_l + n_0 (\gamma - l)$ if l runs from 0 to γ. The calculation up to now does not involve the frequency and deals only with elementary real arithmetics.

(d) The Frequency Dependent Secular Equation

The input of this part consists of the frequency ω, the coefficients a_{lp} and the integers p_l, m_0, n_0, γ. The dynamical matrix is no longer necessary unless an amplitude-weighted frequency distribution is to be calculated, but the present paper deals only with the unweighted distribution. The frequency dependent secular equation is the polynomial of the order $n_1 = \gamma n_0 - m_0$

$$c_0(\omega^2) + c_1(\omega^2) z + c_2(\omega^2) z^2 + \cdots + c_{n_1}(\omega^2) z^{n_1} = 0 \quad [12]$$

with the real coefficients

$$c_q(\omega^2) = \sum_{m=0}^{\gamma} a_{\gamma-m, q+m_0-n_0 m} \, \omega^{2m} . \quad [13]$$

Terms with $m_0 + q < n_0 m + p_{\gamma-m}$ give no contribution to this sum.

(e) Output: The Inverse Phase Frequency Relation

The central computational step is the determination of the n complex roots $z_1, z_2, \ldots, z_n$ of the frequency dependent secular equation for a given frequency ω. Any real solution $z_i > 1$ corresponds to a phonon with energy ω and real phase $\varphi = 2 \arctan \sqrt{z_i - 1}$. The whole set of complex roots is the necessary information (on the sheet of constant energy in Fig. 1) to calculate exactly the frequency distribution at energy ω.

(f) Output: The Frequency Distribution

The exact value of the frequency distribution at frequency ω is the following rational function of the complex roots z_i

$$g(\omega^2) = \frac{1}{\pi \gamma \, c_{n_1}(\omega^2)} \sum_{l=1}^{n_1} \sum_{q=0}^{n_1} c_q'(\omega^2) \, \mathrm{Re} \, \frac{-z_l^{q-1}}{\sqrt{z_l - 1} \prod_{p \neq l}^{n_1} (z_l - z_p)} \quad [14]$$

where $c_q'(\omega^2)$ is given by

$$c_q'(\omega^2) = \sum_{m=1}^{\gamma} m \, a_{\gamma-m, \, q+m_0-n_0 m} \, \omega^{2(m-1)} . \quad [15]$$

The sign of the square root has to be determined such that $\mathrm{Im} \sqrt{z_l - 1} > 0$. If z_l is real and > 1, then the sign is given by

$$\mathrm{sgn} \sqrt{z_l - 1} = -\mathrm{sgn} \left(\sum_{p=1}^{n_1} p \, c_p(\omega^2) z_l^{p-1} \right)$$

$$\times \mathrm{sgn} \left(\sum_{p=0}^{n_1} c_p'(\omega^2) z_l^p \right). \quad [16]$$

3. The Frequency Distribution as Rational Integral

We now begin the theoretical discussion on the basis of eqs. [1] to [8] in order to prove the statements of the last section.

We start with the one-dimensional version of eq. [1] transforming the sum over φ into an integral ($N \to \infty$, periodic boundary conditions)

$$g(\omega^2) = \frac{1}{\gamma N} \sum_{\beta=1}^{\gamma} \sum_{\varphi=-\pi}^{\pi} \delta(\omega^2 - \omega_\beta^2(\varphi)) \quad [17]$$

$$= \frac{1}{2\pi\gamma} \sum_{\beta=1}^{\gamma} \int_{-\pi}^{\pi} \delta(\omega^2 - \omega_\beta^2(\varphi)) \, \mathrm{d}\varphi . \quad [18]$$

Choosing a representation of the delta function as infinitely narrow *Lorentz* curve

$$\delta(\omega^2 - \omega_\beta^2(\varphi)) = -\frac{1}{\pi} \lim_{\varepsilon \to 0+}$$

$$Im \frac{\partial}{\partial \omega^2} \ln(\omega_\beta^2(\varphi) - \omega^2 - i\,\varepsilon) \quad [19]$$

we obtain a representation of the frequency distribution in terms of the reduced secular determination (eq. [6])

$$g(\omega^2) = \frac{1}{2\pi^2\gamma} \lim_{\varepsilon \to 0+}$$

$$Im \int_{-\pi}^{\pi} d\varphi \frac{\partial}{\partial \omega^2} \ln S(\varphi, \omega^2 + i\,\varepsilon). \quad [20]$$

Due to eq. [6] the nonanalytic functions $\omega_\beta^2(\varphi)$ have been eliminated. The reduced secular determinant is a polynomial in $\cos\varphi$ and $\sin\varphi$. In order to apply *Cauchy*'s theorem we need infinite boundaries of the integral. So we substitute $t = \tan\varphi/2$ and get

$$g(\omega^2) = \frac{1}{\pi^2\gamma} \lim_{\varepsilon \to 0+}$$

$$Im \int_{-\infty}^{\infty} \frac{dt}{1+t^2} \frac{\partial}{\partial \omega^2} \ln S(\varphi(t), \omega^2 + i\,\varepsilon). \quad [21]$$

The reduced secular determinant $S(\varphi(t), \omega^2)$ is a rational function of t which can be transformed into a polynomial function by means of

$$P(t, \omega^2) \equiv (1+t^2)^{n_0\gamma} S(\varphi(t), \omega^2). \quad [22]$$

To see this we insert the identity $\exp(i\varphi) = (1+it)/(1-it)$ into eq. [6]

$$S(\varphi(t), \omega^2) = \det \left| \sum_{l=-n_0}^{n_0} D_{ij}^{l0} \left(\frac{1-i\,t}{1+i\,t} \right)^l - \omega^2 \delta_{ij} \right| \quad [23]$$

and multiply each element of the determinant by $(1+t^2)^{n_0}$. The result is

$$P(t, \omega^2) = \det \left| \Lambda_{ij}(t) - \omega^2(1+t^2)^{n_0} \delta_{ij} \right|, \quad [24]$$

where $\Lambda_{ij}(t)$ is the polynomial matrix

$$\Lambda_{ij}(t) = \sum_{l=-n_0}^{n_0} D_{ij}^{l0}(1+i\,t)^{n_0-l}(1-i\,t)^{n_0+l}. \quad [25]$$

Thus $P(t, \omega^2)$ is a polynomial in t and the frequency distribution can be written as

$$g(\omega^2) = \frac{1}{\pi^2\gamma} \lim_{\varepsilon \to 0+} \quad [26]$$

$$Im \int_{-\infty}^{\infty} \frac{dt}{1+t^2} \frac{1}{P(t, \omega^2 + i\,\varepsilon)} \frac{\partial}{\partial \omega^2} P(t, \omega^2 + i\,\varepsilon).$$

This is an infinite integral over a rational integrant, hence *Cauchy*'s theorem can in principle be applied.

4. The Invariants

In order to apply *Cauchy*'s theorem in practice we have to determine the coefficients of the polynomial $P(t, \omega^2)$. First we expand the determinant in power of $(\omega^2 + i\,\varepsilon)(1+t^2)^{n_0}$

$$P(t, \omega^2 + i\,\varepsilon) \quad [27]$$

$$= (-1)^{\gamma+1} \sum_{l=0}^{\gamma} A_l(t) \{(\omega^2 + i\,\varepsilon)(1+t^2)^{n_0}\}^{\gamma-l}.$$

The coefficients $A_l(t)$ are the invariants of the polynomial matrix $\Lambda_{ij}(t)$. Obviously

$$A_0(t) = -1,$$
$$A_\gamma(t) = (-1)^{\gamma+1} \det \mathbf{\Lambda}(t). \quad [28]$$

The other invariants can be calculated, for example by the method of *Leverrier* as modified by *Souriau* (19), which is based on the following algorithm:

$$\mathbf{B}_{l-1} = \mathbf{C}_{l-1} - A_{l-1} \cdot \mathbf{1},$$
$$\mathbf{C}_l = \mathbf{B}_{l-1} \cdot \mathbf{\Lambda}, \quad [29]$$
$$A_l = \frac{1}{l} \operatorname{tr} \mathbf{C}_l,$$

with the initial condition $\mathbf{C}_0 = 0$. The algorithm ends up with $\mathbf{C}_\gamma = \mathbf{1} \cdot A_\gamma$. This last equation can be used as a check. The results begin with

$$A_1 = \operatorname{tr} \Lambda$$
$$A_2 = \tfrac{1}{2} \operatorname{tr}(\Lambda^2) - \tfrac{1}{2}(\operatorname{tr}\Lambda)^2 \quad [30]$$
$$A_3 = \tfrac{1}{3} \operatorname{tr}(\Lambda^3) - \tfrac{1}{2}\operatorname{tr}\Lambda \operatorname{tr}(\Lambda^2) + \tfrac{1}{6}(\operatorname{tr}\Lambda)^3$$

and become more and more complicated. For the polyatomic chain with nearest neighbour interaction it is possible to write down these invariants in a compact form (see below) but for a general polymer the algebra becomes too involved. Therefore we decide to calculate numerically the coefficients of the various powers of t, i.e. the derivatives

$$A_l^{(k)} = (\partial^k/\partial t^k) A_l (t=0)$$

of the invariants. For this purpose the algorithm of *Leverrier* (eq. [29]) is very useful because it can easily be differentiated ($l = 1, 2, \ldots, \gamma$):

$$\boldsymbol{B}_{l-1}^{(k)} = \boldsymbol{C}_{l-1}^{(k)} - A_{l-1}^{(k)} \boldsymbol{1} \quad k = 0, 1, \ldots, 2(l-1) n_0$$

$$\boldsymbol{C}_l^{(k)} = \sum_{p=0}^{k} \binom{k}{p} \boldsymbol{B}_{l-1}^{(p)} \boldsymbol{\Lambda}^{(k-p)}$$
$$k = 0, 1, \ldots, 2l n_0 \quad [31]$$

$$A_l^{(k)} = \frac{1}{l} \operatorname{tr} \boldsymbol{C}_l^{(k)} \quad k = 0, 1, \ldots, 2l n_0 .$$

The initial conditions are

$$\boldsymbol{C}_0^{(k)} = 0 \quad \text{and} \quad A_0^{(k)} = - \delta_{k0}.$$

This algorithm needs as input the derivatives of the polynomial matrix eq. [24] which can be written as

$$\Lambda_{ij}^{(n)} \equiv \frac{\partial^n}{\partial t^n} \Lambda_{ij} (t = 0)$$
$$= c_0^n D_{ij}^{00} + c_1^n D_{ij}^{10} + c_1^{n*} D_{ij}^{-10} + \cdots \quad [32]$$
$$+ c_{n_0}^n D_{ij}^{n_0 0} + c_{n_0}^{n*} D_{ij}^{-n_0 0} \quad n = 0, 1, \ldots, 2 n_0 .$$

The coefficients c_l^n are related to the coefficients of Table 1 by ($i = \sqrt{-1}$)

$$c_l^n = n! \, i^n \, \xi_l^n \quad [33]$$

and depend on the number n_0 of neighbours. They are listed for $n_0 = 1, 2, 3$. Eqs. [31] and [32] are identical with eqs. [9] and [10] if we put

$$\boldsymbol{\Lambda}^{(n)} = n! \, i^n \, \boldsymbol{\alpha}^n$$
$$\boldsymbol{B}_l^{(n)} = n! \, i^n \, \boldsymbol{\beta}_l^n$$
$$\boldsymbol{C}_l^{(n)} = n! \, i^n \, \boldsymbol{\gamma}_l^n \quad [34]$$
$$A_l^{(n)} = n! \, i^n \, \eta_l^n .$$

Eqs. [9] and [10] are more convenient for the computer than the equivalent eqs. [31] and [32] since, by the factor i^n, complex arithmetic is avoided. Furthermore, the factor $n!$ helps in obtaining uniform orders of magnitude.

Once the derivatives of the invariants are known, the coefficients of the polynomial $P(t, \omega^2)$ are in principle determined by inserting the *Taylor* expansion

$$A_l(t) = \sum_{k=0}^{2l n_0} \frac{1}{k!} A_l^{(k)} t^k \quad [35]$$

into eq. [27].

5. The Frequency Dependent Secular Equation and the Complex Integration

The execution of the integral eq. [26] by *Cauchy*'s theorem consists mainly in the determination of the roots $t(\omega)$ of the equation

$$P(t, \omega^2 + i \varepsilon) = 0 \quad [36]$$

which we call the frequency dependent secular equation. It is seen from eqs. [35] and [27] that there are $2\gamma n_0$ roots $t(\omega)$ to be determined. Since we have assumed that the dynamical matrix is symmetric we are able to reduce the order of this algebraic equation from $2\gamma n_0$ to γn_0. From the symmetry of the dynamical matrix it follows that $\boldsymbol{\Lambda}(t)$ is hermitian and consequently the invariants are real. Since $\boldsymbol{\Lambda}(t)$ depends on powers of it ($i = \sqrt{-1}$), the invariants can depend only on even powers of it. Therefore t^2 can be chosen as new variable. Due to the special form of eq. [27] we prefer to introduce $z = 1 + t^2$ instead of t^2. Then eq. [35] takes the form

$$A_l(t) = \sum_{p=0}^{l n_0} a_{lp} z^p . \quad [37]$$

The coefficients are determined by the binomial theorem

$$a_{lp} = (-1)^p \sum_{k=p}^{l n_0} \binom{k}{p} A_l^{(2k)} \frac{(-1)^k}{(2k)!} \quad \begin{array}{l} l = 0, \ldots, \gamma \\ p = 0, \ldots, l \, n_0 . \end{array}$$
$$[38]$$

The last equation is equivalent to eq. [11] (which is more convenient for computing). Eq. [27] can now be written as

$$P(t, \omega^2 + i \varepsilon) \quad [39]$$
$$= (-1)^{\gamma+1} \sum_{l=0}^{\gamma} \sum_{p=p_l}^{l n_0} a_{lp} (\omega^2 + i \varepsilon)^{\gamma-l} z^{n_0(\gamma-l)+p} .$$

The index p_l has been introduced to account for the fact that in most cases some lower terms vanish such that $a_{l p_l} \neq 0$, $a_{l,p_l-1} = 0, \ldots a_{l_0} = 0$. (In the example of the polyatomic linear chain with nearest neighbour interaction the indices p_l are given by $p_l = l$ for $l < \gamma$ and $p_\gamma = \gamma - 1$.) In these cases the polynomial $P(t, \omega^2 + i \varepsilon)$ has a multiplicative factor z^{m_0} where m_0 is the minimum (with respect to l) of $n_0(\gamma-l)+p_l$. m_0 is the multiplicity of the trivial pole $z = 0$ (or $t = i$, or $\varphi = \pi + i\infty$). *Kirkwood*'s model (Fig. 1) has only one trivial pole ($m_0 = 1$), the γ-atomic linear chain with nearest neighbour interaction has only one *nontrivial* pole ($m_0 = \gamma - 1$).

The final expression for the polynomial P is thus

$$P(t, \omega^2 + i \varepsilon) = (-1)^{\gamma+1} z^{m_0} N(z, \omega^2 + i \varepsilon). \quad [40]$$

The polynomial

$$N(z, \omega^2) = \sum_{q=0}^{n_0 \gamma - m_0} c_q(\omega^2) z^q \quad [41]$$

is identical with the left hand side of eq. [12] and the coefficients $c_q(\omega^2)$ are given by eq. [13]. The final expression for the frequency distribution before using *Cauchy*'s theorem is obtained by inserting eq. [40] into eq. [26]

$$g(\omega^2) = \frac{1}{\pi^2 \gamma} \lim_{\varepsilon \to 0+}$$

$$Im \int\limits_{-\infty}^{\infty} \frac{\frac{\partial}{\partial \omega^2} N(1 + t^2, \omega^2 + i\,\varepsilon)}{(1 + t^2)\, N(1 + t^2, \omega^2 + i\,\varepsilon)}\, dt . \qquad [42]$$

In the case of the γ-atomic linear chain the polynomial $N(z, \omega^2)$ is very simple

$$N(z, \omega^2) = c_{\gamma-1}(\omega^2) + c_\gamma(\omega^2)\, z \qquad [43]$$

so that the roots are given analytically. In more complicated cases the zeros can easily be calculated numerically, for any frequency one is interested in.

Due to *Cauchy*'s theorem the integral in eq. [43] is a sum of residues at the poles $t_i = \pm \sqrt{z_i - 1}$. The sign of the square root has to be determined such that the t_i lies in the upper half plane. For frequencies inside a band at least one of the z_i is real and ≥ 1. Thus the corresponding poles $\pm \sqrt{z_i - 1}$ lie on the real axis, and the infinitesimally small imaginary part ε of the energy becomes important because it shifts the poles away from the real axis. The sign of that pole t_j which is shifted into the *upper* half plane can be determined from the equations

$$N(z_j + \delta z, \omega^2 + i\,\varepsilon) = 0 \qquad [44]$$

and from this the sign of the pole

$$\mathrm{sgn}\, t_j = -\,\mathrm{sgn}\!\left(\frac{\partial N}{\partial z}\right) \mathrm{sgn}\!\left(\frac{\partial N}{\partial \omega^2}\right). \qquad [47]$$

Eq. [16] follows immediately. The sum over residues, eq. [14] is obtained in the same way as eq. [22] of Ref. 14.

6. An Analytically Soluble Example: The Poly-Atomic Linear Chain with Nearest Neighbour Interaction

For a linear chain with p particles per unit cell and nearest neighbour interaction ($n_0 = 1$) it is possible to derive an exact analytical expression for the density of states.

We adopt all notations with the only exception $\gamma = p$ (instead of $\gamma = 3p$, i.e. we have pN instead of $3pN$ degrees of freedom) because we consider a motion in only one cartesian direction. The G-matrix of kinetic energy is

$$G_{ik}^{nn'} = \frac{1}{m_i} \delta_{nn'}\, \delta_{ik} \quad i, k = 1, \ldots, p. \qquad [48]$$

The m_i are the masses of the p different particles in the unit cell. The F-matrix of potential energy has the following nonvanishing elements (for arbitrary n)

$$\begin{aligned}
F_{ii}^{nn} &= 2\alpha & i &= 1, \ldots, p \\
F_{p1}^{n-1,n} &= F_{1p}^{n+1,n} = -\alpha & & \\
F_{i,i+1}^{nn} &= -\alpha & i &= 1, \ldots, p
\end{aligned} \qquad [49]$$

α is the coupling constant of the chain.

We start with the matrix $\Lambda_{ij}(t)$ of eq. [25], which can be written as

$$\Lambda(t) = \begin{pmatrix}
2z\omega_1^2 & -z\omega_1\omega_2 & 0 & & & & 0 & 0 & (t-i)^2\omega_1\omega_\gamma \\
-z\omega_1\omega_2 & 2z\omega_2^2 & -z\omega_2\omega_3 & 0 & & & & 0 & 0 \\
0 & -z\omega_2\omega_3 & 2z\omega_3^2 & & & & 0 & & 0 \\
& & 0 & & & & & & \\
& & & & 0 & & & & \\
0 & & & & & & 2z\omega_{\gamma-2}^2 & -z\omega_{\gamma-2}\omega_{\gamma-1} & 0 \\
0 & 0 & & & & 0 & -z\omega_{\gamma-2}\omega_{\gamma-1} & 2z\omega_{\gamma-1}^2 & -z\omega_{\gamma-1}\omega_\gamma \\
(t+i)^2\omega_1\omega_\gamma & 0 & 0 & & & & 0 & -z\omega_{\gamma-1}\omega_\gamma & 2z\omega_\gamma^2
\end{pmatrix}$$

and

$$z_j + \delta z = 1 + (t_j + i\,\tau)^2 = 1 + t_j^2 + 2 i t_j \tau \quad [45]$$

with infinitesimal $\tau > 0$ and $\varepsilon > 0$. Expanding eq. [44] linearly we get (the first term vanishes)

$$\frac{\partial N}{\partial z}\, 2 i t_j \tau + \frac{\partial N}{\partial \omega^2}\, i\,\varepsilon = 0 \qquad [46]$$

where we have introduced the abbreviations

$$z = 1 + t^2 \qquad [51]$$

$$\omega_l = \sqrt{\alpha/m_l} \quad l = 1, \ldots, \gamma . \qquad [52]$$

ω_l is identical with the maximal frequency of a fictitious monatomic chain with masses m_l.

The result of the application of the algorithm [29] is

$$A_l(t) = (1+t^2)^l \, \sigma_l \quad \text{for} \quad l = 1, 2, \ldots, \gamma-1 \qquad [53]$$

$$A_\gamma(t) = t^2 (1+t^2)^{\gamma-1} \, \sigma_\gamma \,. \qquad [54]$$

The first and the last invariant are given by

$$\sigma_1 = 2 \sum_{i=1}^{\gamma} \omega_i^2 \qquad [55]$$

$$\sigma_\gamma = (-1)^{\gamma+1} \, 4 \prod_{i=1}^{\gamma} \omega_i^2 \,. \qquad [56]$$

Introducing the abbreviations

$$S_n \;\;= \sum_{i=1}^{\gamma} \omega_i^2 \, \omega_{i+n}^2$$

$$S_{nm} \;= \sum_{i=1}^{\gamma} \omega_i^2 \, \omega_{i+n}^2 \, \omega_{i+m}^2 \qquad [57]$$

$$S_{nmp} = \sum_{i=1}^{\gamma} \omega_i^2 \, \omega_{i+n}^2 \, \omega_{i+m}^2 \, \omega_{i+p}^2$$

thereby adopting the convention

$$\omega_{i+\gamma}^2 = \omega_i^2 \quad \text{(for all } i) \qquad [58]$$

the other A_l' can be written as

(a) for $\gamma = 3$: $\sigma_2 = -3 S_1$ $\qquad [59]$

(b) for $\gamma = 4$: $\sigma_2 = -3 S_1 - 2 S_2$

$\qquad\qquad\qquad\;\; \sigma_3 = 4 S_{12} ,$ $\qquad [60]$

(c) for $\gamma = 5$: $\sigma_2 = -3 S_1 - 4 S_2$

$\qquad\qquad\qquad\;\; \sigma_3 = 4 S_{12} + 6 S_{13}$ $\qquad [61]$

$\qquad\qquad\qquad\;\; \sigma_4 = -5 S_{123} ,$

(d) for $\gamma = 6$: $\sigma_2 = -3 S_1 - 4 S_2 - 2 S_3$

$\qquad\qquad\qquad\;\; \sigma_3 = 4 S_{12} + 6 S_{13} + 6 S_{14}$

$\qquad\qquad\qquad\qquad\qquad\quad + \tfrac{8}{3} S_{35}$ $\qquad [62]$

$\qquad\qquad\qquad\;\; \sigma_4 = -5 S_{123} - 8 S_{124} - 3 S_{134}$

$\qquad\qquad\qquad\;\; \sigma_5 = 6 S_{123} \,.$

All the A_l' are functions of the ω_i^2 alone and do not depend on t. (The above formulae have been checked by computer algebra.)

By means of the remarkably simple t-dependence of the invariants, the integral in eq. [26] can be done easily. By inserting eqs. [53], [54] and [27] into eq. [26] we get $(n_0 = 1)$

where the $P_i(\omega^2)$ are the following polynomials in ω^2 $(\sigma_0 = -1)$

$$P_1(\omega^2) = \sum_{j=0}^{\gamma-1} (j+1) \, \omega^{2j} \, \sigma_{\gamma-j-1} = \sum_{l=1}^{\gamma} l \, \omega^{2(l-1)} \, \sigma_{\gamma-l}$$

$$P_2(\omega^2) = \sum_{j=1}^{\gamma} \omega^{2j} \, \sigma_{\gamma-j} \qquad [64]$$

$$P_3(\omega^2) = \sum_{j=0}^{\gamma} \omega^{2j} \, \sigma_{\gamma-j} = P_2(\omega^2) + \sigma_\gamma \,.$$

The result of the integration is

$$g(\omega^2) = (\pm) \frac{1}{\gamma \pi} \, P_1(\omega^2) \, \mathrm{Im} \, \frac{1}{\sqrt{P_2(\omega^2) \, P_3(\omega^2)}} \,. \qquad [65]$$

The sign can be determined from the fact that the density of states must be positive.

For very small ω we have

$$P_1(\omega^2) \approx \sigma_{\gamma-1}$$

$$P_2(\omega^2) \approx \omega^2 \, \sigma_{\gamma-1} \qquad [66]$$

$$P_3(\omega^2) \approx \sigma_\gamma \,.$$

Since σ_γ and $\sigma_{\gamma-1}$ are different in sign,

$$g(\omega^2) \approx \frac{1}{\pi \gamma} \sqrt{\left| \frac{\sigma_{\gamma-1}}{\sigma_\gamma} \right|} \, \frac{1}{\omega} \quad (\omega \to 0) \,. \qquad [67]$$

Furthermore, from eqs. [64]–[67] it follows

$$\left| \frac{\sigma_{\gamma-1}}{\sigma_\gamma} \right| = \frac{\gamma}{4\alpha} \, (m_1 + m_2 + \cdots + m_\gamma) \qquad [68]$$

thus

$$g(\omega^2) \approx \frac{1}{2\pi} \sqrt{\frac{1}{\alpha \gamma} (m_1 + m_2 + \cdots + m_\gamma)} \, \frac{1}{\omega}$$
$$(\omega \to 0). \qquad [69]$$

On the other hand, for very large ω, the polynomials $P_i(\omega^2)$ behave like

$$P_1(\omega^2) \approx \gamma \, \sigma_0 \, \omega^{2(\gamma-1)} = -\gamma \, \omega^{2(\gamma-1)}$$

$$P_2(\omega^2) \approx \omega^{2\gamma} \, \sigma_0 \qquad\quad = -\omega^{2\gamma} \qquad [70]$$

$$P_3(\omega^2) \approx -\omega^{2\gamma}$$

so that $g(\omega^2) = 0$ as it should be.

Finally, we consider the 3-atomic linear chain $(\gamma = 3)$ explicitly. The invariants are

$$\sigma_1 = 2\alpha \left(\frac{1}{m_1} + \frac{1}{m_2} + \frac{1}{m_3} \right)$$

$$g(\omega^2) = \frac{-1}{\gamma \pi^2} \lim_{\varepsilon \to 0+} \mathrm{Im} \, P_1(\omega^2 + i\,\varepsilon) \int_{-\infty}^{\infty} \frac{\mathrm{d}t}{P_2(\omega^2 + i\,\varepsilon) + t^2 P_3(\omega^2 + i\,\varepsilon)} \qquad [63]$$

$$\sigma_2 = -3\alpha^2 \left(\frac{1}{m_1 m_2} + \frac{1}{m_2 m_3} + \frac{1}{m_3 m_1} \right) \qquad [71]$$

$$\sigma_3 = \frac{4\alpha^3}{m_1 m_2 m_3}$$

The density of states inside the bands is

$$g(\omega^2) = \frac{1}{3\pi\omega} \qquad [72]$$

$$\frac{\sigma_2 + 2\omega^2 \sigma_1 - 3\omega^4}{\sqrt{(-\sigma_2 - \omega^2 \sigma_1 + \omega^4)(\sigma_3 + \omega^2 \sigma_2 + \omega^4 \sigma_1 - \omega^6)}}$$

Frequencies for which the radicand becomes negative lie outside the bands, i.e. $g(\omega^2) = 0$.

7. Conclusion

The method of complex phases presented in this paper bridges the gap between purely analytical methods on the one side and purely numerical methods on the other. Polymers are such complicated systems that purely analytical methods break down, and the purely numerical methods are so general that they cannot take full advantage of the polymer's short range interaction and one-dimensionality. The method of complex phases requires in general a computer, but is not purely numerical in so far as it does not suffer from typical numerical phenomena such as virtual statistical peaks. The limit $\varepsilon \to 0$ in eq. [19] does not require infinite computing time (as in the histogram method which, however, uses a rectangular function of width ε instead of a *Lorentz* curve). The execution of the limit $\varepsilon \to 0$ means in practical terms that the direction of the infinitesimal shift of complex poles must be determined. However, this can be computed exactly by first order *Taylor* expansion (see eq. [46]).

The method of complex phases is designed such that no tedious work has to be done when changing from one polymer to the other. A computer program can easily be written following the formulae of Section 2. From the dynamical matrix, the program has first to determine the coefficients (eq. [11]) of the frequency dependent secular equation (using the theory of invariants). Then it has to find all complex roots of this higher order algebraic equation (by standard subroutines) for any frequency ω desired. The result is of the type of Fig. 1. All points of the dispersion curve having this frequency ω are

among these roots (the real phases). Finally it will calculate extremely fast and exactly the density of states as rational function of these roots (eq. [14]).

The method is particularly useful for the study of small frequency ranges. There is no obligation to calculate the whole frequency band; any particular frequency can be selected, regardless of whether there are gaps. In contrast, the conventional methods (20) permit a separate approximative treatment of an energy region only if this region is well separated from the rest by a gap in the frequency distribution, such as the gap in polyethylene between the low- and high-frequency vibrations. Of course, a refinement in the frequency resolution does not mean a loss of earlier results in our method, in contrast to conventional methods.

If a large number of frequencies has to be studied, interpolation techniques can in principle be applied to the calculation of the inverse complex phase-frequency relation (e.g. Fig. 1) in the same way as *Gilat* and *Dolling* (8) did for the real dispersion relation. But in general it will not be worthwhile since the frequency distribution itself can be interpolated between exactly calculated points.

We have not yet done calculations for three-dimensional chain-polymer crystals, but we believe that the method of complex phases will be useful here too. The total frequency distribution can be written as

$$g(\omega^2) = \iint g_c(\omega^2, \varphi_a, \varphi_b) \, d\varphi_a \, d\varphi_b , \qquad [73]$$

where $g_c(\omega^2, \varphi_a, \varphi_b)$ is the frequency distribution at frequency ω^2 due to all phonons along the chain direction for fixed perpendicular phases φ_a, φ_b. g_c can be calculated in the framework of this paper. Since the dispersion relation changes little with perpendicular phase differences (except in the very low frequency region), the integral [73] can be approximated e.g. by a sum over a rather coarse mesh of values φ_a, φ_b.

Finally we should like to mention that the technique developed in this paper is useful not only for the calculation of the frequency distribution. If the amplitude weighted frequency distributions (such as neutron scattering cross-sections) and, more generally, the lattice *Green* function (which is fundamental for defect calculations) are to be computed, the derivation and solution of the frequency dependent secular equation is a necessary first step which has to be complemented by further steps. This problem has in principle been solved recently (21) and will be considered in detail in a forthcoming paper.

Summary

The frequency distribution $g(\omega)$ (density of states at frequency ω) of an isolated polymer chain (with arbitrary number γ of degrees of freedom per unit cell) is usually calculated *approximately* as a histogram from the phase frequency curve $\omega_\beta(\varphi)$ $(\beta = 1, \ldots, \gamma)$ with a *large* set of *real* phases φ. In this paper a method is presented for calculating $g(\omega)$ *exactly* as a rational function of a *small* set of *complex* phases $\varphi_i(\omega)$ $(i = 1, \ldots, n_0\gamma, n_0 =$ range of interaction). The $\varphi_i(\omega)$ are the roots of a frequency dependent secular equation the coefficients of which are obtained by the theory of invariants (differentiation of the algorithm of *Leverrier-Souriau*). The method is of basic importance also for the computation of the *Green* function in space energy representation for realistic polymers. Analytical results are obtained for a polyatomic linear chain with arbitrary γ and $n_0 = 1$.

Zusammenfassung

Es ist üblich, die Frequenzverteilung $g(\omega)$ (Zustandsdichte bei der Frequenz ω) einer isolierten Polymerkette (mit beliebiger Anzahl γ von Freiheitsgraden pro Einheitszelle) durch ein Histogramm *anzunähern*, das aus *vielen* Punkten (reelle Phasen φ) der Dispersionsrelation $\omega_\beta(\varphi)$ $(\beta = 1, \ldots \gamma)$ konstruiert wird. In der vorliegenden Arbeit wird eine Methode entwickelt, die es erlaubt, $g(\omega)$ *exakt* auszurechnen als rationale Funktion von *wenigen* Phasen $\varphi_i(\omega)$, die jedoch *komplex* sind $(i = 1, \ldots n_0\gamma, n_0 =$ Reichweite der Wechselwirkung). Die $\varphi_i(\omega)$ sind die Lösungen einer frequenzabhängigen Säkulargleichung, deren Koeffizienten mit der Theorie der Invarianten (Ableitung des Algorithmus von *Leverrier-Souriau*) gewonnen werden. Die Methode ist darüber hinaus von grundlegender Bedeutung für die Berechnung der *Green*schen Funktion eines realistischen Polymers in Ort-Energie-Darstellung. Für das Beispiel einer mehratomigen linearen Kette (mit γ beliebig und $n_0 = 1$) werden analytische Ergebnisse abgeleitet.

References

1) *Kirkwood, J. G.*, J. Chem. Phys. **7**, 506 (1939).
2) *Pitzer, S. J.*, J. Chem. Phys. **8**, 711 (1940).
3) *Liang, C. Y.*, *S. Krimm* and *G. B. B. M. Sutherland*, J. Chem. Phys. **25**, 543 (1956).
4) *Bowers, W.* and *H. B. Rosenstock*, J. Chem. Phys. **18**, 1056 (1950).
5) *van Hove, L.*, Phys. Rev. **89**, 1189 (1953).
6) *Maradudin, A. A.*, *E. W. Montroll, G. H. Weiss* and *I. P. Ipatova*, Solid State Phys. Suppl. **3**, 1 (1971)
7) *Wunderlich, B.*, J. Chem. Phys. **37**, 1207 (1962).
8) *Gilat, G.* and *G. Dolling*, Phys. Lett. 8, 304 (1964).
9) *Piseri, L.* and *G. Zerbi*, J. Chem. Phys. **48**, 3561 (1968).
10) *Kitagawa, T.* and *T. Miyazawa*, Bull. Chem. Soc. Japan **43**, 372 (1970).
11) *Gilat, G.* and *L. J. Raubenheimer*, Phys. Rev. **144**, 320 (1966).
12) *Schmid, C.* and *K. Hölzl*, J. Phys. C: Solid State Phys. **6**, 2401 (1973).
13) *Hölzl, K.* and *C. Schmid*, J. Phys. C: Solid State Phys. **8**, 2235 (1975).
14) *Schmid, C.* and *K. Hölzl*, J. Polym. Sci. A2 **10**, 1881 (1972).
15) *Zerbi, G.*, Conference of the European Physical Society on "Radiation Scattering of Bulk Polymers", Strasbourg, 1972.
16) *Zbinden, R.*, Infrared Spectroscopy of High Polymers. (New York, 1964.)
17) *Hölzl, K.* and *C. Schmid*, to be published.
18) *Zerbi, G.*, Applied Spectroscopy Reviews (New York, 1969.)
19) *Faddeev, D. K.* and *V. N. Faddeeva*, Computational Methods of Linear Algebra. (San Francisco-London, 1963.)
20) *Kitagawa, T.* and *T. Miyazawa*, Adv. in Polymer Sci. **9**, 335 (1972).
21) *Schmid, C.*, J. Phys. C: Solid State Phys. **6**, L458 (1973).

Author's address:

Privatdozent Dr. *Christhard Schmid*
Bayer AG, FS-A
D 4047 Dormagen

Progr. Colloid & Polymer Sci. **58**, 30—35 (1975)

Sonderforschungsbereich „Makromoleküle", Mainz

Untersuchung der Dichteschwankungen in amorphen Polymeren durch Licht- und Röntgenkleinwinkelstreuung *)

H. J. Hölle, R. G. Kirste, B. R. Lehnen und *M. Steinbach*

Mit 3 Abbildungen und 2 Tabellen

(Eingegangen am 26. September 1974)

1. Einleitung

Gefragt sei, inwieweit die Struktur von amorphen Polymeren mit der von niedermolekularen Flüssigkeiten vergleichbar ist. Wir beschränken uns auf die Licht- und die Röntgenkleinwinkelstreuung als Untersuchungsmethode. Dann ist unser Problem gleichbedeutend mit der Frage, ob das Streuverhalten der amorphen Polymeren in Übereinstimmung mit der Schwankungstheorie (1, 2) für Flüssigkeiten ist, deren Gültigkeit für niedermolekulare isotrope Substanzen hinreichend sichergestellt ist. Es gibt bereits etliche experimentelle Untersuchungen zu dieser Frage. Dabei wurden stets mehr oder weniger ausgeprägte Abweichungen von den aus der Schwankungstheorie ableitbaren Daten gefunden (3—10). Berichtet wurden insbesondere exzessive und winkelabhängige Werte der Rayleighstreuung (3, 10) und unterschiedliche Werte für V_h und H_h beim Streuwinkel $\theta = 90°$ (7). Hierbei ist V_h die Vertikalkomponente und H_h die Horizontalkomponente des gesamten Streulichts bei horizontal polarisiertem Primärlicht. Bei der Röntgenkleinwinkelstreuung wird allgemein ein Anstieg der Streuintensität zu kleineren Winkeln hin für $\varkappa < 0,03$ gefunden, wobei $\varkappa = (4\pi/\lambda) \sin \theta/2$ der Streuvektor und λ die Wellenlänge ist.

Wir haben die referierten Diskrepanzen zwischen Theorie und Experiment ebenfalls gefunden. Das Ausmaß der Diskrepanz hängt aber stark von der Präparation ab. Bei besonders guter Präparationstechnik wird im Grenzfall das theoretische Streuverhalten gefunden.

---

*) *Herrn Prof. Dr. G. V. Schulz zum 70. Geburtstag gewidmet*

2. Theoretische Grundlagen

a) Lichtstreuung

Betrachtet wird das *Rayleigh*-Verhältnis R, das ist die Intensität der gesamten, d.h. über alle Frequenzen integrierten Streustrahlung bezogen auf die Einheit der Primärintensität, die Einheit des Streuvolumens und die Einheit des Abstandes zwischen Streuer und Detektor. Nach der Schwankungstheorie gilt für reine isotrope Substanzen (2, 11)

$$R_u = \frac{2\pi^2 kT \beta}{\lambda_0^4} \left(\varrho n \frac{dn}{d\varrho} \right)^2 \frac{6 + 6\sigma_u}{6 - 7\sigma_u} \text{ bei } \theta = 90°$$

und

$$R_v = \frac{4\pi^2 kT \beta}{\lambda_0^4} \left(\varrho n \frac{dn}{d\varrho} \right)^2 \frac{3 + 3\sigma_v}{3 - 4\sigma_v}$$

[1]

unabhängig vom Streuwinkel. [2]

Dabei ist β die isotherme Kompressibilität, λ_0 die Wellenlänge des Lichtes im Vakuum, ϱ die Dichte und n der Brechungsindex der Substanz. Die Indizes u und v beziehen sich auf unpolarisiertes bzw. vertikal polarisiertes Primärlicht, σ_u und σ_v sind die zugehörigen Depolarisationsgrade bei $\theta = 90°$.

$$\sigma_u \equiv \frac{H_u}{V_u} \quad \text{und} \quad \sigma_v \equiv \frac{H_v}{V_v},$$

[3]

H_u, V_u, H_h, H_v, V_h und V_v sind die Horizontal- bzw. Vertikalkomponenten des Streulichts. R_v, V_v und $H_v = V_h$ sind nach der Theorie unabhängig vom Streuwinkel. Für H_h gilt

$$H_h = H_v + (V_v - H_v) \cos^2 \theta.$$

[4]

b) Röntgenkleinwinkelstreuung

Wir gehen aus von der Gleichung (12)

$$R(\varkappa) = \left(\frac{e^2}{mc^2} \right)^2 \cdot \frac{1 + \cos^2 \theta}{2} \varrho_N |F(\varkappa)|^2 S(\varkappa)$$

[5]

e und m sind Ladung und Masse eines Elektrons, c ist die Lichtgeschwindigkeit, $\bar{\varrho}_N$ ist die mittlere Teilchendichte, F ist der über alle Raumrichtungen gemittelte Formfaktor der einzelnen Flüssigkeitsmoleküle und S ist die Interferenzfunktion.

Mit der Ortskorrelationsfunktion $g(x)$ ist

$$S(\varkappa) = 1 + \int\limits_0^\infty [g(x) - \bar{\varrho}_N] \frac{\sin \varkappa x}{x \varkappa} 4\pi x^2 \, dx \ . \quad [6]$$

Für die Streuung unter hinreichend kleinen Winkeln wird der Polarisationsfaktor $(1 + \cos^2\theta)/2$ gleich 1, $F(\varkappa)$ wird gleich der Zahl der Elektronen Z im Molekül, und $S(\varkappa)$ läßt sich durch das Schwankungsquadrat der Teilchendichte (13) und dieses durch die isotherme Kompressibilität (14) ausdrücken, so daß für die Streuintensität

$$R(0) = \left(\frac{e^2}{m\,c^2}\right)^2 Z^2 (\varrho\,N_{L/}M)^2 \, kT \, \beta \quad [7]$$

erhalten wird, wobei N_L die *Loschmidt*sche Zahl ist. Diese Gleichung wird mit der experimentell bestimmten Streuintensität verglichen.

3. Durchführung der Experimente

a) Herstellung und Reinigung der verwendeten Substanzen

Untersucht wurden Polydimethylsiloxan (PDMS) und Polymethylmethacrylat (PMMA). Folgende Präparate wurden verwendet:

PDMS — W 1 ist ein käufliches Präparat der Firma Wacker. Das Gewichtsmittel des Polymerisationsgrades P_w ist 380 (Molekulargewicht $M_w = 28\,000$). Die durch Gelpermeationschromatographie bestimmte Uneinheitlichkeit ist $U = 1,5$.

PDMS — W 2 wurde aus PDMS — W 1 durch einmalige Umfällung mit THF als Lösungs- und Methanol als Fällungsmittel erhalten.

PDMS — W 3: wie PDMS — W 1, jedoch $M_w = 138\,000$.

PDMS — M 1 ist ein selbst hergestelltes Präparat. Es wurde die anionische Technik nach der Arbeitsweise von *Lee* et al. angewendet (15, 15a). P_w ist 50 und $U \leqq 0,13$.

PDMS — 5 — das lineare Pentamere — wurde uns dankenswerterweise von der Fa. Wacker zur Verfügung gestellt. Zur Reinigung wurde 2mal über eine 35 cm Drehbandkolonne destilliert; danach war konstanter Siedepunkt und Brechungsindex erreicht.

PDMS — 50 und PDMS — 900 sind hochgereinigte selbst hergestellte Präparate mit ausschließlich Trimethylsilylendgruppen. Für die Herstellung gilt das bei PDMS — M 1 Gesagte. Der gute Reinigungseffekt wurde durch mehrmalige sorgfältige Umfällungs- und Filtrationsoperationen erreicht. P_w ist 50 bzw. 900 (genauer 49 bzw. 930), und für die Uneinheitlichkeit gilt $U \leqq 0,13$.

Die PMMA-Präparate wurden durch radikalische Polymerisation hergestellt (16). Das Präparat R 240 ist ein Plexiglas der Firma Röhm GmbH ($M_w = 5 \cdot 10^6$). Die Präparate K$_1$ und K$_7$ wurden von *W. Kruse* nach der in Zitat 17 beschriebenen Methode hergestellt ($M_w = 1 \cdot 10^5$ bei K$_1$ und $1 \cdot 10^6$ bei K$_7$). Die Uneinheitlichkeit liegt bei allen drei Präparaten nahe 1. Wesentlich bei der Präparation von K$_1$ und K$_7$ war, daß nach Entgasung des Monomeren unter dem Dampfdruck des Monomeren polymerisiert wurde. Ferner wurde bei dieser Präparation besonders auf Staubfreiheit geachtet. Alle Präparate wurden durch Tempern oberhalb der Glastemperatur und langsames Abkühlen weitgehend spannungsfrei gemacht.

Tetrahydrofuran (THF), Benzol und Toluol waren p. A. Substanzen der Firma Merck, die vor der Verwendung sorgfältig durch Destillation über Kolonnen mit großer Bödenzahl gereinigt wurden.

b) Streulichtexperimente

Gemessen wurden die PDMS-Präparate, Benzol, Toluol und THF. Benzol und Toluol dienten der Eichung. THF wurde gemessen, weil es eine niedermolekulare Substanz ist, deren Brechungsindex mit dem des PDMS bis auf 0,2% übereinstimmt. Dadurch konnten Fehler bei der Brechungskorrektur eliminiert werden. Über die Lichtstreuung an den PMMA-Präparaten ist in einer früheren Arbeit berichtet worden (10).

Es wurde ein Streulichtphotometer vom Typ Fica 50 der Firma Sofica benutzt. Als Badflüssigkeit diente o-Xylol ($n_{25}^{436} = 1,52$). Die Küvetten waren zylindrisch und aus Quarz ($n = 1,47$), der Durchmesser betrug 25 mm. Nach Justierung des Polarisationsfilters wurde für Benzol, Toluol und THF die Relation [4] und $V_h = H_v$ verifiziert. Die gemessenen Depolarisationsgrade stimmen mit Literaturwerten überein: Bei $\lambda = 5460$ Å und $T = 25\,°C$ wurde für Toluol $\sigma_u = 0,49$ gemessen, der Literaturwert (18) ist 0,48. Dto Benzol: gemessen $\sigma_u = 0,42$, Literaturwert (18) ebenfalls 0,42.

Für die Absolutbestimmung von R_v und R_u wurden die entsprechenden Literaturwerte von Benzol verwendet:

$$R_v = \left(\frac{R_v}{U_v}\right)_{\text{Benzol}} \times U_v \times \left(\frac{n}{n_{\text{Benzol}}}\right)^2 \quad [8]$$

mit $R_{v\,(\text{Benzol})} = 62,4 \cdot 10^{-6}$ (cm^{-1}) bei $\lambda = 4360$ Å und $22,0 \cdot 10^{-6}$ (cm^{-1}) bei $\lambda = 5460$ Å, beide Werte bei $25\,°C$ (19). U_v bezeichnet den unkorrigierten Meßwert zu R_v in Skalenteilen. Analog wurde bei R_u verfahren mit $R_{u\,(\text{Benzol})} = 44,5 \cdot 10^{-6}$ (cm^{-1}) (20) bei $\lambda = 4360$ Å und $15,8 \cdot 10^{-6}$ (cm^{-1}) (19, 20) bei $\lambda = 5460$ Å, beide Werte bei $25\,°C$.

c) Röntgenkleinwinkelstreuung

Gemessen wurden PMMA und PDMS-Präparate, sowie Benzol (21). Die allgemeine Meßtechnik ist an anderer Stelle beschrieben (22, 23). Da die interessierenden Streuintensitäten (bei kleinen $\varkappa$-Werten) sehr gering sind, oberhalb $\varkappa = 0,3$ Å^{-1} (im Bereich des Halos) aber stark ansteigen, rührt das gemessene Intensität bei Verwendung langer Strahlquerschnitte überwiegend von größeren Streuwinkeln her. Deshalb wurde mit relativ kurzen Strahlquerschnitten gearbeitet. Die effektive Länge des Strahlquerschnitts (nach Faltung mit dem

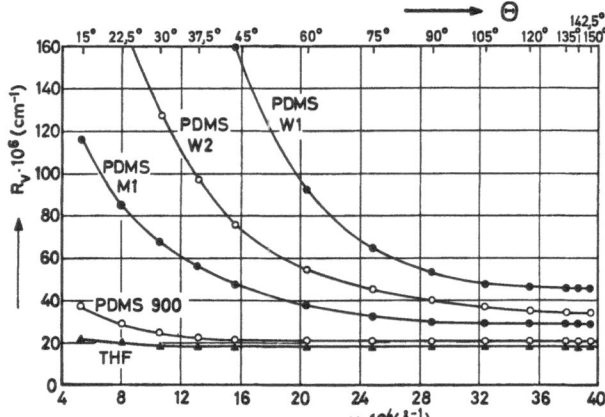

Abb. 1. Lichtstreuung an PDMS-Präparaten. Zum Vergleich die Streuung an THF

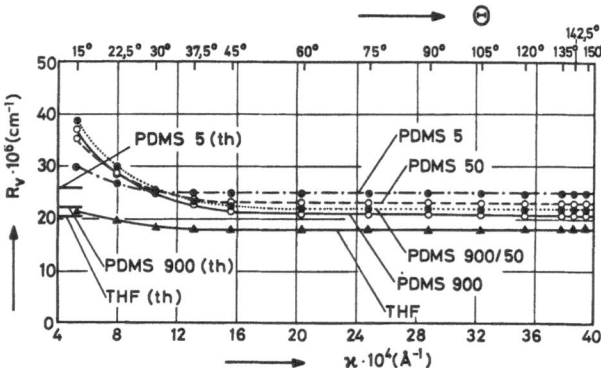

Abb. 2. Lichtstreuung an hochgereinigten molekular einheitlichen PDMS-Präparaten. Zum Vergleich THF und die theoretischen Intensitätswerte

Zählrohrspalt) betrug 1,9 cm, das sind 0,2 Å⁻¹ beim Streuvektor. Die effektive Breite des Strahlquerschnitts betrug 0,018 cm oder 0,0017 Å⁻¹ bei Streuvektor.

Die PMMA-Präparate waren Platten von 0,13 cm Dicke, während PDMS und Benzol in Markkapillaren mit einem Durchmesser von 0,035 bzw. 0,2 cm („optimale Dicken") gemessen wurden. Die Blindstreuung wurde unter Berücksichtigung der Absorption durch das Präparat eliminiert. Die gemessenen Streuintensitäten der Präparate wurden durch Multiplikation mit $e^{\mu D}/D$ normiert: $\mu =$ Absorptionskoeffizient, $D =$ Präparatdicke. Analog zu Gl. [8] wurde

$$R(\varkappa) = \left(\frac{R(0)}{I(0)}\right)_{\text{Benzol}} \times I(\varkappa) \qquad [9]$$

zur Berechnung des *Rayleigh*-Verhältnisses aus der gemessenen relativen Intensität I verwendet.
$R(0)_{\text{Benzol}} = 3{,}1 \cdot 10^{28}$ Elektronen/cm³ wurde gemäß Gl. [7] unter Weglassung von $(e^2/mc^2)^2$ berechnet.

In einer besonderen Meßreihe wurde nachgewiesen, daß Gl. [7] für Benzol erfüllt ist. Hierzu wurde die Streuintensität bei kleinen Streuwinkeln ($\varkappa < 0{,}3$) für

etliche Lösungsmittel, u.a. Benzol und Äthyläther in ihrem relativen Verhältnis zueinander gemessen¹). Dieses Verhältnis stimmte außer bei Wasser mit dem aus Gl. [7] ableitbaren Wert

$$\frac{R_A(0)}{R_B(0)} = \left(\frac{\varrho_A Z_A/M_A}{\varrho_B Z_B/M_B}\right)^2 \frac{\beta_A}{\beta_B} \qquad [10]$$

überein. Die Indizes A und B bezeichnen zwei Substanzen. Bei Wasser wird, wie *Hendricks* nachgewiesen hat (32), das Ergebnis bei üblichen Arbeitsbedingungen durch Mehrfachstreuung verfälscht. Für Äthyläther ist die Gültigkeit von Gl. [7] bereits nachgewiesen worden (33), so daß sie damit allgemein für niedermolekulare Flüssigkeiten als erfüllt angesehen werden kann.

Bei den Messungen wurden die direkt gemessenen Intensitäten und nicht durch Entschmierungsprozeduren veränderte Werte eingesetzt. Der Halo ist deshalb durch Verschmierungseinfluß zu kleineren $\varkappa$-Werten verschoben, und zwar gilt

$$\varkappa^* = \sqrt{\varkappa_0^{*2} - 0{,}1^2}\,, \qquad [11]$$

wenn $\varkappa^*$ der beobachtete Beginn des Halos und $\varkappa_0^*$ der wahre Beginn des Halos ist. Bei PMMA ist $\varkappa^* \approx 0{,}3$ (Abb. 3), der Halo ist also in diesem Fall nicht wesentlich nach links verschoben. Der konstante Teil wird nicht verfälscht, während der bei $\varkappa < 0{,}03$ auftretende Anstieg der Intensität zu kleineren $\varkappa$-Werten hin durch den Verschmierungseinfluß abgeschwächt ist.

4. Ergebnisse

a) Lichtstreuung

Die Abb. 1 und 2, sowie Tabelle 1 zeigen die wesentlichen Ergebnisse der Streulichtmessungen an PDMS. Bei $\theta = 90°$ wird $H_v = V_h = H_h$ gefunden.

Zur Berechnung von R_v und R_u nach Gl. [1] und [2] wurden die Materialdaten von Tabelle 2 verwendet. Die isotherme Kompressibilität ist

Tabelle 1. Streulichtmessungen bei $\theta = 90°$, $\lambda = 4360$ Å und $T = 25\,°C$. R_u und R_v in 10^{-6} cm⁻¹

Substanz	R_v	R_v nach Gl. [2]	R_u	R_u nach Gl. [1]	σ_v	σ_u
Benzol	62,4*)	62,3	44,5*)	44,3	0,28	0,43
Toluol	69,4	72,1	52,8	54,4	0,34⁵	0,51
THF	18,2	20,6	10,1	11,4	0,056	0,107
PDMS 900	20,7	22,3	10,9	11,7	0,029	0,055
PDMS 5	24,9	26,0	12,8	13,6	0,029	0,050

*) Literaturwert (19, 20).

¹) Diese Messung wurde von R. *Oberthür* in unserem Labor durchgeführt.

Tabelle 2. Verwendete Daten zur Berechnung von R_u und R_v nach Gl. [1] bzw. [2] und von $R(0)$ nach Gl. [6]

Substanz	T (°C)	β (cm^2 dyn^{-1}) $\cdot 10^{11}$	n	ϱ
Benzol	25	9,65 (24, 25)	1,519	0,871
Toluol	25	9,2 (26)	1,515	
THF	25	10,5 (27)	1,413	
PDMS 900	25	11,8 (28)	1,416 (29)	0,97 (28)
PDMS 5	25	14 (28, 29)	1,403 (28, 29)	
PMMA	25	2,5 (30)		1,19
PMMA	110	5,0 (30)		1,18

bei THF und PDMS aus der adiabatischen Kompressibilität gemäß

$$\beta_{\text{isoth}} = \beta_{\text{ad}} + \frac{\alpha^2 T}{\varrho \, c_p} \qquad [12]$$

berechnet worden. Dabei ist α der isobare Ausdehnungskoeffizient und c_p die spez. Wärme bei konstantem Druck.

Für die Berechnung von $\varrho n \cdot \mathrm{d}n/\mathrm{d}\varrho$ wurde die *Eykmannsche* Beziehung (31) zwischen n und ϱ verwendet, welche bessere Resultate liefert als die analogen Gleichungen von *Clausius-Mosotti*, *Onsager*, *Gladstone-Dale* und *Laplace-Ramanathan* (20). Sie führt auf die Gleichung

$$\varrho \, n \cdot \frac{\mathrm{d}n}{\mathrm{d}\varrho} = \frac{n(n^2 - 1)(n + 0,4)}{n^2 + 0,8\,n + 1} . \qquad [13]$$

Beim Brechungsindex von THF und PDMS ($\approx 1,41$) sollte die Beziehung nur zu Fehlern von etwa 2% führen (20).

Die Streulichtmessungen an PDMS wurden bei einer zweiten Temperatur (50 °C) und bei einer weiteren Wellenlänge ($\lambda = 5460$ Å) wiederholt. Die Ergebnisse sind nicht wesentlich anders als die hier gezeigten. Die Temperaturabhängigkeit der Streuintensität ist in Übereinstimmung mit der Theorie.

Bezüglich der Streulichtmessungen an den PMMA-Präparaten verweisen wir auf die Arbeit von *Dettenmaier* und *Fischer* (10).

b) Röntgenkleinwinkelstreuung

Abb. 3 zeigt die Röntgenkleinwinkelmessungen dieser Arbeit (21). Die PMMA-Werte wurden an den Präparaten K_1 und K_7 erhalten, zwischen denen kein Unterschied im Streuverhalten festgestellt werden konnte. Ebenso stammen die PDMS-Werte von Messungen an PDMS W 1 und W 3. Es gibt also keinen Einfluß des

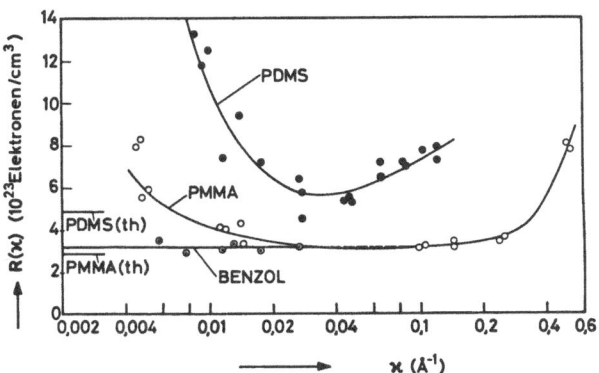

Abb. 3. Röntgenkleinwinkelstreuung an PMMA und PDMS. Zum Vergleich Benzol und die theoretischen $R(0)$-Werte. Bei PMMA wurde zur Berechnung von $R(0)$ nach Gl. [7] die Kompressibilität am Glaspunkt verwendet

Molekulargewichts auf das Streuverhalten. Dagegen besteht ein deutlicher Unterschied zwischen den PMMA-Präparaten K_1 und K_7 einerseits und dem Präparat R 240 andererseits: siehe hierzu Fig. 1 der zitierten Arbeit von *Wendorff* und *Fischer* (8): Obere Kurve = PMMA R 240, untere Kurve = PMMA K_7. Fig. 1 in der Arbeit unter Zitat 9 zeigt das Ergebnis an K_7 in Relation zur Röntgenkleinwinkelstreuung an anderen Polymeren. Daraus ergibt sich, daß die hier untersuchten Präparate (Abb. 3) außerordentlich schwach streuen. Leider standen bei der Durchführung der Röntgenexperimente die hochgereinigten PDMS-Präparate noch nicht zur Verfügung. Da die PDMS-Kurve in Abb. 3 keinen horizontalen Teil erhält, ist sie im gesamten Verlauf durch den Verschmierungseinfluß verfälscht.

Es wurde nach Gl. [7] $R(0)$ für Benzol, PMMA und PDMS berechnet (= theoretische Werte in Abb. 3). Der theoretische Benzolwert wurde weiter nach Gl. [9] als Eichwert für die experimentellen $R(\varkappa)$-Daten von Benzol, PMMA und PDMS verwendet. Hierdurch wird bei Benzol $R(0)_{exp}$ eo ipso identisch mit $R(0)_{th}$.

5. Diskussion

Die gezeigten Ergebnisse legen die Annahme nahe, daß rein amorphe Polymere im thermodynamischen Gleichgewicht Licht- und Röntgenlicht so beugen, wie es von der Schwankungstheorie für Flüssigkeiten vorhergesagt wird. In Abb. 1 und 2 sind die Abweichungen von der Theorie bei den bestgereinigten Präparaten geringfügig und leicht verständlich. Bei $\theta = 90°$ beträgt die Diskrepanz in R_v 13% bei THF, 8% bei PDMS 900 und 4% bei PDMS 5. Wesentlich ist hierbei der Unterschied in der Diskrepanz zwischen PDMS und THF. Ungenauigkeiten in den verwendeten Stoffkonstanten und in der Brechungskorrektur für das Streuvolumen, sowie Störreflexe können für die Diskrepanz bei THF verantwortlich sein. Der Unterschied von PDMS zu THF könnte dann auf einer restlichen Verunreinigung beruhen; die zähflüssigen Polysiloxane können ja nicht wie die niedermolekularen Lösungsmittel direkt vor der Messung durch Zentrifugation gereinigt werden. Ebenso muß der Anstieg zu kleinen Winkeln hin bei $\theta < 30°$ gedeutet werden.

Bemerkenswert ist, daß eine Mischung von PDMS-Präparaten sehr verschiedenen Molekulargewichts keine deutliche zusätzliche Exzeßstreuung verursacht (vgl. Abb. 2). Das gleiche Ergebnis ist für kleine Beimengungen von Monomeren und anderen niedermolekularen Flüssigkeiten gefunden worden. Daß die Exzeßstreuung bei PMMA ebenfalls stark von der Präparation abhängt, ist in vorhergehenden Arbeiten sowohl für die Licht- als auch für die Röntgenkleinwinkelstreuung gezeigt worden (8, 10). Allerdings ist es bei PMMA bisher nicht gelungen, Präparate mit sehr geringer Exzeßstreuung bei $\varkappa < 0,03$ herzustellen. Der feste Aggregatzustand bedeutet für die Präparation offenbar eine Erschwerung gegenüber PDMS. Eine detaillierte Deutung der Exzeßstreuung ist von *Dettenmaier* und *Fischer* (8) versucht worden. Im Bereich $0,03 < \varkappa < 0,2$ gehorcht die Röntgenkleinwinkelstreuung von PMMA der Theorie, falls unterhalb des Glaspunktes die Kompressibilität am Glaspunkt eingesetzt wird (Abb. 3 und Tabelle 2). Das Röntgenkleinwinkelstreuexperiment an PDMS hat nur orientierenden Charakter. Bei $\varkappa \approx 0,03$ ist die Streuintensität nicht deutlich höher als der theoretische Wert. Es ist deswegen zu vermuten, daß bei einem guten Präparat Übereinstimmung mit der Theorie gefunden worden wäre.

Zusammenfassung

Die Intensität der Licht- und der Röntgenkleinwinkelstreuung an amorphen Polymeren ist in der Regel höher, als nach der Schwankungstheorie für Flüssigkeiten zu erwarten ist. Der Exzeßwert der Streuung ist außerdem winkelabhängig. Durch besondere Sorgfalt bei der Präparation kann aber eine Annäherung an den theoretischen Wert erreicht werden. Bei PDMS, das wegen der flüssigen Konsistenz leichter als andere Polymere zu reinigen ist, wurde bei den bestgereinigten Präparaten noch eine Diskrepanz von etwa 5% gefunden. Deshalb liegt die Vermutung nahe, daß die Struktur von reinen amorphen Polymeren im thermodynamischen Gleichgewicht in Übereinstimmung mit der Schwankungstheorie ist. Bei glasförmigem PMMA wird nur dann eine Annäherung an die Theorie gefunden, wenn zur Berechnung der theoretischen Streuintensität die Kompressibilität am Glaspunkt eingesetzt wird.

Summary

The intensity of light and of small angle X-ray scattering by amorphous polymers is in general larger than can be estimated from the fluctuation theory for liquids. Moreover the excess value of scattering depends on the scattering angle. However, by careful preparation of the samples an approximation to the theoretical scattering intensity can be achieved. In PDMS which can be purified more easy on account of its liquid state, in one case only a discrepancy of 5% has been found. One can assume therefore that the structure of pure amorphous polymers in thermodynamical equilibrium is in agreement with the fluctuation theory. In glassy PMMA an approximation to the theory can be found only if for the calculation of the theoretical scattering power the compressibility at the glas transition point is taken.

Literatur

1) *Smoluchowski, M. V.*, Ann. Physik **25**, 205 (1908).
2) *Einstein, A.*, Ann. Physik **33**, 1275 (1910).
3) *Debye, P.* und *A. M. Bueche*, J. Appl. Phys. **20**, 518 (1949).
4) *Bueche, F.*, J. Appl. Phys. **23**, 1280 (1952).
5) *Peticolas, W. L., G. I. A. Stegeman* und *B. P. Stoicheff*, Phys. Rev. Letters **18**, 1130 (1967).
6) *Friedman, E. A., A. Y. Ritger* und *R. D. Andrews*, J. Appl. Phys. **40**, 4243 (1968).
7) *Romberger, A. B., D. P. Eastman* und *J. L. Hunt*, J. Chem. Phys. **51**, 3723 (1969).
8) *Wendorff, J. H.* und *E. W. Fischer*, Kolloid-Z. u. Z. Polymere **251**, 876 (1973).
9) *Wendorff, J. H.* und *E. W. Fischer*, ibid **251**, 884 (1973).

10) *Dettenmaier, M.* und *E. W. Fischer*, ibid **251**, 922 (1973).
11) *Carr, C. I.*, Jr. und *B. H. Zimm*, J. Chem. Phys. **18**, 1616 (1950).
12) *Guinier, A.*, X-ray diffraction (San Francisco-London, 1963).
13) *Guinier, A.* und *G. Fournet*, Small angle scattering of X-rays (New York-London, 1955).
14) *Hill, T. L.*, Statistical Mechanics (New York, 1956).
15) *Lee, C. L, C. L. Frye* und *O. K. Johannson*, Polymer Preprints **10**, 2, 1361 (1969).
(15a) *H. J. Hölle* und *B. R. Lehnen*, Europ. Polym. J. im Druck.
16) *Küchler, L.*, Polymerisationskinetik (Heidelberg, 1951).
17) *Kirste, R. G., W. A. Kruse* und *K. Ibel*, Polymer **16**, 120 (1975).
18) *Common, D. J., E. L. Mackor* und *J. Hijmans*, Trans. Faraday. Soc. **60**, 1539 (1964).
19) *Cantow, H. J.*, Makromol. Chem. **18/19**, 367 (1956).
20) *Kerker, M.*, The Scattering of Light (New York, London, 1969).
21) *Steinbach, M.*, Dissertation Mainz (1973).
22) *Kirste, R. G.* und *O. Kratky*, Z. physik. Chem. (Frankfurt) **31**, 363 (1962).
23) *Wunderlich, W.* und *R. G. Kirste*, Z. Elektrochem., Ber. Bunsenges. physik. Chem. **68**, 646 (1964).
24) *Holder, G. A* und *E. Whalley*, Trans. Faraday Soc. **58**, 2095 (1962).
25) *Deželić, G.*, J. Chem. Physics **45**, 185 (1966).
26) *Freyer, E. B., J. C. Hubbard* und *D. H. Andrews*, J. Am. Chem. Soc. **51**, 759 (1929).
27) *Weissler, A.*, JACS **71**, 419 (1949).
28) *Noll, W.*, Chemie und Technologie der Silicone (Weinheim, 1960).
29) *Barry, A. J.* und *H. N. Beck* in *Stone, F. G. A.* und *W. A. G. Graham*, Inorganic Polymers (New York-London, 1962).
30) *Hellwege, K. H., W. Knappe* und *P. Lehmann*, Colloid & Polymer Sci. **183**, 110 (1962).
31) *Eykman, J. F.*, Rec. Trav. Chim. **14**, 177 (1895).
32) *Hendricks, R. W., P. G. Mardon* und *L. B. Shaffer*, J. Chem. Physics **61**, 319 (1974).
33) *Thomas, P.*, Z. Physik **208**, 338 (1968).

Für die Verfasser:

Prof. *R. G. Kirste*
Chemisches Institut der Universität Mainz
6500 Mainz

Progr. Colloid & Polymer Sci. **58**, 36—43 (1975)

Institut für Nichtmetallische Werkstoffe der Technischen Universität Berlin

Lineare Parakristalle mit bimodaler Koordinationsstatistik

R. Bonart

Mit 8 Abbildungen und 1 Tabelle

(Eingegangen am 15. Juni 1974)

I.

Bei der elastischen Dehnung teilkristalliner orientierter Polymerer weitet sich die Langperiode teils mehr, teils weniger auf, als der makroskopischen Probendehnung entspricht (1). Experimentelle Basis hierfür sind entsprechende Röntgenkleinwinkelinterferenzen, aus deren Winkellage mit Hilfe der *Bragg*-Gleichung auf die Langperiode geschlossen wird. Dies ist jedoch problematisch, da die *Bragg*-Gleichung zunächst nur für idealperiodische Strukturen abgeleitet worden ist, die betrachteten Kolloidstrukturen aber keinesfalls idealperiodisch sind, so daß mit Abweichungen von der *Bragg*-Gleichung zu rechnen ist.

Entsprechende Abweichungen sind in der Literatur u. a. bei sehr kleinen Kristalliten diskutiert worden (2), wo der Strukturfaktor zu Reflexverschiebungen Anlaß geben kann. Ferner bei sehr stark gestörten Strukturen (3), deren Interferenzdiagramm nur noch Interferenzen 1. Ordnung zeigen, und schließlich bei Strukturen mit unsymmetrischen Koordinationsstatistiken (4). Darüber hinaus sollen im folgenden Strukturen mit bimodalen Koordinationsstatistiken betrachtet werden.

Spezieller Anlaß hierzu sind besonders krasse Abweichungen von der *Bragg*-Gleichung im Kleinwinkeldiagramm von gedehntem α-Keratin (s. Abb. 1) sowie bei der chemischen Modifizierung von α-Keratinen (Tabelle 1) (5). In beiden Fällen treten Reflexverschiebungen auf, die an Hand der *Bragg*-Gleichung keine befriedigende Deutung zulassen. Dies zumindest dann nicht, wenn man davon ausgeht, daß alle Reflexe von der gleichen Grundperiode herrühren, sofern man nicht eine unrealistisch große Grundperiode und dementsprechend unrealistisch hohe Indizierungen der beobachtbaren Reflexe annehmen

will (6). *Spei* postuliert deshalb ein Matrixmodell und identifiziert die Kleinwinkelreflexe an Hand von drei verschiedenen idealen Perioden teils als Fibrillen-, teils als Matrixreflexe (5). Er nimmt an, daß neben den Fibrillen auch die

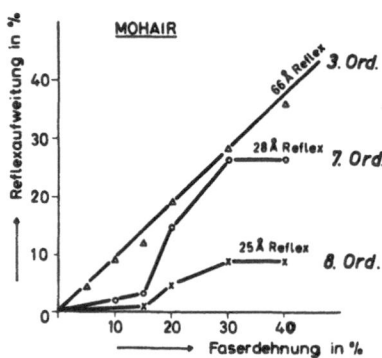

Abb. 1. Aufweitung der *Bragg*-Abstände im Röntgenkleinwinkeldiagramm von gedehntem α-Keratin (Mohairfasern in 2.2.2-Trifluoräthanol) [nach *Spei* (5)]

Tabelle 1. Meridionale Netzebenenabstände im Röntgenkleinwinkeldiagramm von chemisch behandeltem Mohair [nach *Spei* (5)]

Ordnungszahl des Reflexes n	Nitriertes Mohair behandelt mit Natrium-Dodecylsulfat $D\,[\text{Å}]$	n. D. [Å]	DNP-Mohair behandelt mit Natrium-Decylsulfat $D\,[\text{Å}]$	n. D. [Å]
1	200	200	— nicht beobachtet —	
2	86,6	173,2	92,5	185,0
3	61,2	183,6	64,9	194,7
5	36,5	182,5	38,3	191,5
6			32,3	193,8
7			28,0	196,0
8			24,5	196,0
10			19,6	196,0

D = Netzebenenabstand.

n. D. = Grundperiode.

Matrixsubstanz orientiert und periodisch aufgebaut ist und daß Matrix und Fibrillen sowohl auf mechanische Belastungen wie auch auf chemische Markierungen unterschiedlich reagieren, so daß man es tatsächlich nicht mit nur einer, sondern mit mehreren Grundperioden zu tun hat. Bei genauerer Betrachtungsweise führt jedoch auch das Matrixmodell zu erheblichen Schwierigkeiten. Unter anderem sollte man erwarten, daß jede der drei beteiligten Grundperioden zu einem mehr oder weniger vollständigen Reflexsystem mit höheren Ordnungen Anlaß gibt, so daß drei sich überlagernde Reflexsysteme beobachtbar sein müßten. Tatsächlich beobachtet man nur ein einziges Reflexsystem, das jedoch, wie gesagt, mit der *Bragg*-Gleichung nicht konsistent ist. Es ist deshalb von Interesse, daß entsprechende Abweichungen von der *Bragg*-Gleichung ohne Zuhilfenahme des Matrixmodelles auch auf inhomogene Dehnungen bzw. inhomogene ,,Quellungen" oder ,,Kontraktionen" zurückgeführt werden können. Ohne direkten Bezug auf ein spezielles experimentelles Ergebnis soll dies in der vorliegenden Arbeit näher erläutert werden. Die entsprechenden Befunde am α-Keratin sollen erst in einer späteren Arbeit diskutiert werden.

II.

Inhomogene Dehnungen sind dadurch gekennzeichnet, daß sich belastete Fibrillen nicht gleichmäßig, sondern ungleichmäßig längen, so daß gelängte und ungelängte Fibrillenabschnitte mehr oder weniger statistisch aufeinanderfolgen. Man kann sich dies hypothetisch so vorstellen, daß die Kettenmoleküle anfänglich eine kontrahierte Konformation besitzen, die bei der Belastung gleichsam sprunghaft in eine gestreckte Konformation übergeht, wobei die Konformationsänderung in verschiedenen Fibrillenabschnitten nacheinander, d.h. bei unterschiedlichen Gesamtdehnungen der Fibrillen er-

folgen. Liegen bei einem bestimmten Dehnungsgrad der Fibrillen α-% aller Fibrillenabschnitte in gestreckter bzw. $(1 - \alpha)$-% aller Fibrillenabschnitte in kontrahierter Konformation vor, so kann sich bei fortschreitender Gesamtdehnung der Prozentsatz an gestreckten Konformationen erhöhen, ohne daß sich die bereits gestreckten Fibrillenabschnitte noch weiter längen.

In Abb. 2 ist eine inhomogene aufgeweitete Struktur mit den Periodenlängen c_1 und c_2 grob schematisch wiedergegeben.

Die Konformationsänderungen brauchen sich keinesfalls, wie durch Abb. 2 nahegelegt wird, jeweils nur auf einzelne isolierte Perioden zu erstrecken. Vielmehr können sich auch mehrere zusammenhängende Perioden oder Bruchteile hiervon gemeinsam längen, während andere Bruchteile von Perioden ungelängt bleiben. Sind Bruchteile von Perioden mit im Spiel, so hat man trotz definierter Konformationsänderungen neben den Periodenlängen c_1 und c_2 der voll kontrahierten bzw. der voll gestreckten Konformationen entsprechende Mittelwerte für die Periodenlängen zu berücksichtigen.

In ähnlicher Weise wie bei mechanischen Belastungen können sich inhomogen gestörte Strukturen auch bei chemischen Markierungsversuchen ausbilden, wenn sich beispielsweise einzelne Fibrillenabschnitte in unterschiedlicher Weise durch ,,Quellung" oder ,,Entquellung" längen bzw. gegebenenfalls kontrahieren.

Abb. 2 b stellt eine weitere Schematisierung der Struktur Abb. 2 dar, indem man nur die Anfangs- bzw. Endpunkte der Perioden betrachtet, die im folgenden als Gitterpunkte aufgefaßt werden. Die dazugehörige Koordinationsstatistik Abb. 3 gibt die statistische Verteilung aller Abstände zwischen unmittelbar benachbarten Gitterpunkten wieder. Sie setzt sich aus zwei Teilstatistiken $g_1(x - c_1)$ und $g_2(x - c_2)$ zusammen, die sich auf die ungelängten bzw. gelängten Perioden beziehen, wobei die Parameter β_1 bzw. β_2 eventuelle Schwankungen um c_1 und c_2 beschreiben. Mit den Gewichtsfaktoren

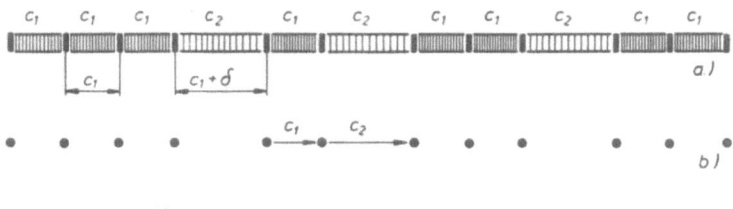

Abb. 2. Schematische Darstellung einer inhomogen aufgeweiteten Struktur mit den Perioden c_1 und c_2

Koordinationsstatistik h(x)

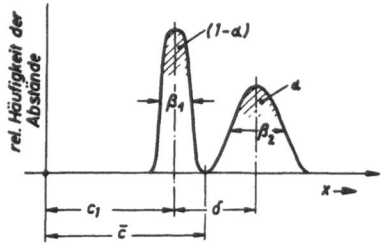

$x \triangleq$ *Abstand zwischen benachbarten*
Gitterpunkten
$\bar{c} = c_1 + \alpha\, d \triangleq$ *mittlerer Koordinations-*
abstand

Abb. 3. Bimodale Koordinationsstatistik einer inhomogen gestörten Struktur. $(1 - \alpha)$ bzw. (α) geben die statistische Häufigkeit an, mit der die Perioden c_1 bzw. $c_2 = c_1 + \delta$ auftreten

α bzw. $(1 - \alpha)$ ergibt sich für die bimodale Koordinationsstatistik

$$h(x) = (1 - \alpha)\, g_1(x - c_1) + \alpha\, g_2(x - c_2), \qquad [1]$$

wobei die $g_i(x)$ im einfachen Fall durch normierte *Gauß*-Kurven gegeben sind

$$g_i(x) = \frac{1}{\sqrt{\pi}\, \beta_i} \cdot \frac{\sqrt{\pi\, \beta_i}}{1}\, e^{-\left(\frac{x}{\beta_i}\right)^2}.$$

In der vorliegenden Arbeit beschränken wir uns einfachheitshalber zunächst auf den Fall $\beta_1 = \beta_2 = 0$. Für die mittlere Periode der bimodal gestörten Struktur gilt dann:

$$\bar{c} = (1 - \alpha)\, c_1 + \alpha\, c_2 = c_1 + \alpha \cdot \delta$$

mit $\delta = c_2 - c_1$,

für die relative Strukturaufweitung gegenüber der mit c_1 idealperiodischen Struktur $(\alpha = 0)$:

$$\varepsilon = \frac{\bar{c} - c_1}{c_1} = \frac{\alpha \cdot \delta}{c_1}$$

bzw. für die relative Gitterstörung

$$g = \frac{\sqrt{\overline{\Delta x}}}{\bar{c}} = \sqrt{\alpha(1 - \alpha)}\, \frac{\delta}{c_1 + \alpha \cdot \delta}\,.$$

III.

Unabhängig von der speziellen Form der Koordinationsstatistik (Gl. [1] bzw. Abb. 3) können die unterschiedlich langen Perioden (beispielsweise c_1 oder c_2) mehr oder weniger statistisch oder aber in Blöcken aufeinanderfolgen. Falls ausreichend lange Blöcke vorliegen,

wirken diese wie aneinandergereihte, homogen aufgebaute „Kristallite" beispielsweise mit den Gitterkonstanten c_1 oder c_2. In gleicher Weise wie beim Matrixmodell müßten in diesem Fall im Interferenzdiagramm voneinander unabhängige Reflexsysteme auftreten, die sich mit Hilfe der *Bragg*-Gleichung auf die betreffenden Gitterkonstanten zurückführen lassen. Folgen die unterschiedlich langen Perioden dagegen statistisch aufeinander, so hat man nur ein einziges Reflexsystem zu erwarten, für das sich allerdings mit Hilfe der *Bragg*-Gleichung im allgemeinen keine einheitliche Grundperiode angeben läßt.

Im Falle des α-Keratins sprechen die bisherigen Befunde für eine ungeordnete, statistische Aufeinanderfolge wechselnder Gitterabstände. Voneinander unabhängige Reflexsysteme, die jeweils in sich mit der *Bragg*-Gleichung konsistent sind, werden nicht beobachtet, wohl aber ein einziges Reflexsystem, das nicht exakt, sondern nur näherungsweise mit der *Bragg*-Gleichung verträglich ist. Wir konzentrieren uns deshalb im folgenden auf eine rein statistische Aufeinanderfolge längerer und kürzerer Gitterabstände, wie es den Bedingungen des idealen Parakristalles entspricht (7).

Der Gitterfaktor des eindimensionalen idealen Parakristalles lautet

$$K(b) = \frac{1 - |H(b)|^2}{1 - 2\,\mathbf{Re}\,H(b) + |H(b)|^2} \qquad [2]$$

mit $b = \dfrac{2 \sin \Theta}{\lambda}$; $\quad 2\,\Theta \triangleq$ Streuwinkel,
$\quad\quad\quad\quad\quad\quad\quad\lambda \triangleq$ Wellenlänge der Röntgen-
$\quad\quad\quad\quad\quad\quad\quad\quad\quad\quad$ strahlung,

wo der sog. Statistikfaktor

$$H(b) = \int\limits_{-\infty}^{+\infty} h(x)\, e^{-2\pi i b x}\, dx$$

durch Fouriertransformation aus der Koordinationsstatistik $h(x)$ folgt.

Bei langsam veränderlichem $|H(b)|^2$ ergeben sich Maxima im Gitterfaktor überall dort, wo der Nenner Minima, d.h. wo $\mathbf{Re}\,H(b)$ Maxima hat. Bei schnell veränderlichem $|H(b)|^2$ gelten dagegen andere Zusammenhänge.

Langsam veränderliche $|H(b)|^2$ treten vor allem bei homogenen Koordinationsstatistiken auf. Mit $\alpha = 0$, $\bar{c} = c_1$ gilt

$$\mathbf{Re}\,H(b) = e^{-(\pi b \delta)^2} \cos 2\pi b\, \bar{c}. \qquad [3]$$

Falls δ genügend klein ist, kann der e-Term als langsam veränderlich gegenüber $\cos 2\pi b\, \bar{c}$ angesehen werden, so daß die Maxima in Re $H(b)$ und damit im Gitterfaktor direkt durch

$$\cos 2\pi b\,\bar{c} = 1,$$

also durch

$$b\,\bar{c} = l \quad (l \triangleq \text{ganzzahlig}) \qquad [4]$$

gegeben sind, was mit der *Bragg*-Gleichung identisch ist. Bis zu einem Störungsgrad von etwa 30% ($g \approx 0{,}3$; $\beta \approx 0{,}3 \cdot \sqrt{2} \cdot \bar{c}$) haben homogene Gitterstörungen praktisch keinen Einfluß auf die Reflexlagen, so daß die *Bragg*-Gleichung gültig bleibt. Sie wirken sich ausschließlich in den Reflexbreiten aus, die mit der Ordnungszahl der Reflexe monoton anwachsen, wodurch die Zahl der separierbaren Maxima begrenzt ist

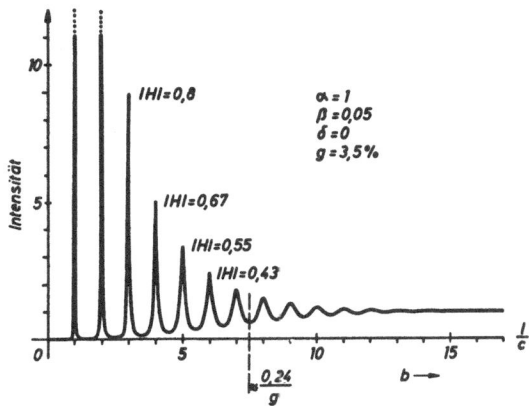

Abb. 4. Gitterfaktor eines inhomogen gestörten Parakristalles

(s. Abb. 4). Als Richtwert für den Bereich der separierbaren Maxima gibt *Hosemann* (7) die Beziehung

$$l \leqq 0{,}24/g \qquad [5]$$

an, die selbstverständlich nur qualitative Bedeutung haben kann. Falls $g > 0{,}3$ ist, fällt der e-Term in Gl. [3] so schnell ab, daß die Oszillationen des cos-Termes nicht mehr zum Tragen kommen und damit weder diskrete Maxima in Re $H(b)$ noch im Gitterfaktor auftreten, so daß auch die *Bragg*-Gleichung nicht mehr sinnvoll anwendbar ist.

Im Falle einer bimodalen symmetrischen Koordinationsstatistik ($\alpha = 0{,}5$) mit $\beta_1 = \beta_2 = 0$ lautet der Realteil des Statistikfaktors

$$\text{Re}\,(b) = \cos \pi b\,\delta \cos 2\pi b\,\bar{c}. \qquad [6]$$

Falls δ genügend klein gegenüber der mittleren Periode $\bar{c}$ ist, kann auch hierbei $|H(b)|^2$ als

langsam veränderlich gelten. Trotzdem ergeben sich gravierende Unterschiede gegenüber der homogenen Störung, da $\cos \pi b\,\delta$ oszillierend positive und negative Werte annimmt, während der e-Term in Gl. [3] monoton auf Null abfällt. Wo $\cos \pi b\,\delta$ negativ ist, sind die Maxima von Re $H(b)$ nicht durch $\cos 2\pi b\,\bar{c} = +1$, sondern durch $\cos 2\pi b\,\bar{c} = -1$ gegeben, so daß die Reflexlage statt durch die *Bragg*-Gleichung Gl. [4] durch

$$b\,\bar{c} = l + 0{,}5$$

festgelegt ist (vgl. Abb. 5). Obwohl die zugrunde liegende Koordinationsstatistik symmetrisch und $|H(b)|^2$ langsam veränderlich ist, treten Reflexverschiebungen um die halbe reziproke Periode auf, so daß weder die Symmetrie noch die langsame Veränderlichkeit von $|H(b)|^2$ ein ausreichendes Kriterium für die Gültigkeit der *Bragg*-Gleichung darstellen. Ferner sind wegen der Oszillation von $\cos \pi b\,\delta$ bis zu beliebig hohen Ordnungen immer wieder scharfe Reflexe beobachtbar, im Gegensatz zu der Relation Gl. [5]. Umgekehrt darf deshalb aus scharfen Reflexen hoher Ordnung keinesfalls direkt auf eine störungsfreie Struktur geschlossen werden. In der Umgebung um $b \approx 0$ kann $\cos \pi b\,\delta$ mit einem geeigneten β durch eine *Gauß*kurve, also durch

$$\cos \pi b\,\delta \approx e^{-(\pi b\beta)^2}; \quad b \approx 0 \qquad [7]$$

angenähert werden. Vergleichbare homogen bzw. inhomogen gestörte Strukturen haben deshalb im sog. Homogenitätsbereich, d.h. im Gültigkeitsbereich der Näherung Gl. [7] praktisch identische Gitterfaktoren, so daß insbesondere auch die *Bragg*-Gleichung in diesem Bereich gilt. Der Homogenitätsbereich erstreckt sich in den Rechenbeispielen Abb. 5a und b bis zur 4. bzw. zur 2. Ordnung.

Der exakte Verlauf des Gitterfaktors ergibt sich in Abhängigkeit vom jeweiligen Statistikfaktor aus Gl. [2]. Die qualitative Diskussion wird jedoch insbesondere in komplizierteren Fällen wesentlich erleichtert, wenn man neben den reziproken Gitterpunkten $l/\bar{c}$ auch die Rasterpunkte l'/c_1 und l''/c_2 in Rechnung stellt. Es zeigt sich nämlich, daß überall dort, wo die Raster l' und l'' koinzidieren, scharfe Gittermaxima auftreten, während die Gittermaxima in den dazwischenliegenden Bereichen breit und diffus sind (vgl. (5)).

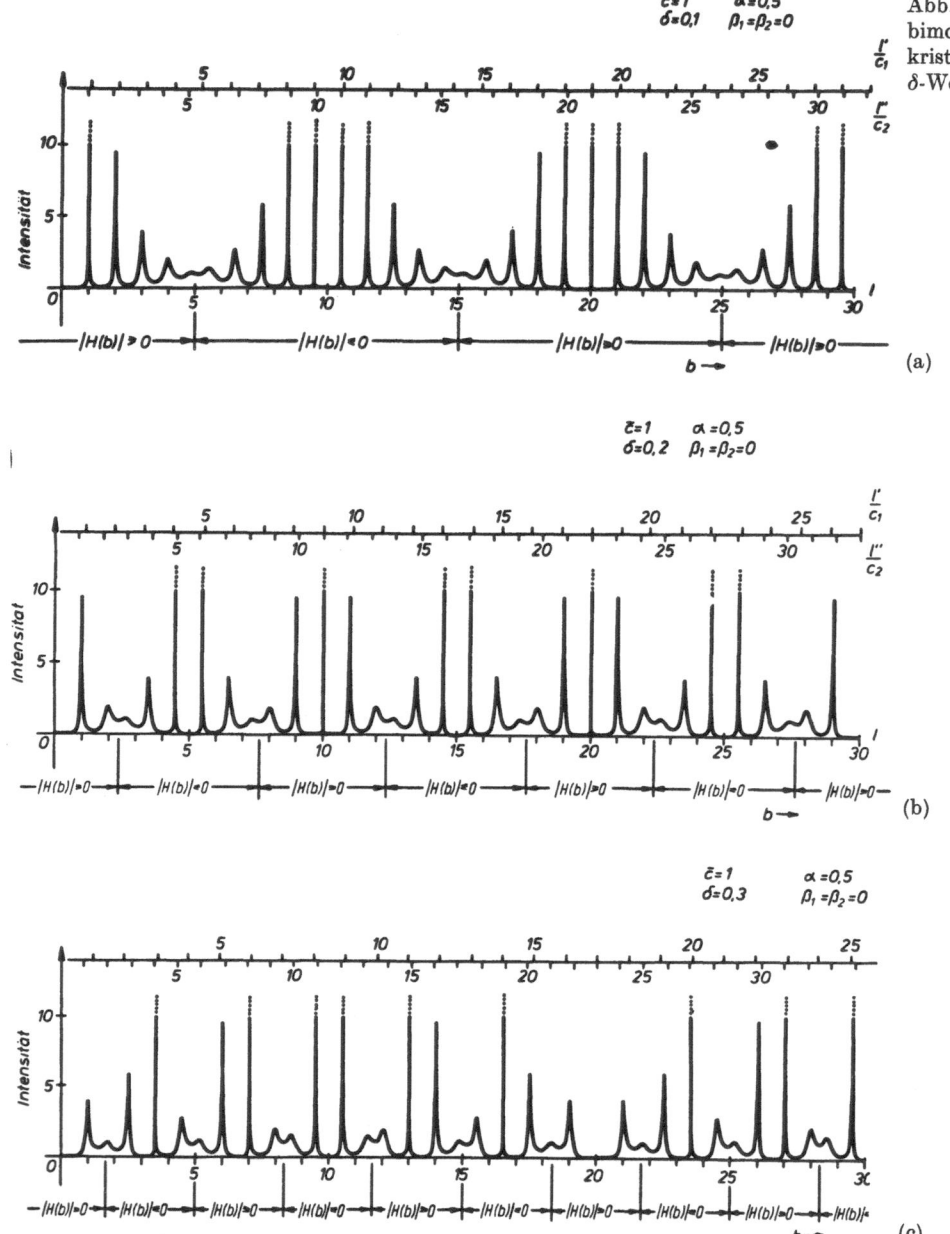

Abb. 5. Gitterfaktoren bimodal gestörter Parakristalle mit „kleinen" δ-Werten.

Die Aufeinanderfolge dieser Koinzidenzbereiche ist durch die Differenz $\delta = c_2 - c_1$ gegeben. Mit wachsendem δ werden der Homogenitäts- und die Koinzidenzbereiche immer kürzer, bis sie explizit nicht mehr erkennbar sind (s. Abb. 5c). Nach wie vor treten jedoch scharfe Maxima jeweils nur dort auf, wo sich Koinzidenzen zwischen den Rastern l' und l'' ergeben, während das Raster l, d.h. die reziproken Gitterpunkte vergleichsweise bedeutungslos sind. Mit Hilfe der *Bragg*-Gleichung können

deshalb allenfalls die Perioden c_1 oder c_2 gefunden werden, nicht jedoch die tatsächliche mittlere Periode $\bar{c}$. Aber auch bezüglich der Perioden c_1 und c_2 ergeben sich Abweichungen von der *Bragg*-Gleichung, da die Gittermaxima in den Bereichen unvollständiger Koinzidenz weder exakt in den Rasterpunkten l' noch in den Rasterpunkten l'' liegen.

Die bessere oder schlechtere Koinzidenz der Raster l' und l'' wirkt sich wie ein zusätzlicher Strukturfaktor aus. Dies ist insbesondere bei chemischen Markierungsversuchen zu beachten, da veränderte Reflexintensitäten keinesfalls direkt auf einen veränderten Strukturfaktor, d.h. auf definierte Anlagerungen der Markierungssubstanz an spezielle chemische Gruppen zurückgeführt werden können. Vielmehr können auch lokale Quellungen oder Kontraktionen einzelner Fibrillenabschnitte mit im Spiel sein, die zu gegenseitigen Verschiebungen der Raster l' und l'' und zu entsprechend veränderten Koinzidenzen führen, wobei die mittlere Periode konstant bleibt oder sich auch geringfügig ändern kann. Derartiges ist insbesondere dann zu vermuten, wenn sich mit der Intensität auch die Breiten und die *Bragg*-Abstände der verschiedenen Reflexe in unterschiedlicher Weise ändern, wie es beim α-Keratin offensichtlich der Fall ist (Tabelle 1). Entsprechende Rechenbeispiele, allerdings mit konstantem $\bar{c}$, sind in Abb. 6 wiedergegeben, wo die Änderungen der Reflexintensitäten in keiner Weise durch einen Strukturfaktor, sondern durch unterschiedliche Koinzidenzen der Raster l' und l'' bedingt sind.

Bei symmetrischen Koordinationsstatistiken ($\alpha = 0{,}5$; $\beta_1 = \beta_2$) haben die beiden Raster l' und l'' das gleiche Gewicht, während sie bei unsymmetrischen Statistiken ($\alpha \neq 0{,}5$; oder/und $\beta_1 \neq \beta_2$) mit unterschiedlichen Gewichten in

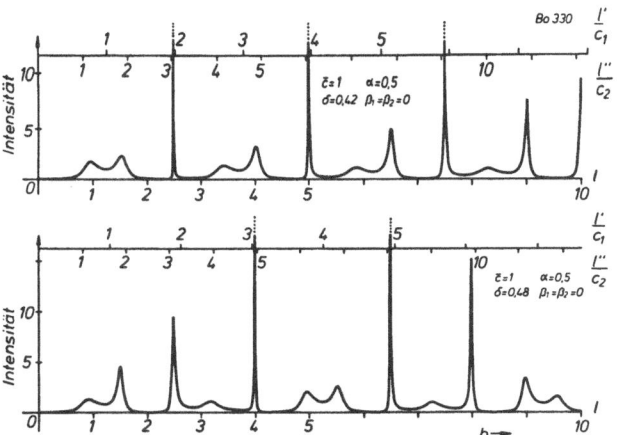

Abb. 6. Gitterfaktoren bimodal gestörter Parakristalle mit „großen" δ-Werten. Bei konstanter mittlerer Periode hat eine Änderung von δ markante Änderungen der Reflexintensitäten und teilweise auch Reflexverschiebungen zur Folge

den Interferenzeffekt eingehen. Falls $\beta_1 = \beta_2 = 0$ ist, setzt sich im Gitterfaktor entweder das Raster l' oder l'' durch, je nachdem, ob α kleiner oder größer als 0,5 ist. Als Beispiel hierfür mögen die Abb. 7 dienen, die sich auf die gleiche mittlere Periode beziehen, trotzdem aber charakteristisch unterschiedliche Reflexlagen beschreiben.

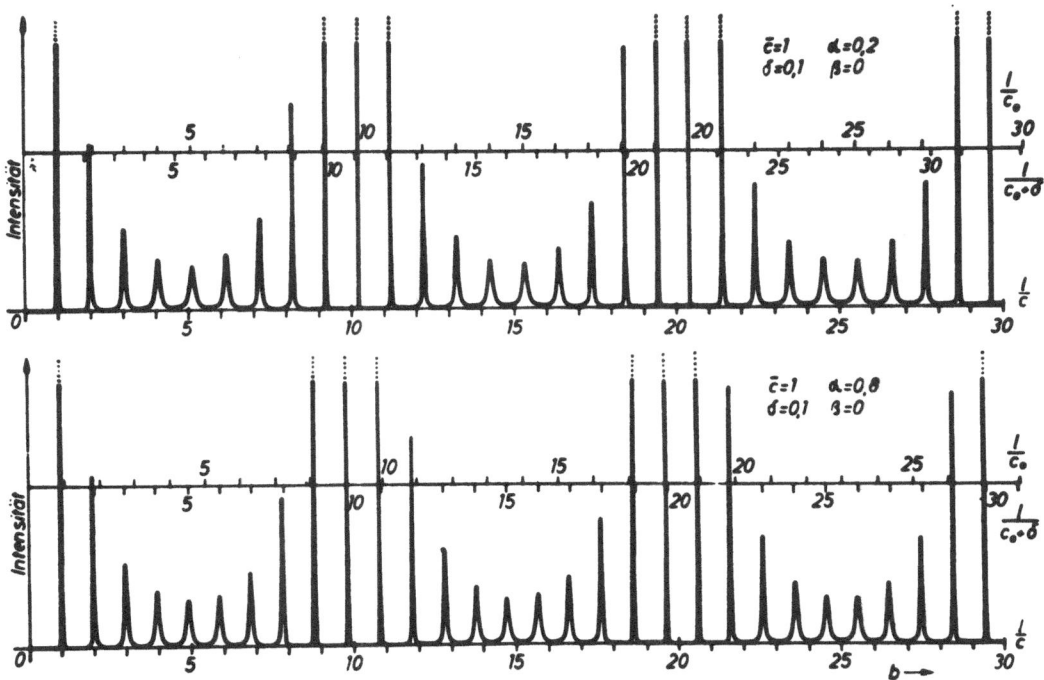

Abb. 7. Gitterfaktoren bimodal gestörter Parakristalle mit unsymmetrischer Koordinationsstatistik. Der jeweils dominierende Mode setzt sich in der Reflexlage durch

In Abb. 7a ist die mittlere Periode $\bar{c} = 1$ durch $c_1 = 0,98$ und $c_2 = 1,08$ mit den Gewichtsfaktoren $(1 - \alpha) = 0,8$ bzw. $\alpha = 0,2$ realisiert, so daß sich im Gitterfaktor die Periode $c_1 = 0,98$ durchsetzt. Alle Gittermaxima liegen rechts von den dazugehörigen reziproken Gitterpunkten.

In Abb. 7b ist die gleiche mittlere Periode $\bar{c} = 1$ durch $c_1 = 0,92$ und $c_2 = 1,02$ mit den Gewichtsfaktoren $(1 - \alpha) = 0,2$ bzw. $\alpha = 0,8$ realisiert. Im Gitterfaktor dominiert jetzt die Periode $c_2 = 1,02$. Trotz der gleichen mittleren Periode wie in Abb. 7a liegen die Reflexe nun links von den dazugehörigen reziproken Gitterpunkten.

Im ersten Fall setzt sich die Periode $c_1 = 0,98$, im zweiten Fall dagegen die Periode $c_2 = 1,02$ durch, obwohl die mittlere Periode in beiden Fällen gleich eins ist. Die unterschiedlichen Reflexlagen rühren also ausschließlich von einer unterschiedlichen Realisierung ein und derselben mittleren Periode her.

Wie man bei genauerer Betrachtung erkennt, bilden sich die Reflexverschiebungen jeweils in den Übergängen zwischen benachbarten Koinzidenzbereichen aus. In den Homogenitäts- und den Koinzidenzbereichen selber ist der gegenseitige Abstand zwischen benachbarten scharfen Maxima nicht durch die dominierende, sondern durch die mittlere Periode, d.h. durch $1/\bar{c}$ gegeben. Sofern unmittelbar benachbarte scharfe Gittermaxima beobachtbar sind, kann deshalb

zwar nicht aus ihrer Lage, wohl aber aus ihrem gegenseitigen Abstand auf die mittlere Periode zurückgeschlossen werden, während ihre Lage die dominierende Periode anzeigt.

In den Gitterfaktoren Abb. 7 sind die Homogenitäts- und die Koinzidenzbereiche klar voneinander getrennt. Geht man zu größeren δ-Werten über (Abb. 8), so treten statt zusammenhängender Bereiche unregelmäßig aufeinanderfolgende Einzelkoinzidenzen, wie in Abb. 6, auf, wobei sich je nachdem c_1 bzw. c_2 durchsetzt.

Im Unterschied zu den bisherigen Rechenbeispielen beziehen sich die Gitterfaktoren Abb. 8 nicht auf eine konstante mittlere Periode, sondern auf konstante Perioden c_1 und c_2, wobei die mittleren Perioden $\bar{c} = 1,244$ bzw. $\bar{c} = 1,366$ sind. Gegenüber einer hypothetischen idealperiodischen Ausgangsstruktur $(\alpha = 0)$ ist die der Abb. 8 zugrunde liegende Struktur um ca. 24%, die der Abb. 8b zugrunde liegende Struktur um ca. 37% aufgeweitet. Wegen des inhomogenen Charakters der Aufweitung macht sich diese jedoch nicht in kontinuierlichen Reflexverschiebungen bemerkbar, sondern in einer Verbreiterung einzelner Reflexe (Abb. 8) bzw. im Auftreten zusätzlicher neuer Intensitätsmaxima (Abb. 8). Insbesondere durch Abb. 8 wird deutlich, daß inhomogene Dehnungen (im vorliegenden Fall bis ca. 20%) völlig verborgen bleiben können, wenn man nur die Lage der scharfen Reflexe beobachtet.

Ähnlich wie unterschiedliche Gewichtsfaktoren wirken sich auch unterschiedliche Parameter $\beta_1 = \beta_2$ aus. Hierauf soll jedoch erst später in Zusammenhang mit speziellen experimentellen Ergebnissen eingegangen werden.

Die Rechnungen wurden auf einer Hewlett-Packard 9820 A durchgeführt. Frau *Scanelli* danke ich für die Erstellung des Programmes und die Ausführung der Rechnung.

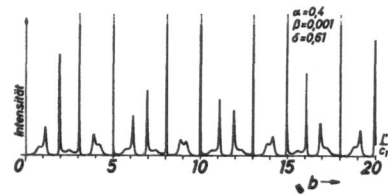

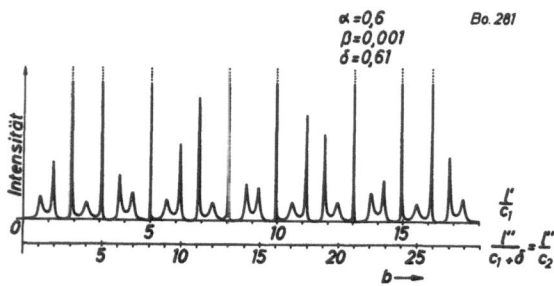

Abb. 8. Gitterfaktoren, die im Verlauf einer inhomogenen Dehnung auftreten. Die Strukturaufweitung kommt praktisch ausschließlich in den Reflexbreiten, nicht aber in den Reflexlagen zum Ausdruck

Zusammenfassung

Im Röntgenkleinwinkeldiagramm von gedehntem α-Keratin beobachtet man Reflexverschiebungen, die weder der makroskopischen Probendehnung entsprechen noch vergleichbare Werte für die verschiedenen Reflexordnungen zeigen. Um zu klären, ob dies eventuell auf inhomogene Dehnungen der Fibrillen zurückzuführen ist, wird in der vorliegenden Arbeit die Theorie des idealen Parakristalles für bimodale Koordinationsstatistiken diskutiert. Dabei zeigt sich, daß teilweise recht erhebliche Abweichungen von der *Bragg*-Gleichung auftreten können. Neben den reziproken Gitterpunkten werden reziproke Rasterpunkte eingeführt, die den beiden Moden der Koordinationsstatistik entsprechen. Scharfe und intensive Gittermaxima treten überall dort auf, wo die reziproken Rasterpunkte der Moden miteinander koinzidieren (Koinzidenzbereiche), weitgehend unabhängig von der Lage der reziproken Gitterpunkte. Bei unsymmetrischen Koordinationsstatistiken setzt sich das Raster der dominierenden Mode verstärkt durch.

Die in der vorliegenden Arbeit allgemein dargestellten Zusammenhänge sollen in einer folgenden Arbeit auf α-Keratin angewendet werden.

Summary

Under load the *Bragg*-spacings in the small angle X-ray diagram of α-Keratin don't increase neither proportional to the macroscopic strain nor even by a constant ratio for different orders. To clearify whether this could be due to an inhomogeneous strain along the fibrils, the theory of ideal paracrystals is discussed in this paper for the case of bimodal statistics. Rather remarkable deviations from *Bragg*'s Law are found. Besides the well known Reciprocal Lattice Points so called Reciprocal Elementary Points are introduced corresponding to the different modes of the statistics. Sharp and intense maxima in the interference diagram are found, where the Reciprocal Elementary Points coincide with each other (coincidence regions). The maxima are the sharper the better the coincidence is. Their position is almost independent of the position of the Reciprocal Lattice Points. With unsymmetric bimodal statistics the two sets of Reciprocal Elementary Points gets differently pronounced.

The more general relationships given in this paper will be specialized to the X-ray findings with α-Keratin in a following one.

Literatur

1. *Zahn, H.,* und *U. Winter,* Kolloid-Z. u. Z. Polymere **128**, 142 (1952).
   D. R. Beresford und *H. Bevan,* Polymer **5**, 247 (1964).
   Bonart, R., Kolloid-Z. u. Z. Polymere **231**, 438 (1969).
   Tuichev, Sh., N. Sutanov, B. M. Ginzburg und *S. Ya. Frenkel,* Pol. Sci. USSR **12**, 2298 (1970).
   Brestkin, Yu. V., B. M. Ginzburg und *K. B. Kurbanov,* Pol. Sci. USSR **13**, 1966 (1971).
2. *Wallner, L. G.,* Mh. f. Chem. **79**, 271 (1947); zitiert in *Stuart* Physik d. Hochpolymeren III, S. 67 (Berlin-Heidelberg-New York 1955).
3. *Bonart, R.,* Kolloid-Z. u. Z. Polymere **211**, 14 (1966).
4. *Reinhold, Ch., E. W. Fischer* und *A. Peterlin,* J. Appl. Phys. **35**, 71 (1964).
5. *Spei, M.,* Kolloid-Z. u. Z. Polymere **250**, 214 (1972).
6. *Fraser, R. D. B.* und *T. P. McRae,* Nature **233**, 138 (1971); Polymer **14**, 61 (1973).
7. *Hosemann, R.* und *S. N. Bagchi,* Direct Analysis of Diffraction by Matter (1962).

Anschrift des Verfassers:

Dr. *R. Bonart*
Institut für nichtmetallische Werkstoffe
der Technischen Universität Berlin
1 Berlin 12, Englische Str. 20

Progr. Colloid & Polymer Sci. **58**, 44–52 (1975)

C.N.R.S.
Centre de Recherches sur les Macromolécules, Strasbourg (France)

Melting behaviour of low molecular weight
poly (ethylene-oxide) fractions
I. Extended chain crystals

C. P. Buckley *) and *A. J. Kovacs*

With 6 figures and 1 table

(Received April 8, 1974)

Introduction

Low molecular weight fractions of poly (ethylene-oxide) (PEO) are known to crystallise under atmospheric pressure with chains either fully extended or folded only a small number of times. In both cases the OH-terminated chain ends are rejected onto the surface layer of the crystalline lamellae, such that the number n of folds per molecule is zero or a small integer (1, 2). It has been shown also that n can be controlled in a systematic manner by crystallisation (or annealing) temperature (3, 4) and that the crystal growth process involves sharp transitions at well defined critical temperatures (5) at which n suddenly varies to $n \pm 1$. Chain folded crystals are metastable with respect to extended chain crystals: they thicken, more or less rapidly, even during their growth (5) by stepwise decrease of n.

PEO is therefore a model polymer for studying the influence of chain folding on the physical properties of polymer crystals. The present series of papers concerns the effect of varying n on the melting behaviour of PEO.

Central to this problem is the question: to what extent does the surface free energy of crystalline lamellae, σ_e, depend on chain folding? Recent measurements of crystal growth rate (5) suggest that chain folding significantly increases σ_e. On the other hand, melting point studies have led to the opposite conclusion (6, 7, 8). As a first step towards resolving the paradox,

this paper considers in some detail the melting of extended chain crystals of PEO.

The melting point T_m of a thin lamellar crystal, of thickness d, of a simple pure substance may be expressed by the equation of *Tammann* (9):

$$T_m(d) = T_m(\infty)\left(1 - \frac{2\,\sigma_e}{\Delta H \cdot d}\right), \qquad [1]$$

where ΔH is the enthalpy of fusion and σ_e the *Gibbs'* free energy of the large surfaces of the lamellar crystal. Eq. [1] has been applied to the melting of extended-chain lamellar crystals of PEO to obtain: $\sigma_e \simeq 70\ \text{ergs/cm}^2$ and $T_m(\infty) \simeq 69\ °C$ (5, 10). This latter value is in excellent agreement with $T_m(\infty)$ as estimated by the self seeding technique for high molecular weight PEO (11), and also with the convergence point obtained when the newly discovered critical temperatures (for the crystal growth rate) are plotted versus reciprocal molecular weight and extrapolated to $1/M_n = 0$ (5, 12).

Recently, however, *Beech* and *Booth* (13), *Afifi-Effat* and *Hay* (14) claimed that eq. [1] is invalid for lamellar polymer crystals composed of extended chains. These authors favour the theory of *Flory* and *Vrij* (15) derived for linear chain homologs. They also present some evidence, based on a $T_m(T_c)$ extrapolation rule (T_c being the crystallisation temperature) proposed for high molecular weight polymers (16), which leads to $T_m(\infty) \simeq 76\ °C$ and hence, for reasons to be discussed below, to quite different values for the important parameter σ_e.

The aim of the present paper is to resolve the conflict. Eq. [1] and that of *Flory* and *Vrij* (15)

---

*) Present address: Department of Textile Technology, UMIST, P. o. Box 88, Manchester, U. K.

will be briefly rederived on a common basis to underline the different assumptions involved in each. A new set of values of ΔH and T_m for extended-chain crystals of PEO, covering a wide range of molecular weight (M.w.), will be reported. They will be compared with both theoretical expressions, in order to select the most appropriate one for evaluating $T_m(\infty)$ and σ_e in a consistent manner.

Theoretical

Melting of Extended Chain Polymer Crystals

Consider a model polymer crystal (Fig. 1), in which chains are packed parallel onto a simple square lattice[1]) to form a lamella. All

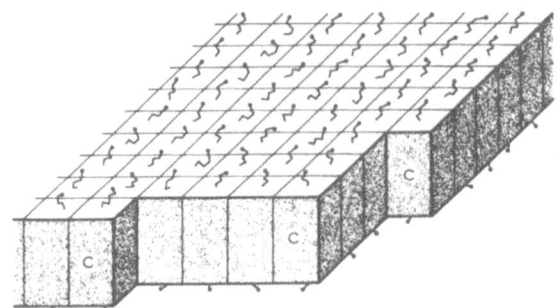

Fig. 1. Model for monolayer extended-chain polymer crystal

the chain ends, including a few monomer units, are located in the surface layers, in which they are disorganised. If melting is treated as a simple reverse of crystallisation, the kinetic theory of polymer crystal growth based on coherent surface nucleation (see for example ref. 17) leads to the following conclusions concerning T_m.

1) $T < T_m$: Crystals grow by deposition of surface nuclei onto the growth faces, followed by more or less rapid advance of the surface steps along the face to complete a new monomolecular layer.

2) $T > T_m$: Mature crystals contract by more or less rapid recession of surface steps which always exist on the prism faces, at least at the

---

[1]) This simple model applies to PEO, by assuming that the growth faces are {120} planes (27); the same treatment can, however, be easily extended to other *two dimensional crystal lattices*.

corners, and may possibly be nucleated by lattice defects.

Consequently, T_m is the temperature at which steps on the growth face neither advance nor recede, i.e. at which there is thermodynamic equilibrium between molecules in sites of type c (Fig. 1) and molecules in the surrounding liquid (l) phase. In terms of chemical potentials μ, therefore, T_m is that value of T which satisfies

$$\Delta\mu(T) = \mu_l(T) - \mu_c(T) = 0 . \qquad [2]$$

This phenomenological model of the melting process of polymer crystals excludes the possibility of melting along the chain direction ("Boundary melting") proposed by *Fischer* (31). Direct observations on PEO single crystals — fully extended, or folded — have shown that lamellar crystals embedded in the liquid, melt effectively by progressive lateral shrinkage, involving, in isothermal conditions, a constant rate of recession, starting often at the corners (32). Thus, the model of melting put forward above has strong experimental support, at least in the case of PEO crystals.

The important fact to be emphasised here is that according to the model adopted the expression for $\Delta\mu(T)$ depends upon the immediate environment of chain ends at the large crystal surface. Two simple possibilities may be distinguished which will now be discussed.

A. Crystal Surrounded by Melt

In this case the environment of chain ends located in the disordered surface layer of the crystal might be assumed identical to that of chain ends in the melt, whatever, the degree of polymerisation p. $\Delta\mu$ would then take the form

$$\Delta\mu(T) = p[\Delta H^*(T) - T\Delta S^*(T)] - 2\sigma_e'(T) \qquad [3]$$

where ΔH^* and ΔS^* are respectively the molar enthalpy and entropy of fusion, per monomer unit, of an infinitely thick and wide single crystal of PEO, while the surface free energy σ_e' refers to one mole of chain ends which may include several monomer units.

Eq. [3] can be expanded about a reference temperature T_0. By assuming the enthalpic and entropic contributions to $\sigma_e'(\Delta H_e'/2$ and $\Delta S_e'/2$ in the notation of *Flory* and *Vrij* (15)) to be independent of T, one can write:

$$\Delta\mu(T) = p[\Delta H^*(T_0) + \Delta C_p(T - T_0)]$$
$$- pT[\Delta S^*(T_0) + \Delta C_p \ln(T/T_0)] \qquad [4]$$
$$- (\Delta H_e' - T\Delta S_e') ,$$

where ΔC_p, the difference in specific heat between melt and crystal, is also assumed temperature independent.

Equating $\Delta\mu(T)$ to zero (i.e. putting $T = T_m$) and choosing for the reference temperature $T_m(\infty)$, the equilibrium melting temperature of an extended-chain crystal composed of infinitely long chains:

$$T_0 = T_m(\infty) = \frac{\Delta H^*[T_m(\infty)]}{\Delta S^*[T_m(\infty)]} \qquad [5]$$

yields the following expression:

$$\frac{1}{T_m(p)}\left\{1 - \frac{\Delta H'_e}{p\,\Delta H^*[T_m(\infty)]}\right\}$$
$$+ \frac{\Delta C_p}{\Delta H^*[T_m(\infty)]}\left[1 - \frac{T_m(\infty)}{T_m(p)} + \ln\frac{T_m(\infty)}{T_m(p)}\right]$$
$$= \frac{1}{T_m(\infty)} - \frac{\Delta S'_e}{p\,\Delta H^*[T_m(\infty)]} \equiv X(1/p). \qquad [6]$$

Eq. [6] can be recognised as a modified form of eq. [1] allowing for temperature dependences of ΔH^*, ΔS^* and $\sigma'_e = [\Delta H'_e - T_m(p)\,\Delta S'_e]/2$.

B. Closely Stacked Lamellar Crystals

In this case the environment of chain ends is assumed to be provided solely by other chain ends protruding from the surface of the same lamella and/or adjacent lamellae. Obviously chain end pairing (15) would be a special but not necessary case. In any event the environment might reasonably be assumed to be independent of p. However, one must account for the entropy change of chain ends on melting which does vary with p [2]). Following *Flory* and *Vrij* (15) this entropy change ("entropy of mixing") amounts to $R\ln Cp$ per mole, where C is a constant, the value of which depends on the chain flexibility. Taking this term into account eq. [3] must be modified to:

$$\Delta\mu(T) = p[\Delta H^*(T) - T\,\Delta S^*]$$
$$- 2\,\sigma_e(T) - RT\ln Cp. \qquad [7]$$

Proceeding as before, i.e. expanding about the reference temperature $T_m(\infty)$ and assuming the enthalpic and entropic contributions to σ_e ($\Delta H_e/2$ and $\Delta S_e/2$ respectively) and ΔC_p to be temperature independent, one obtains:

$$\frac{1}{T_m(p)}\left\{1 - \frac{\Delta H_e}{p\,\Delta H^*[T_m(\infty)]}\right\}$$

---

[2]) We do not consider here the possibility of end pairing in the melt.

$$+ \frac{\Delta C_p}{\Delta H^*[T_m(\infty)]}\left[1 - \frac{T_m(\infty)}{T_m(p)} + \ln\frac{T_m(\infty)}{T_m(p)}\right]$$
$$- \frac{R}{\Delta H^*[T_m(\infty)]}\frac{\ln p}{p}$$
$$= \frac{1}{T_m(\infty)} - \frac{\Delta S_e - R\ln C}{p\,\Delta H^*[T_m(\infty)]} \equiv Y(1/p) \qquad [8]$$

which includes the terms due to the randomisation of chain ends after melting.

The left hand sides of eqs. [6] and [8] define respectively two functions $X(1/p)$ and $Y(1/p)$, in which the molecular weight dependence is represented by the degree of polymerisation p. Clearly, models A and B predict a linear dependence on p^{-1} for X and Y respectively, both having $T_m(\infty)^{-1}$ as the ordinate intercept at $p^{-1} = 0$. This provides a simple means for comparing models A and B with experiment, although both cases can be envisaged as occurring in appropriate experimental conditions. In fact, isothermal melting of individual single crystals embedded in the melt has recently been observed and will be described elsewhere (32).

Experimental

T_m and ΔH were measured using a Perkin-Elmer DSC-1 B Differential Scanning Calorimeter (DSC). T_m was determined from the peak temperature of the melting endotherm at a heating rate (scan speed) of 0.5 °C/min, corrected for thermal lag (18). This correction amounted to approximately -0.2 °C. Precision of temperature measurement was found to be better than ± 0.2 °C. For heat flux calibration the latent heat of melting of benzoic acid was used (19). Heat flux noise level was always less than ± 0.001 cal/min. and the scatter in measurements of ΔH was found to be $\pm 1.5\%$.

Polymers studied were all commercial OH — terminated low molecular weight fractions of PEO from Union Carbide (prefix C), Hoechst (H) or Fluka (F).

Table 1. Molecular weights of PEO samples used. Average values of melting points and enthalpy of fusion of extended chain crystals

Polymer	M_n	$\dfrac{M_w}{M_n}$	$100/p$ $(4400/M_n)$	$T_m(p)$ (°C)	$\Delta H(p)$ (cal/g)
H 400	375	—	11.74	2.8	21.7
H 600	640	—	6.87	19.9	31.5
C 1000	1110	1.04	3.96	43.3	37.5
C 1500	1350	1.26	3.26	46.0	41.4
H 2000	1890	1.27	2.33	52.7	41.7
F 3000	2780	1.08	1.58	57.6	42.7
H 4000	3900	1.11	1.13	60.4	44.9
H 6000	5970	1.06	0.74	63.3	—
C 8000	7760	1.19	0.57	64.3	—

Number average molecular weight, M_n, was determined by OH end-group analysis and weight average, M_w, by light scattering from solution in methanol. Characteristic parameters of the samples used are listed in Table 1.

Polymers solid at room temperature ($M_n > 1000$) were stored as powder in vacuo over P_2O_5 for several months prior to use. They were melted at 70 °C in the DSC capsules before being crystallised isothermally at different temperatures T_c.

Liquid polymers were freeze-dried from benzene solution and then sealed in the DSC capsules under argon. These samples were crystallised in the DSC below room temperature. For all polymers crystallisation conditions were chosen such that complete crystallisation could be assumed.

Polymers with $M_n < 3000$ always gave a single peak (at least at scan speeds less than 32 °C/min), corresponding to the melting of extended-chain crystals. H 4000 showed complex melting behaviour, strongly dependent on scan speed, which will be described in detail elsewhere (20). At a scan speed of 0.5 °C/min, however, it consistently gave a prominent melting peak corresponding to melting of extended-chain crystals and except for the highest value of T_c, a small peak at lower temperatures, corresponding to melting of a small fraction ($< 15\%$) of chains crystallised in the once-folded form.

Samples H 6000 and C 8000 could only be partially crystallised in extended-chain form even at the greatest crystallisation times and temperatures which could reasonably be employed. As a result they always gave multiple but clearly separable endotherm peaks. In order to reduce the time necessary for complete crystallisation, polymers H 6000 and C 8000 were crystallised using the self-seeding procedure (5, 11).

Results

The shape of the melting endotherm peak for those polymers which crystallised purely in extended-chain form was found to vary considerably with M.w. Representative results are shown

in Fig. 2. Clearly with decreasing M.w. the melting endotherms shift to low temperatures and significantly broaden. Most of the broadening takes the form of a low temperature tail, which in the case of H 400 extends to about -25 °C, thus considerably below -13 °C, the T_m value of ethylene glycol monomer (21). The origin of this low temperature tail might be assigned to differences in the crystal structure

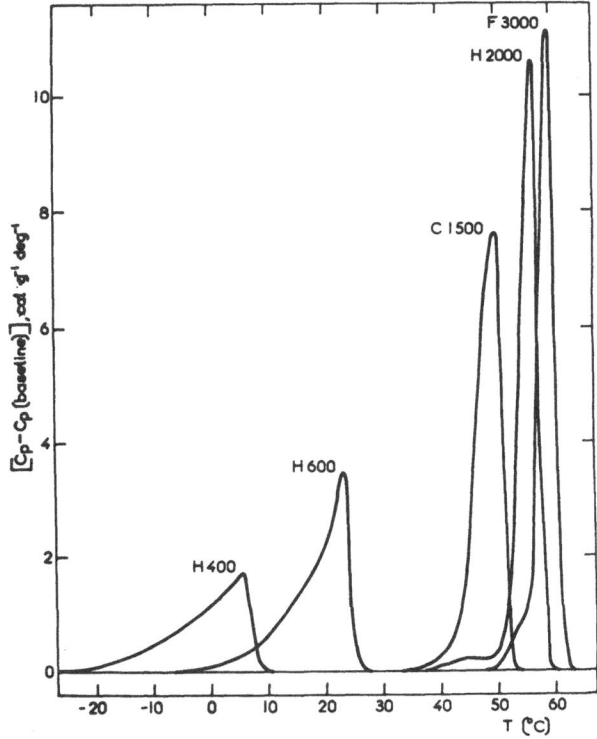

Fig. 2. Representative melting endotherms of extended-chain crystals for various PEO samples obtained at a scan speed of 8 °C/min

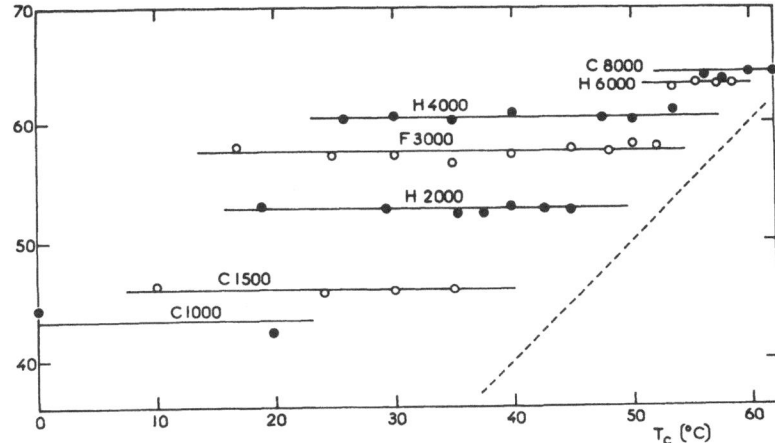

Fig. 3. Melting point T_m of extended chain PEO crystals isothermally crystallised at temperatures T_c (°C). Broken line $T_c = T_m$

of polymolecular samples and that of pure oligo(ethylene oxide)s. Obviously oligomers with $p < 7$ cannot adopt the usual 7/2 helix conformation (27) in the lattice.

The variation of T_m (of extended-chain crystals) with T_c for some of the polymers studied is shown in Fig. 3. As expected, the value of T_m for each polymer investigated was found to be essentially independent of T_c. Furthermore, the enthalpy of fusion ΔH for fully extended-chain samples was also independent of crystallisation temperature.

Average values of T_m and ΔH of extended-chain crystals of the various polymers studied, collected in Table 1, were used in the analysis which follows. All ΔH values refer to samples crystallised entirely in extended-chain form, as do the T_m values for polymers of M.w. less than 3000. The value of T_m for H 4000 is an average over results for specimens in which the fraction in extended-chain form was greater than ca. 0.85. The T_m value for H 6000 is an average over results for samples in which the extended chain fraction was greater than 0.5. Finally, for C 8000, the value of T_m refers to one sample fully crystallised at 42.8 °C and annealed at 60.15 °C for 172 h., in which about 32% of chains were extended. (Simple isothermal crystallisation of this polymer yielded at most 2% in extended-chain form).

Analysis and Discussion

Analysis of T_m measurements using eqs. [6] and [8] requires prior knowledge of $\Delta H^*[T_m(\infty)]$, ΔH_e and ΔC_p in order to be able to construct the functions $X(1/p)$ and $Y(1/p)$. For an approximate $T_m(\infty)$ of 70 °C the data of *Beaumont* et al. (22) allow an estimate of $\Delta C_p = 2.07$ cal/mole deg., although this value must be considered approximate as it refers to CH_3 terminated PEO.

If all molecules are crystallised the enthalpy of melting $\Delta H(p)$ per mole of repeating unit is given by (15):

$$\Delta H(p) = \Delta H^*[T_m(\infty)] - \Delta C_p \Delta T(p) - (\Delta H_e/p) \quad [9]$$

where $\Delta T(p) = T_m(\infty) - T_m(p)$. A plot of $\Delta H(p) + \Delta C_p \Delta T(p)$ versus $1/p$ should therefore yield a straight line with $\Delta H^*[T_m(\infty)]$ as the ordinate intercept at $(1/p) = 0$ and $-\Delta H_e$ as slope.

The values of $\Delta H(p)$ given in Table 1 are plotted in this form in Fig. 4, taking $T_m(\infty)$ to be approximately 70 °C. Also shown are the results obtained by *Braun* et al. (23) for three polymers which may be assumed to be crystallised in extended-chain form. The reader will note the satisfactory agreement between eq. [9] and the present data. The least squares fitted straight line shown in Fig. 4 leads to:

$$\Delta H^*[T_m(\infty)] = (2.07 \pm 0.03) \text{ Kcal/mole, and}$$
$$\Delta H_e/2 \quad\quad = (4.18 \pm 0.22) \text{ Kcal/mole.} \quad [10]$$

These values were used in the calculations described below.

As already noted by *Afifi-Effat* and *Hay* (14) the data of *Braun* et al. (cf. Fig. 4) also appear to obey eq. [9], but lead to greater values:

$$\Delta H^*[T_m(\infty)] = 2.30 \text{ Kcal/mole} \quad \text{and}$$
$$\Delta H_e/2 \quad\quad = 5.85 \text{ Kcal/mole.} \quad [11]$$

There is no obvious explanation for this difference. The results derived below, however, do not depend critically on the choice of one or other set of constants.

The values of $\Delta H^*[T_m(\infty)]$ and ΔH_e obtained above (eqs. [10]) and the assumed value for ΔC_p were used, in conjunction with the T_m data of Table 1, to evaluate the $X(1/p)$ and $Y(1/p)$ functions, taking in turn two trial values for $T_m(\infty)$: 70 °C and 75 °C. Straight lines were fitted to $X(1/p)$ and $Y(1/p)$ using the method of least squares.

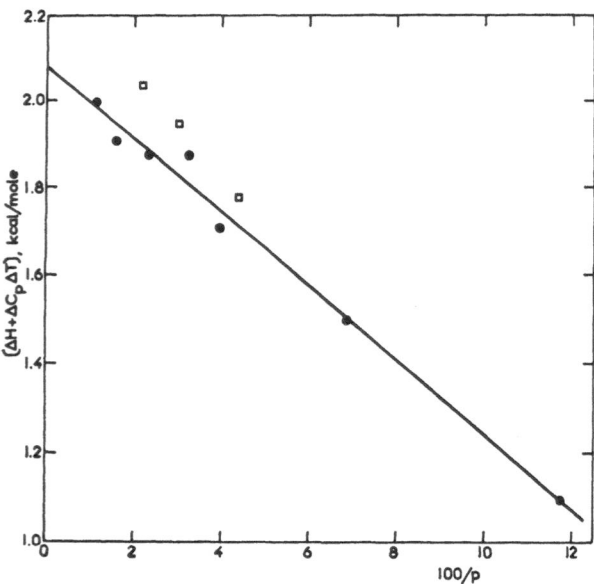

Fig. 4. Molar $\Delta H(p)$ values from Table 1, all referred to a reference temperature of 70 °C, $[\Delta T = 70 - T_m(p)]$. The straight line is a least squares fit to eq. [9]. Open squares: data of *Braun* et al. (Ref. 23).

For $M_n > 1000$ a good straight line fit was obtained for $Y(1/p)$ leading, *for both trial values* of $T_m(\infty)$, to:

$T_m(\infty) = (68.9 \pm 0.4)\ °C$ and

$\Delta S_e - R \ln C = (20.3 \pm 0.3)$ cal/deg. mole.   [12]

Over the same range of M.w. a significantly poorer fit was obtained for $X(1/p)$, giving:

$T_m(\infty) = (67.1 \pm 0.5)\ °C$ and
$\Delta S_e' = (14.5 \pm 0.4)$ cal/deg. mole.   [13]

To confirm that these conclusions are not restricted to the samples and calorimetric procedure employed, the same analysis was repeated, using in addition all the values of T_m so far reported in the literature (6, 10, 14, 23—26) for industrial OH-terminated low M.w. fractions of PEO, crystallised in extended-chain form. The corresponding plots are presented in Fig. 5, which shows a more pronounced curvature of $X(1/p)$ compared with $Y(1/p)$. The straight line for $Y(1/p)$ calculated from the present data alone is clearly consistent with the literature data at least for $(1/p) < 0.08$, $(M_n > 550)$.

The constants given in eqs. [10] and [12] were used to calculate from eq. [8] the molecular weight dependence of T_m, which is shown in

Fig. 6, as a $T_m(1/M_n)$ plot, together with available data. This theoretical curve illustrates the relation between eqs. [1], [6] and [8]. Since the thickness of extended-chain lamellae is proportional to M_n, eq. [1] predicts a linear decrease of T_m vs. $1/M_n$, whereas the theoretical curve derived from eq. [8] clearly shows an inflexion in the vicinity of $M_n \simeq 2000$.

The downward curvature in the low M.w. range originates from the temperature dependence of the entropic part $(-T_m \Delta S_e/2)$ of the surface free energy allowed for in both eqs. [6] and [8]. On the other hand, the upward curvature at high M.w. results from the additional $RT_m \ln Cp$ term introduced in eq. [8]. This makes a small but significant contribution to the extrapolated value of $T_m(\infty)$ — about 1.8 °C if compared to that obtained from eq. [6] — as already pointed out by *Flory* and *Vrij* (15).

Surface Free Energy

The present analysis allows an estimation of the surface free energy of extended chain crystals of PEO. As shown above, available evidence points to the applicability of eq. [8] rather than that of eq. [6]. The values of ΔH_e, $T_m(\infty)$ and $\Delta S_e - R \ln C$, derived by fitting

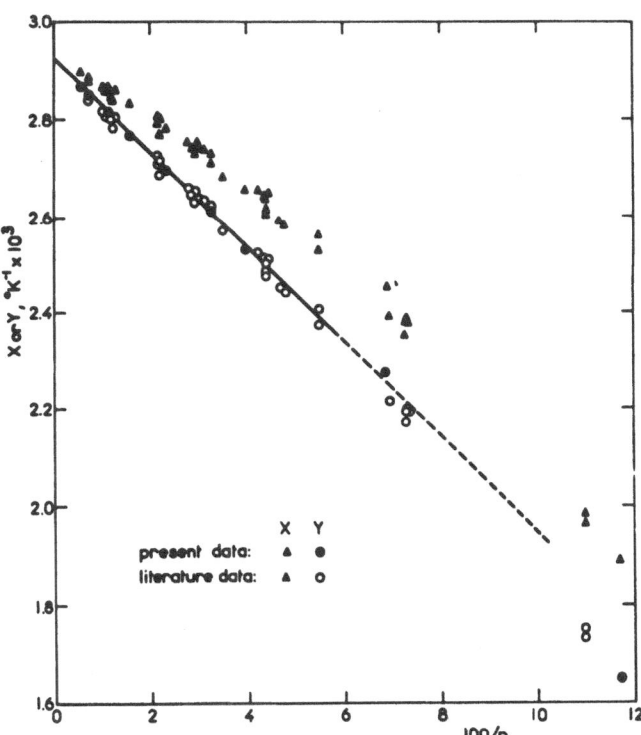

Fig. 5. Values of $X(1/p)$ and $Y(1/p)$ functions (cf. eqs. [6] and [8]) constructed from present and literature data (Ref. 6, 10, 14, 23—26). The straight line is a least squares fit of $Y(1/p)$ based on present data alone over the range $M_n > 1000$

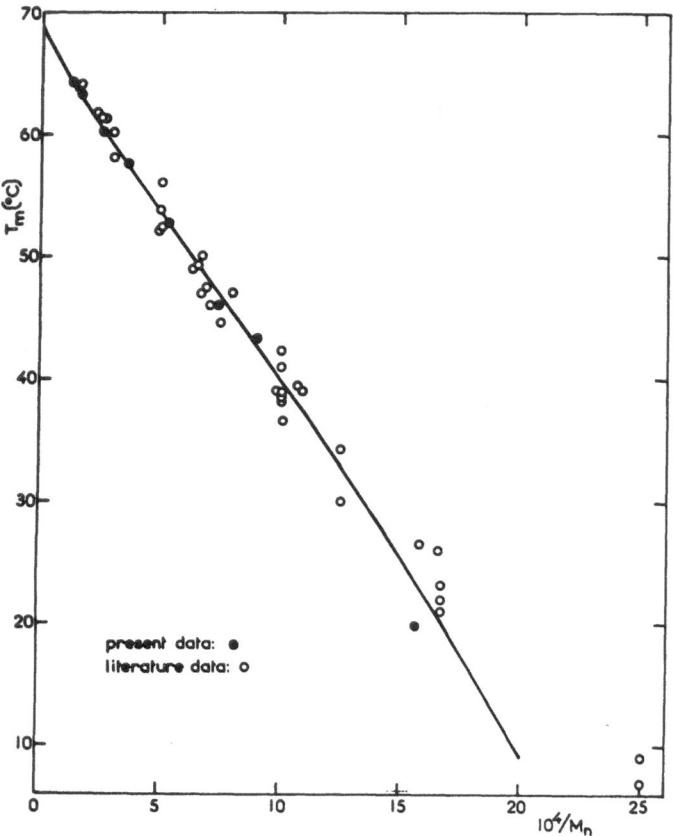

Fig. 6. T_m vs. $(1/M_n)$ plot from present and literature data. The theoretical curve is constructed from eq. [8] using parameters defined by eqs. [10] and [12]

eq. [8] to the data (cf. eqs. [10] and [12]), yield:

$$2\,\sigma_e[T_m(\infty)] + T_m(\infty)\,R\ln C$$
$$= 1.42\,\text{Kcal/mole.}\,[3)] \qquad [14]$$

Unfortunately, the correct value to assign C is not known. It may however be bounded with some confidence by assuming that the dimensionless product Cp in eq. [7] lies between the number of bonds and the number of freely orienting segments, per molecule, derived from the mean square end-to-end distance (28). Accordingly one obtains:

$$3/4 < C < 3. \qquad [15]$$

Combined with the above results this yields:

$$0.34 < \sigma_e[T_m(\infty)] < 0.81 \text{ (Kcal/mole)} \quad \text{or,}$$
$$11 < \sigma_e[T_m(\infty)] < \quad 26 \text{ (erg/cm}^2)\,. \qquad [16]$$

Although σ_e can only be bounded, its value appears to be considerably lower than all previous estimates for extended-chain crystals

of PEO. These were: $\sigma_e = 2.28$ Kcal/mole (10) and $\sigma_e = 2.00$ Kcal/mole (5) both obtained using eq. [1], or $\sigma_e > 1.40$ Kcal/mole (6) (the actual value depending on M.w.), obtained using essentially eq. [8], but assuming a priori $T_m(\infty) = 76\,°$C. Another value: $\sigma_e = 2.18$ Kcal/mole (14) was also derived from eq. [8], but applying it to data limited to low M.w. fractions ($2000 > M_n > 100$) extending down to a range where its applicability is questionable.

It should be noted that the value of σ_e determined above (eq. [16]) for extended chain crystals of PEO is significantly lower than that reported (15, 29) for *n*-alcanes: about 1.6 Kcal/mole (assuming $C = 1$). The difference should be attributed to the possibility, in PEO, of hydrogen bonding between the OH end-groups concentrated in the disordered surface layer of extended chain crystals.

It must be pointed out also that the foregoing analysis assumes implicitly the value of $\sigma_e[T_m(\infty)]$ to be independent of M.w. *Booth* and co-workers (6, 8) have taken a different view. Adopting for $T_m(\infty)$ a value of 76 °C (13), they reach the conclusion that $T_m(M_n)$ data for

---

[3]) The cross-sectional area of one molecule in the PEO crystal (27) is 21.4 Å^2: hence the molar units of σ_e may be converted using 1 Kcal/mole $\equiv 32.5$ erg/cm^2.

extended chain PEO crystals lead to σ_e values which strongly depend upon M.w. In fact, since the data closely approximate eq. [1] at high M.w. (cf. Fig. 6), the reported linear dependence of σ_e on M_n (8) can easily be predicted from eq. [1] if $T_m(\infty)$ is given any value other than 69.0 °C. This apparent variation of σ_e was attributed (7) to an increase in breadth of the M.w. distribution with increasing M_n. Such a variation cannot a priori be ruled out. The evidence in its favour, however, is not compelling. Firstly, the straight line fit of Fig. 4 strongly suggests that the enthalpic part of σ_e is independent of M.w. Secondly, the straight line obtained for $Y(1/p)$ in Fig. 5 shows that the entropic part of σ_e is also independent of M.w. Thirdly, low M.w. fractions of PEO are known to segregate according to M.w. during extended-chain crystallisation rather than to form mixed crystals (1, 30). Finally, the value for $T_m(\infty)$ of 76 °C, derived from $T_m(T_c)$ measurements (13, 14) is highly questionable: the data are in some cases ambiguous (13), and it may be questioned whether the empirical method of linear extrapolation employed (16) can be safely applied to PEO. Note that the applicability of this extrapolation procedure implies that the ratio of the final thickness of the lamellae and the initial thickness of the surface nucleus is temperature independent (16, 17). This should result in a linear plot of T_m vs. T_c, whereas the critical data of *Beech* and *Booth* (13) (on partially crystallised samples) show a rather abrupt change of slope.

In the present work no correction has been made to allow for breadth of M.w. distribution. Since molecular weight determinations are subject to up to 5% uncertainty, large possible error in the important parameter: $[(M_w/M_n) - 1]$ which characterises the width of M.w. distribution (7, 8) precludes any such attempt.

Conclusions

The present work has shown that it is possible to make an objective choice between the available theories of melting in extended chain polymer crystals. The two theories commonly used embody different assumptions concerning the environment of a crystal as it melts. Both the data reported here and all other data in the literature on melting of extended-chain PEO crystals support the use of a model in which crystalline lamellae are stacked closely together without an intervening layer of molecules in the liquid state.

This model which leads to the theory of *Flory* and *Vrij* (15), when applied to the present results yields $T_m(\infty) = (68.9 \pm 0.4)$ °C which is about 7 °C lower than the recent estimates by *Beech* and *Booth* (13) and *Afifi-Effat* and *Hay* (14), who used a less well-founded extrapolation procedure.

In addition, this model suggests a value for σ_e within the bounds: $0.34 < \sigma_e[T_m(\infty)] < 0.81$ (Kcal/mole). Such a remarkably low value for σ_e, hitherto unsuspected, may be attributed to hydrogen bonding of OH end-groups localised in the disorganised surface layers between adjacent lamellae.

Acknowledgements

One of us (C. P. B.) was supported during the course of this work by a NATO Research Fellowship. Dr. *A. Gonthier* kindly performed some of the measurements shown in Fig. 3. We are grateful to Dr. *P. Calmé* and Dr. *Cl. Strazielle* for M_n and M_w determinations respectively.

Summary

Two available theories of melting in extended chain polymer crystals are critically examined and compared with calorimetric data. Melting points, T_m, and enthalpy of fusion, ΔH, have been measured for extended-chain crystals of poly(ethylene-oxide) fractions, covering a wide range of molecular weights. It appears that the theory of *Flory* and *Vrij* is the most appropriate to describe the present results and literature data. Application of this theory to experiments yields $T_m(\infty) = (68.9 \pm 0.4)$ °C and suggests that the value of the surface free energy, σ_e, lies within 0.34 and 0.81 Kcal/mole, its enthalpic part ($\Delta H_e/2$) being (4.18 ± 0.22) Kcal/mole. Such a low value of σ_e may be attributed to hydrogen bonding of OH end-groups located in the disordered surface layer of closely stacked crystalline lamellae.

Résumé

Les deux théories actuelles de la fusion de cristaux de polymères, à chaines étirées, sont examinées d'une manière critique. Leurs prédictions sont comparées avec des mesures calorimétriques portant sur la température, T_m et de l'enthalpie, ΔH, de fusion de cristaux à chaines étirées de fractions de polyoxyéthylène, couvrant une large gamme de masses moléculaires. On constate que la théorie de *Flory* et *Vrij* est celle qui décrit le mieux les résultats obtenus et ceux de la littérature. L'application de cette théorie conduit à une valeur de $T_m(\infty) = 68,9$ °C et suggère que la valeur de l'énergie libre superficielle, σ_e, est comprise entre 0,34 et 0,81 Kcal/mole, sa partie enthalpique étant de (4,18 $\pm$ 0,22) Kcal/mole. Une valeur aussi faible de σ_e peut être interprétée par la formation de liaisons hydrogènes entre les groupements terminaux OH des chaines, localisés dans les couches superficielles désordonnées séparant les cristaux lamellaires empilés.

Zusammenfassung

Zwei bestehende Theorien des Schmelzens in Polymerkristallen mit gestreckten Ketten werden kritisch geprüft und mit kalorimetrischen Daten verglichen. Die Schmelzpunkte T_m und die Schmelzenthalpie ΔH wurden für „extended-chain"-Kristalle von Polyäthylenoxyd-Fraktionen gemessen, die einen weiten Bereich unterschiedlicher Molekulargewichte überspannen. Es scheint, daß die Theorie von *Flory* und *Vrij* am besten die vorliegenden Ergebnisse und die Literaturdaten beschreibt. Die Anwendung der Theorie auf die Experimente liefert $T_m(\infty) = (68,9 \pm 0,4)$ °C und legt einen Wert für die freie Oberflächenenergie σ_e innerhalb 0,34 und 0,81 kcal/Mol nahe. Sein Enthalpieanteil ($\Delta H_e/2$) beträgt (4,18 $\pm$ 0,22) kcal/Mol. Ein solcher niedriger Wert für σ_e ist den lokalisierten Wasserstoffbindungen der OH-Endgruppen in der ungeordneten Oberflächenschicht von dicht ineinander gesteckten kristallinen Lamellen zuzuschreiben.

References

1) *Arlie, J. P., P. A. Spegt* and *A. E. Skoulios*, Makromol. Chem. **99**, 160 (1966).
2) *Spegt, P.*, Makromol. Chem. **140**, 167 (1970).
3) *Arlie, J. P., P. Spegt* and *A. Skoulios*, Makromol. Chem. **104**, 212 (1967).
4) *Spegt, P.*, Makromol. Chem **139**, 139 (1970).
5) *Kovacs, A. J.* and *A. Gonthier*, Kolloid-Z. u. Z. Polymere **250**, 530 (1972).
6) *Ashman, P. C.* and *C. Booth*, Polymer **13**, 459 (1972).
7) *Beech, D. R., C. Booth, D. V. Dodgson, R. R. Sharpe* and *J. R. S. Waring*, Polymer **13**, 73 (1972).
8) *Beech, D. R., C. Booth, C. J. Pickles, R. R. Sharpe* and *J. R. S. Waring*, Polymer **13**, 246 (1972).
9) *Tammann, G.*, Z. Anorg. Chem. **110**, 166 (1920).
10) *Spegt, P., J. Terrisse, B. Gilg* and *A. Skoulios*, Makromol. Chem. **107**, 29 (1967).
11) *Vidotto, G., D. Levy* and *A. J. Kovacs*, Kolloid-Z. u. Z. Polymere **230**, 289 (1969).
12) *Gonthier, A.*, Thèse, Université Louis Pasteur (Strasbourg, 1973).
13) *Beech, D. R.* and *C. Booth*, J. Polymer Sci., Part B, **8**, 731 (1970).
14) *Afifi-Effat, A. M.* and *J. N. Hay*, J.C.S. Faraday II, **68**, 656 (1972).
15) *Flory, P. J.* and *A. Vrij*, J. Amer. Chem. Soc. **85**, 3548 (1963).
16) *Hoffman, J. D.* and *J. J. Weeks*, J. Res. Nat. Bur. Stds. **66 A**, 13 (1962).
17) *Hoffman, J. D.*, S.P.E. Trans. **4**, 1 (1964).
18) *Perkin-Elmer Corp*, Thermal Analysis News Letter N° 5.
19) *Ginnings, D. C.* and *G. T. Furukawa*, J. Amer. Chem. Soc. **75**, 522 (1953).
20) *Buckley, C. P.* and *A. J. Kovacs*, Prog. Colloid & Polymer Sci., (in press).
21) *Gallaugher, A. F.* and *H. Hibbert*, J. Amer. Chem. Soc. **58**, 813 (1936).
22) *Beaumont, R. H., B. Clegg, G. Gee, J. B. M. Herbert, D. J. Marks, R. C. Roberts* and *D. Sims*, Polymer **7**, 401 (1966).
23) *Braun, W., K. H. Hellwege* and *W. Knappe*, Kolloid-Z. u. Z. Polymere **215**, 10 (1967).
24) *Koizumi, K.* and *T. Hanai*, Bull. Inst. Chem. Res. (Kyoto) **42**, 115 (1964).
25) *Hay, J. N., M. Sabir* and *R. L. T. Steven*, Polymer **10**, 187 (1969).
26) *Godovsky, Yu. K., G. L. Slonimsky* and *N. M. Garber*, J. Polymer Sci., Part C, **38**, 1 (1972).
27) *Takahashi, Y.* and *H. Tadokoro*, Macromolecules **6**, 672 (1973).
28) *Brandrup, J.* and *E. H. Immergut* (Eds.), Polymer Handbook (New York, 1966).
29) *Broadhurst, M. G.*, J. Res. Nat. Bur. Stds. **70 A**, 481 (1966).
30) *Gilg, B.* and *A. Skoulios*, Makromol. Chem. **140**, 149 (1971).
31) *Fischer, E. W.*, Kolloid-Z. u. Z. Polymere **231**, 458 (1969).
32) *Kovacs, A. J., A. Gonthier* and *C. Straupe*, J. Polymer Sci. Part C, **50**, 1975, (in press).

Authors' address:

A. J. Kovacs
Centre de Recherches sur les Macromolécules
67083, Strasbourg, France

Progr. Colloid & Polymer Sci. **58**, 53—60 (1975)

Abteilung für Experimentelle Physik I der Universität Ulm — Federal Republic of Germany

Thermodynamic characterization of metastable organic liquids at the quasi-static glass-transition-temperature

H. G. Kilian

With 11 figures and 1 table

(Received July 26, 1974)

Introduction

From the *van der Waals* state equation it can be demonstrated that there are fundamental similarities between the fluid and the gas state. Therefore a general phenomenological description of the fluid equilibrium states might be thought to be possible from this point of view.

But on account of the peculiarities of the molecular structure and interactions (1—14) such an universal characterization might be questionable. Form anisotropy of the molecules should perhaps lead to cluster structures as

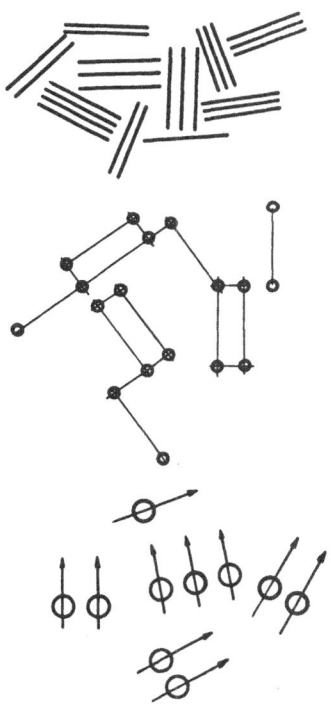

Fig. 1. Various models (a, b, c)

shown in the sketch of Fig. 1a. Localized interactions should influence the spatial arrangements of the molecules distinctly (Figs. 1b and 1c). Moreover linear macromolecules are expected to build up bundlelike superstructure elements within certain regions. Therefore a general phenomenological characterization of fluid systems might be thought to become problematic essentially at relative low temperatures.

We want to demonstrate in this paper that a generalized thermodynamic description of metastable organic fluid systems can nevertheless be established at the quasi-static glass transition temperature. This approach is based on a model proposed by *Frenkel* (15—17): *The fluid will be looked at as a mixture of molecules and voids.* The mixture should be *saturated* with respect to voids.

The Isobar State Diagram

It is known that if crystallization does not occur undercooled metastable fluid systems become glassy at a certain temperature, $T_{g\infty}$. Without going into kinetic aspects in this paper, the glass transition is considered to take place under isobaric "quasi-static" conditions, i.e. the cooling rate of the supercooled liquid must be extremely slow (10^{-3} degree a minute). We assume that in an "idealized system" the cooperative glass transition of the total system occurs at $T_{g\infty}$. Since in real systems the volume change occurs continuously within a certain temperature range near $T_{g\infty}$ (see sketch in Fig. 2), $T_{g\infty}$ is obtained as suggested by the intersection of lines extended from the linear portions of the cooling curve (Fig. 2).

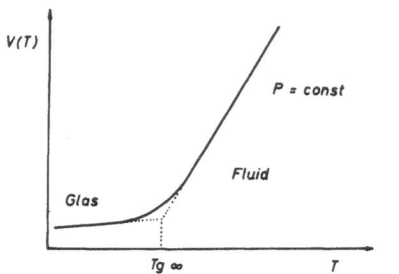

Fig. 2. Sketch of a $V - T$ ($P =$ constant) diagram for a liquid system undergoing glass transition at $T = T_{g\infty}$. $T_{g\infty}$ refers to the "quasi"-static-glass-transition temperature

We want to describe the metastable fluid system at the temperature $T^* = T_{g\infty} + \Delta T$ with $\Delta T/T_{g\infty} \ll 1$ by means of thermodynamics.

The Frenkel Model

The fluid should represent a mixture out of molecules and voids. *A size distribution of voids* is assumed to exist (19). Voids of different size are regarded to represent different particles within the mixture. At constant temperature and pressure, the condition of saturation of the voids of various sizes demands that:

$$\Delta \mu_v = \mu_v(T, P, a_v) - \mu_{v0} = 0 \quad \text{(all } v\text{)} \quad [1]$$

$\mu_v(T, P, a_v)$: chemical potential of the voids of size v within the liquid at (T, P)

a_v: activity of the voids;

$\mu_{v0} = 0$: hypothetical chemical potential of the pure void reservoir.

Using

$$\mu_v = h_v - T s_v \qquad [2]$$

h_v: partial molar enthalpy of the voids;

s_v: partial molar entropy of the voids.

We get from [1] the relations

$$\Delta h_v = T \cdot \Delta s_v; \quad \text{all } v \qquad [3]$$

$\Delta h_v = h_v(T, P)$: mixing enthalpy of the voids;

$\Delta s_v = s_v(T, P)$: mixing entropy of the voids.

The thermodynamic description is related to the hypothetical standard states of the single component systems of molecules with no voids and with voids of uniform size resp. ("void-reservoirs v") at the same (T, P) as in the liquid mixture.

We have to develop a more detailed formulation of the general relations [3].

T* of Solutions

For the purpose of application later we need to take into consideration binary liquid mixtures of molecules. The molar volume fraction of the molecules within the liquid will be

$$\varphi_k = n_k y_k / \sum n_i y_i \qquad [4]$$

n_k, n_i: mole numbers of the particles of the liquid, voids included;

y_k, y_i: size parameter of the particles related to the volume y_0 of the hypothetical lattice cell.

The molar mixing enthalpy of the solution can be written as

$$\Delta H = \{ \sum_v (A_{pv}^h \varphi_p \varphi_v + A_{sv}^h \varphi_s \varphi_v) \\ + A_{ps}^h \varphi_p \varphi_s \} \quad (\sum n_i y_i) . \qquad [5]$$

Here we confine ourselves to the description of the excess contributions by means of the second virial coefficients only. A_{pv}^h, A_{sv}^h and A_{ps}^h are the enthalpy terms of the second virial coefficients for p-molecule-void-, s-molecule-void- and for the p-molecule-s-molecule-contacts resp. The total free enthalpy of void formation related to the parameter y_0 is assumed to obey the empirical relation (29, 30, 31)

$$A = A^h - T A^s . \qquad [6]$$

A^s is the entropy term of the excess function A related to y_0.

The volume fraction for the p- and s-molecules and for the voids v are normalized:

$$\varphi_s + \varphi_p + \sum_v \varphi_v = 1 . \qquad [7]$$

From [5] we derive for the partial molar enthalpy of mixing of the voids y_i adopting

$$A_{py_k}^h \cong A_{pv}^h; \quad A_{sy_k}^h \cong A_{sv}^h \qquad [8]$$

for all voids k

$$\partial \Delta H / \partial k_v |_{T, P, n_k \neq n_v} = \Delta h_v$$

$$\Delta h_v = y_v \{ A_{pv}^h \varphi_p^2 + A_{sv}^h \varphi_s^2 \\ + (A_{pv}^h + A_{sv}^h - A_{ps}^h) \varphi_p \varphi_s \} . \qquad [9]$$

Assumption [8] implies that the enthalpy of void formation for example should be proportional to y_v. This is a somewhat sophisticated supposition introduced for the sake of simplicity and because of our lack of knowledge concerning the molecular structure of the liquid mixtures.

Employing the lattice model of fluids we derive for the partial molar entropy of mixing of the voids y_v using the analogous approximations as before

$$\Delta s_v = - R \{\ln \varphi_v + 1 - y_v/\bar{y}\} + y_v A^s ,$$
$$A^s = A^s_{pv} \varphi_p^2 + A^s_{sv} \varphi_s^2$$
$$+ (A^s_{pv} + A^s_{sv} - A^s_{ps}) \varphi_p \varphi_s , \quad [10]$$

where

$$\bar{y} = 1/\sum_i (\varphi_i/y_i) \qquad [11]$$

represents the average size of the particles within the liquid (voids included!).

From [3], [9] and [10] we get

$$\frac{1}{T} = \frac{1}{A^h} \left\{ \frac{R}{y_v} \left(-1 - \ln \varphi_v + \frac{y_v}{\bar{y}} \right) + A^s \right\} \qquad [12]$$

where $A^h = \Delta h_v/y_v$.

With the help of

$$\frac{1}{\bar{y}} = \frac{\varphi_p}{y_p} + \frac{\varphi_s}{y_s} + \frac{(1 - \varphi_p - \varphi_s)}{\bar{y}_L}$$
$$\sum (\varphi_v/y_v) = \frac{1 - \varphi_p - \varphi_s}{\bar{y}_L} . \qquad [13]$$

We have instead of [12]

$$\frac{1}{T} = \frac{R}{A^h} \left\{ \frac{\varphi_p}{y_p} + \frac{\varphi_s}{y_s} - \frac{1 + \ln \varphi_v}{y_v} \right.$$
$$\left. + \frac{1 - \varphi_p - \varphi_s}{\bar{y}_L} + \frac{A^s}{R} \right\} . \qquad [14]$$

oligomer and polymer systems seem to possess *the same relative free volume at $T_{g\infty}$* ("iso-free-volume-hypothesis"). The vibrational parts of the volume of a molecule are not taken as a contribution to the free volume but should be added to the "occupied volume" of the molecules. There have been objections to this hypothesis too (25—27).

Accepting the iso-free-volume-hypothesis we have at T^*

$$1 - \varphi^* = \sum_v \varphi_v ; \qquad [15]$$
$$\varphi_p = \varphi^* \cdot \varphi_p^0 ; \quad \varphi_s = \varphi^* \cdot \varphi_s^0 ; \quad \varphi_p^0 + \varphi_s^0 = 1 .$$

φ^* represents the "occupied volume" at T^*; φ_p^0, φ_s^0 the volume fractions for molecules p and s resp. within the hypothetical standard state with no voids at all. These parameters can in practice be evaluated from the weight fraction of the molecules within the liquid mixture.

Adopting a *size-distribution with rational volume* ratio for the various voids ($v = 1, 2, \ldots, \infty$) the following relations have been verified elsewhere (19)

$$\bar{y}_L = (1 - \varphi^*) \cdot e/\ln \{(1 - \varphi^*) \cdot e + 1\} \qquad [16]$$

and

$$\varphi_v = \frac{1}{e} \{(1 - \varphi^*) \cdot e/[(1 - \varphi^*) \cdot e + 1]\}^{y_v} . \qquad [17]$$

Using the above eqs. [15]—[17] we derive from [14] the basic equation of this treatment

$$\boxed{ \frac{1}{T^*} = \frac{R}{A^{k0} \varphi^*} \cdot \left\{ \frac{\varphi_p^0}{y_p} + \frac{\varphi_s^0}{y_s} + \frac{1}{\varphi^*} \left[\frac{e+1}{e} \cdot \ln \{(1 - \varphi^*) \cdot e + 1\} - \ln(1 - \varphi^*) - 1 \right] \right\} + \frac{A^{s0}}{A^{k0}} }$$

$$[18]$$

where

$$A^{k0} = A^h_{pv} \varphi_p^{02} + A^h_{sv} \varphi_s^{02} + (A^h_{pv} + A^h_{sv} - A^h_{ps}) \varphi_p^0 \varphi_s^0$$
$$A^{s0} = A^s_{pv} \varphi_p^{02} + A^s_{sv} \varphi_s^{02} + (A^s_{pv} + A^s_{sv} - A^s_{ps}) \varphi_p^0 \varphi_s^0 . \qquad [19]$$

Describing polymer-solvent systems with $y_p \gg 1$ we have the approximative eq. [20] instead of [18]

$$\boxed{ \frac{1}{T^*} \cong \frac{R}{A^{k0} \varphi^*} \left\{ \frac{\varphi_s^0}{y_s} + \frac{1}{\varphi^*} \left[\frac{e+1}{e} \ln \{(1 - \varphi^*) \cdot e + 1\} - \ln(1 - \varphi^*) - 1 \right] \right\} + \frac{A^{s0}}{A^{k0}} . }$$

$$[20]$$

Relation [14] must be fulfilled for voids of each size y_v.

The Situation at $T^* = T_g \infty + \triangle T$

Within a certain limit of accuracy various authors (18, 20—24) have shown that liquid

The relations [18] and [20] suggest the following conclusions:

a) Within the limits of our approximations, especially on account of the saturation conditions [3], T^* depends on the total free volume fraction $\sum_v \varphi_v = \varphi_v^{tot}$ only. *No influ-*

ence of the size distribution of the voids is apparent.

b) The temperature T^* will be depressed the more, the smaller the molecules will be. This effect is well known from measurements for the $T_{g\infty}$ of oligomer-polymer-homologues (5, 7).

The "Iso-Excess-Entropy-Hypothesis"

Describing the quasi-static glass-transition temperatures of oligomer-polymer-homologues of polystyrene with the help of [18] ($\varphi_s^0 = 0$) we discussed the following general conclusions (19):

If the "iso-free-volume-hypothesis" holds, the specific *enthalpy of void formation A_{pv}^h stays constant for* all homologues (18, 19). If we further introduce the above size distribution of voids we need a pronounced negative specific excess entropy of mixing A_{pv}^s to fit the experimental results. *A_{pv}^s turns out to be independent on y_p for the total description, too.* Therefore we stated that *probably all homologues show up the same fundamental deviations of the ideal athermical mixing behaviour at T^* ("iso-excess-entropy-hypothesis").* *The volume fractions of the voids of various sizes reveals to be a monotonous function* becoming rapidly smaller the larger y_v will be (19).

Polymer-Solvent-Mixtures

We intend to examine the above characteristics. Therefore we add small solvent molecules of very different structure to polystyrol as the polymer component. If the above statement will be of general validity we should be able to *evaluate the T^* from* [20] *with invariant* φ^*, A_{pv}^h *and* $A^{s0} \equiv A_{pv}^s$. From [19] we learn that we have to know additionally y_s, A_{sv}^h and A_{ps}^h only.

A_{sv}^h we evaluate from the glass transition temperatures, $T_{g\infty}$, of the pure solvent systems employing the same assumption for the characterization of undercooled liquid state at $T_{g\infty}$ as above. For this purpose we use [20] with $\varphi_s^0 = 1$. If the $T_{g\infty}$ is not known we estimate this parameter from the so-called 2/3-rule:

$$T_{g\infty} \equiv 0.66 \cdot T_{M\infty} . \qquad [21]$$

$T_{M\infty}$: melting temperature of the „infinite extended" crystals of the pure solvent system.

The following discussion is referred to measurements of $T_{g\infty}$ on various polystyrene-solvent mixing systems published by *Jenckel* and *Heusch* (28). The χ-parameters of the systems we are interested in have been determined from appropriate measurements (29). The solution-polymer-excess-term is usually defined by (29, 30, 31).

$$\Delta G_S^E = A^* \cdot \varphi_p^2 = R \cdot T \cdot \chi \cdot \varphi_p^2 . \qquad [22]$$

Using our formalism we get instead of [22]

$$\Delta G_S^E = y_s \cdot A \cdot \varphi_p^2 \qquad [23]$$

indicating that our phenomenological quantities are related to the "unit cell" of the hypothetical lattice y_0. Very often the empirical relation has been employed (29, 30)

$$\chi = \alpha + \beta/T , \qquad [24]$$

where α and β design the entropy- and the enthalpy part of the second virial coefficient. The appropriate relation of these parameters to those used within this paper will be

$$A_{ps}^h = R \cdot \beta/y_s; \quad A_{ps}^s = - R \cdot \alpha/y_s . \qquad [25]$$

Unknown values of α and β may be interpolated from the plot from α resp. β versus χ as shown in Fig. 3 provided the total χ-parameter itself be known.

The size of the solvent molecules will be evaluated from atomic radii r_{at} (32) of the atoms i of the molecules:

$$v_s = \frac{4\pi}{3} \sum_i r_{ai}^3 . \qquad [26]$$

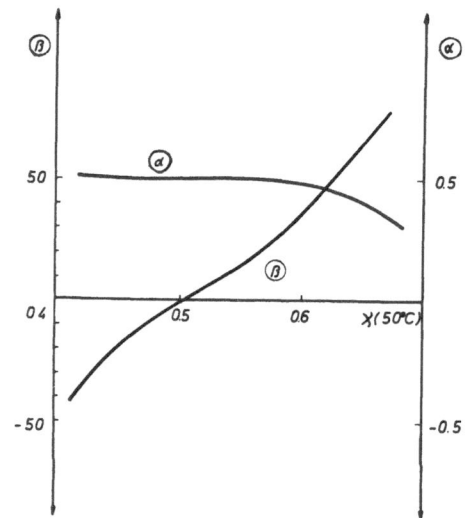

Fig. 3. The parameters α, β as functions of χ using data published by *R. Rehage* (29)

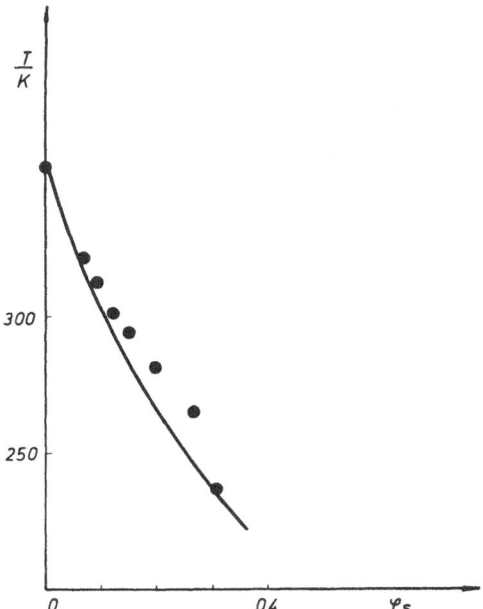

Fig. 4. Quasi-static-glass transition temperatures of the system polystyrene-*n*-butylacetone using data of *Jenckel* and *Heusch* (28). The drawn curve represents the dependence of $T_{g\infty}(\varphi_0)$ calculated from eq. [20] employing the parameters listed in Table 1

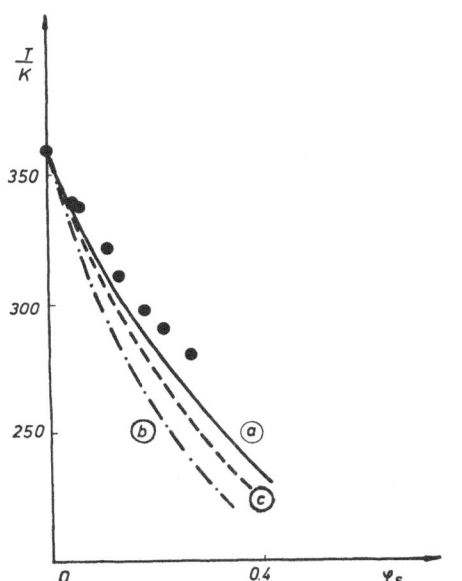

Fig. 5. System: polystyrene-chloroform curves (a), (b) and (c) are theoretical

a) iso-excess-entropy hypothesis: $A'_{ps} \equiv 0$, $A^{s0} = A'_{ph}$ of the polystyrene homologues

b) taking into account α, the entropy-excess contribution of the polymer-solvent-mixture

c) ideal-athermal limit

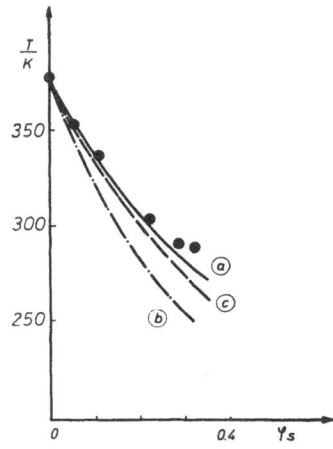

Fig. 6. System: polystyrene – carbon-tetrachloride (a), b) and c): see Fig. 5)

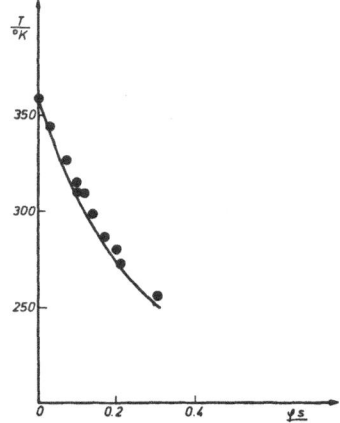

Fig. 7. System: polystyrene-benzene

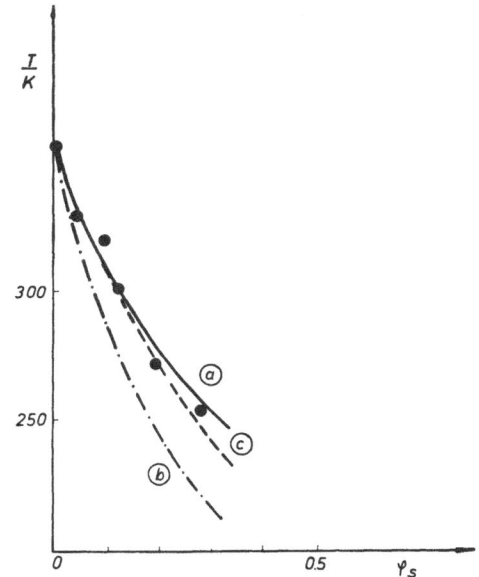

Fig. 8. System: polystyrene – toluene (a), b) and c): see Fig. (5))

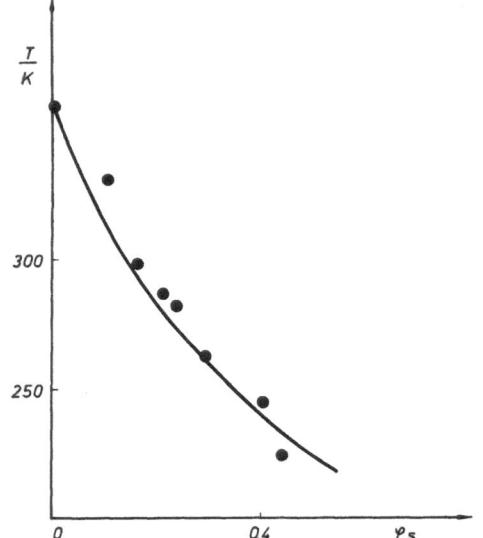

Fig. 9. System: polystyrene – nitro-benzene

y_s the size parameter of the solvent molecules is related to the volume of the periodic unit of polystyrene (y_{st})

$$y_s = v_s/y_{st} .$$ [27]

Parameters used in the calculations are listed in Table 1.

The set of theoretical curves of various mixtures compared with the experiments is shown in the Figs. 4—10. The fit of the experiments is surprisingly good, *if we use energetic excess parameter adopting only an invariant iso-entropy state at* $T_{g\infty}$.

Table 1. $A_{pv}^h = 343$ cal/mol; $T_{g\infty}$ (Polyst. = 358 K, $A_{pv}^s = A^{s0} = -1.44$ cal/mol degree

system polystyrene + solvent	y_S	$\frac{T_{g\infty}^s}{K}$	$\frac{A_{sv}^h}{cal/mol}$	β (29)	α (29)
benzene	0.77	182	720	-18	~ 0.5
nitrobenzene	1.11	178	610	-10	~ 0.5
toluene	0.94	178	600	-16	~ 0.5
carbon tetra- chloride	2.02	167	345	-32	~ 0.55
chloroform	1.55	141	345	-44	~ 0.55
n-butyl acetate	0.89	130	370	-18	~ 0.5
salicyl acid β-naphthyl ester	3.03	243	410	$+40$	$-$
salicyl acid methyl ester	1.54	175	425	~ 0	$-$

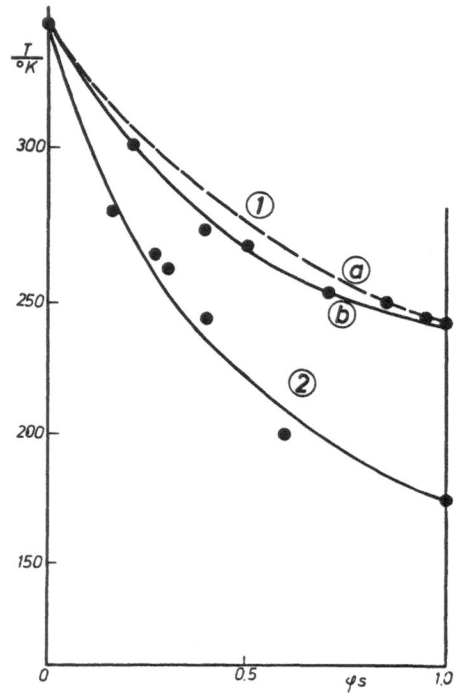

Fig. 10. System (2) salicyl acid – – naphthyl ester,
(1)   salicyl acid methyl ester,
(1a),   athermical mixture,
(1b)   excess enthalpy ($\beta = 40$)

Discussion

From the above agreement between calculations and experiment, the general validity of the iso-free-volume and the iso-excess-entropy hypothesis seems to be examined with some evidence:

The metastable equilibrium state of organic liquids at T^* *reveals general phenomenological characteristics independent of the "molecular-weight distribution" as well as independent of the average size of the molecules.*

It may be interesting to intensify the interpretation:

A. The "Configuration" of the Metastable Liquids at T^*

The basis of the configurational characterization of the liquid is the mathematical procedure of the lattice model supposing athermal conditions. If the molecules would have the same size as the voids the configurational entropy of mixing would be due to randomly distributed molecules and voids. The average excess entropy, which is needed to describe the metastable liquids at T^* (in any case independent of the relative size of all particles), seems to indicate *reliable deviations of this random arrangement of the elements*. Vibrational entropy mixing contributions which might be related to the empty space of the voids should enlarge the entropy. *Therefore the negative excess entropy should be due to considerable short range ordering within the metastable liquids at T^*. There is strong evidence that the molecules should be arranged within certain clusters on account of their mutual interactions.*

It is quite unexpected that the *specific excess entropy does not depend on the molecular weight or mutual interactions within the liquids at T^*. The deviations from the idealized athermical arrangements in spite of reliable differences in size of the participants have been proved to be the same within molecular solutions at the corresponding T^*.*

If there should exist individual arrangements of molecules within clusters (3—14) — especially if for example bundle-like arrangements of the chains are adopted — the distribution of the voids must be correlated to these configurations exactly in this manner that the average specific excess entropy of the mixture stays invariant at T^*.

B. Criteria of Stability

Looking at the sketch of the free enthalpy of a single component system as shown in Fig. 11

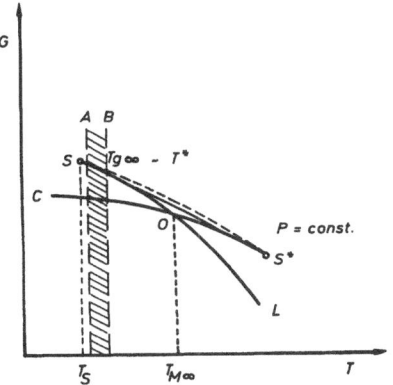

Fig. 11. The G-T plot at constant P for a single-component system. $T_{M\infty}$ refers to the melting temperature of an "infinite" extended system: S, S^* represents the limits of stability which we get from the extrapolation of the metastable states OS resp. OS^*. AB indicates qualitatively the temperature region where glass transition should occur. *Borchers* (35) has studied the stability of polystyrene systems and verified that T_S should be expected below $T_{g\infty}$ only. From fundamental reasons it is expected that $T_S < T_{g\infty}$ holds in any case

we might draw the following conclusion: *The state at the temperature T^* is not far from the limit of stability S* (34, 35). Therefore within the metastable liquids fluctuations may exist which should become infinitely large at the limit of stability. It might be suspected that these fluctuations will produce an "uniform excess entropy" of all metastable organic liquid systems at T^* — independent of the molecular parameters.

From this point of view the occurrence of the glass transition appears to be related in a more fundamental manner to criteria of stability of metastable undercooled liquids — inspite of evident kinetic aspects.

Criticism

The above thermodynamical description implies suppositions some of them we want to discuss shortly.

From the molecular point of view the simple lattice model seems to be not very accurate if we treat dense packed polymer systems. But attempts of refining reveals at least no relevant deviations of the simple description (33).

The assumptions $A^H = y_v A^h$ or $A^S = y_v A^s$ appear to be very synthetical ones as mentioned above. It is completely improbable that the enthalpy of void formation will be enlarged proportional to the volume of the voids themselves. This assumption can only be accepted as a rough approximation if the voids are relatively small. The above results — if they would be correct — would therefore indicate that *small voids should fill up the major portion of the free volume* metastable fluids at T^* — in accordance to our qualitative conclusions drawn elsewhere (19).

It is hard to understand that the enthalpy of void formation within the group of oligomer- and polymer-homologues turns out to be independent on temperature (independent on molecular weight). This might be due to co-operative contributions of a corresponding molecular neighbourhood of single voids which could be estimated — if possible at all — from exact knowledges on the molecular structure of the liquids only.

Acknowledgement

I am very obliged to Prof. Dr. *G. S. Y. Yeh* for useful discussions and for his kindness to improve the English text of this paper. I have to offer many thanks to Prof. Dr. *G. Rehage* for giving me useful thermodynamic data to me part of them are not yet published.

Summary

Describing the metastable equilibrium states at the quasi-static glass transition temperature of organic liquids with the help of the "*Frenkel* model" as "void-saturated-systems" it turns out that these states can be thermodynamically characterized by a common value of the specific excess entropy ("iso-excess-entropy-state") if an "iso-excess-volume state" at this temperature does exist. Therefore pronounced short range ordering producing the same specific entropy content of the total system should be apparent at this temperature independent of the structure of the molecules themselves. This hypothesis can be shown to be valid for various polymer solvent mixtures too. An understanding of this general result may be related to criteria of stability of metastable quasi-static organic liquids at their temperature of glass transition.

Zusammenfassung

Metastabile Gleichgewichtszustände von organischen Flüssigkeiten bei der quasi-statischen Glastemperatur $T_{g\infty}$ werden mit Hilfe des „*Frenkel*-Modells" als „Leerstellen-gesättigte" Mischungen thermodynamisch beschrieben.

Wenn bei dieser Temperatur dasselbe „freie Volumen" besteht, sollten die Systeme auch stets dieselbe spezifische Exzeß-Entropie besitzen. Die spezifische Exzeß-Entropie erweist sich als unabhängig von der molekularen Zusammensetzung der Systeme. Die Leerstellen verteilen sich offenbar nicht statistisch zufällig.

Im Rahmen der Meßgenauigkeit kann dies für verschiedene Polymer-Lösungsmittelmischungen bestätigt werden.

Ein grundlegendes Verständnis könnte aus Betrachtungen der Stabilität von organischen Flüssigkeiten im metastabilen Gleichgewicht bei der quasi-statischen Glastemperatur gewonnen werden.

References

1) *Mackenzie, J. D.*, Modern Aspects of the Vitreous State, Butterworths (1960).
2) *Hosemann, R.* and *S. N. Bagchi*, Direct Analysis of Diffraction of Matter, Worth Holland (1962).
3) *Pechhold, W.*, Kolloid-Z. u. Z. Polymere **251**, 818 (1973).
4) *Ovchinnikov, Yu. K., G. S. Markova* and *V. A. Kargin*, Polymer Sci. USSR **11**, 369 (1969).
5) *Bokhyan, E. B., Yu. K. Ovchinnikov, G. S. Markova* and *V. A. Kargin*, Polymer Sci. USSR **13**, 2026 (1971).
6) *Wilkes, C. E.* and *M. H. Lehr*, J. Macromol. Sci. **B 7**, 225 (1973).
7) *Robertson, R. E.*, J. Phys. Chem. **69**, 1575 (1965).
8) *Bose, E.*, Review Physics 8, 513 (1907).
9) *Ornstein, L. S.* and *W. Kast*, Faraday Soc. **19**, 932 (1933).
10) *Frank, W., H. Goddar* and *H. A. Stuart*, J. Polym. Sci. **B 5**, 711 (1967).
11) *Schoon, Th. G. F.*, Proc. Internat. Rubber Conf. **277** (1967).
12) *Yeh, G. S.* and *P. H. Geil*, J. Macromol. Sci. **B 1**, 235, 251 (1967).
13) *Yeh, G. S. Y.*, Macromol. Sci. **B 6**, 451, 456 (1972).
14) *Klement, J. J.* and *P. H. Geil*, J. Macromol. Sci. **B 6**, 31 (1972).
15) *Frenkel, J. I.*, Koll. **190**, 1 (Berlin, 1957).
16) *Eyrius, H.*, J. Chem. Phys. **4**, 283 (1936).
17) *Hirai, N.* and *W. Eyring*, J. Appl. Phys. **29**, 810 (1958).
18) *Kanig, G.*, Kolloid-Z. u. Z. Polymere **190**, 1 (1963).
19) *Kilian, H. G.*, Colloid & Polymer Sci., **252**, 353 -357 (1974)
20) *Fox, T. G.* and *P. J. Flory*, J. Appl. **21**, 581 (1950).
21) *Kanig, G.*, Kolloid-Z. u. Z. Polymere **233**, 829 (1969).
22) *Williams, M. L., R. F. Landel* and *J. D. Ferry*, J. Far. Soc. **77**, 3701 (1955).
23) *Simha, R.* and *R. F. Boyer*, J. Chem. Phys. **37**, 1003 (1962).
24) *Maocamir, J.* and *R. Simha*, J. Chem. Phys. **45**, 964 (1966).
25) *Miller, A. A.*, J. Pol. Sci. A 1, 1857, 1865 (1963).
26) *Miller, A. A.*, J. Pol. Sci. A 2, 1095 (1964).
27) *Boudi, A.*, J. Pol. Sci. A 2, 3159 (1964).
28) *Jenckel, E.* and *R. Heusch*, Koll.-Z. **130**, 89 (1953).
29) *Rehage, G.*, Koll.-Z. **196**, 97 (1964).
30) *Haase, R.*, Thermodynamik der Mischphasen (1956).
31) *Flory, P. J.*, Chemistry (1953).
32) *Lax, D'Ans*, Taschenbuch für Chemiker (1964).
33) *Tompa, H.*, Polymer Solutions (London, 1956).
34) *Callen*, Thermodynamics (1960).
35) *Borchers, K.*, private communication.

Author's address:

Prof. Dr. *H. G. Kilian*
Universität Ulm
Abt. für Experimentelle Physik I
75 Ulm
Oberer Eselsberg

Progr. Colloid & Polymer Sci. **58**, 61—76 (1975)

Meß- und Prüflaboratorium der BASF Aktiengesellschaft, Ludwigshafen/Rhein

Die Änderung der Dichte und der Schmelzwärme beim Tempern von 6.6-Polyamid

K. H. Illers

Mit 15 Abbildungen und 3 Tabellen

(Eingegangen am 8. August 1974)

I. Einleitung

Zwischen der experimentellen Schmelzwärme ΔH_M und dem bei 20 °C gemessenen spezifischen Volumen v besteht bei vielen Polymeren ein linearer Zusammenhang. Bei 6-Polyamid, das je nach Kristallisationsbedingung in der α- oder γ-Modifikation vorliegen kann, wurde für jede der beiden kristallinen Modifikationen eine andere lineare Beziehung zwischen ΔH_M und v gefunden [1]. In einer Kurzmitteilung haben wir 1971 eine Zusammenfassung unserer Ergebnisse für 6.6-Polyamid veröffentlicht [2]. Auch bei diesem Polyamid erhält man zwei verschiedene Geraden, wenn man die Schmelzwärme als Funktion des spezifischen Volumens aufträgt.

Die 1972 publizierten Untersuchungen von *Hinrichsen* [3] an unverstrecktem und verstrecktem 6.6-Polyamid stimmen mit unseren Ergebnissen nicht überein. Insbesondere wurde in [3] gefunden, daß zwar die Dichte mit zunehmender Tempertemperatur (80 — 260 °C) stark ansteigt, die Schmelzwärme sich dabei jedoch nur geringfügig ändert. Die Dichtezunahme wird in [3] hauptsächlich auf einen Anstieg der Kristalldichte mit steigender Tempertemperatur zurückgeführt, während sich die Kristallinität beim Tempern nur wenig ändern soll.

Im folgenden sind die unserer Kurzmitteilung [2] zugrundeliegenden experimentellen Ergebnisse und weitere inzwischen vorliegenden Resultate ausführlich beschrieben. Es wird versucht, die Ursachen für die Diskrepanzen zwischen [2] und [3] aufzuklären.

II. Experimentelles

1. Probenherstellung

1.1. Temperung abgeschreckter Preßplatten

Trockenes Granulat aus Ultramid A 3® wurde bei 280 °C zwischen 0,02 mm dicken Aluminiumfolien zu 0,5 mm dicken Platten verpreßt und so schnell wie möglich in einem n-Heptanbad von 0 °C abgeschreckt. Die Aluminiumfolien kleben sehr fest auf den Oberflächen der abgeschreckten Platten und verhindern das Eindringen von Luftfeuchtigkeit. Die abgeschreckten Proben wurden bis zur weiteren Verwendung über P_2O_5 gelagert.

Die nach dem Abschrecken nur geringfügig kristallinen Proben (Dichte 1,098 g/cm³) wurden durch Tempern zwischen 47 und 260 °C kristallisiert. Die Temperzeiten lagen zwischen 2 und 10^4 min. Temperungen unter 80 °C wurden im Kalorimeter unter Spülung mit trockenem Stickstoff und bei höheren Temperaturen in thermostatiertem flüssigem *Wood*-Metall durchgeführt. Die auf beiden Probenoberflächen haftende Aluminiumfolie wurde erst unmittelbar vor der Messung entfernt. Auf diese Weise wird außer der Feuchtigkeitsaufnahme auch die Oxidation bei höheren Tempertemperaturen vermieden. Nur an den ungeschützten Stirnflächen der Proben trat während längerer Temperung bei höherer Temperatur eine Verfärbung auf. Diese Teile der Probe wurden deshalb verworfen.

1.2. Temperung langsam abgekühlter Preßplatten

Die bei 280 °C zwischen Aluminiumfolien gepreßten Platten wurden in der Presse durch Ausschalten der elektrischen Heizung sehr langsam abgekühlt und dann 1 Std. bei 200, 220, 230 und 245 °C in flüssigem *Wood*-Metall getempert. Im Anschluß an die Temperung wurde schnell auf Raumtemperatur abgekühlt. Die gleichen Temperungen wurden ferner an Proben vorgenommen, die durch Einschaltung der Wasserkühlung schneller in der Presse abgekühlt waren.

1.3. Isotherme Kristallisation aus der Schmelze

Die Proben wurden einschließlich der Aluminium-folien im Kalorimeter 1 min bei 280 °C aufgeschmolzen, dann schnell auf die Kristallisationstemperatur von 240 °C gebracht und nach einer Kristallisationsdauer von 6 min bis 14 Std. rasch auf Raumtemperatur abgekühlt. Auch diese Proben wurden bis zur Messung über P_2O_5 aufbewahrt.

1.4. Unverstreckte, trocken abgeschreckte Drähte

Die 1,2 mm dicken Drähte wurden aus der Düse unmittelbar in ein CCl_4-Bad von -20 °C gesponnen und nach Entfernen des oberflächlich anhaftenden CCl_4 sofort in einen Exsikkator überführt. 1stündige Temperungen wurden bei 121,5, 160, 200 und 240 °C in Wood-Metall durchgeführt und die Proben anschließend über P_2O_5 gelagert.

1.5. Unverstreckte, feucht gelagerte Drähte

Die Borsten wurden in ein Wasserbad von 20 °C gesponnen und bis zur Temperung 3 Wochen in der Laboratmosphäre gelagert. Die Temperbedingungen stimmen mit denen in 1.4. überein. Zwischen Temperung und Messung lagerten die Proben im Exsikkator.

1.6. Heiß-verstreckte, feucht gelagerte Drähte

Die Drähte wurden in ein Wasserbad von 20 °C gesponnen und anschließend 1 : 4,3 heiß verstreckt. Vor der 1stündigen Temperung in Wood-Metall bei 160, 200 und 240 °C wurden die Proben in der feuchten Laboratmosphäre, nach der Temperung jedoch über P_2O_5 gelagert.

2. Dichtebestimmung

Die Dichte der Proben wurde nach der Schwebemethode in n-Heptan/CCl_4-Mischungen bei 20 °C ermittelt. Die Reproduzierbarkeit an ein und derselben Probe betrug $\pm 0,0005$ g/cm³.

3. Kalorische Messungen

Die Schmelzwärme wurde mit Hilfe des Perkin-Elmer DSC-1 B-Differentialkalorimeters gemessen. Die Probeneinwaage von ca. 15 mg wurde auf $\pm 0,05$ mg genau ermittelt. Die Messungen wurden zwischen 0 und 300 °C mit 16 °C/min bei einer Empfindlichkeit von 32 mcal/sec/inch durchgeführt. Die Eichung des Kalorimeters erfolgte durch je 5 Messungen an folgenden reinen Stoffen mit bekannter Schmelzwärme: Indium (28,4 J/g), Zinn (59,6 J/g) und Harnstoff (243 J/g). Die mit diesen Sub-

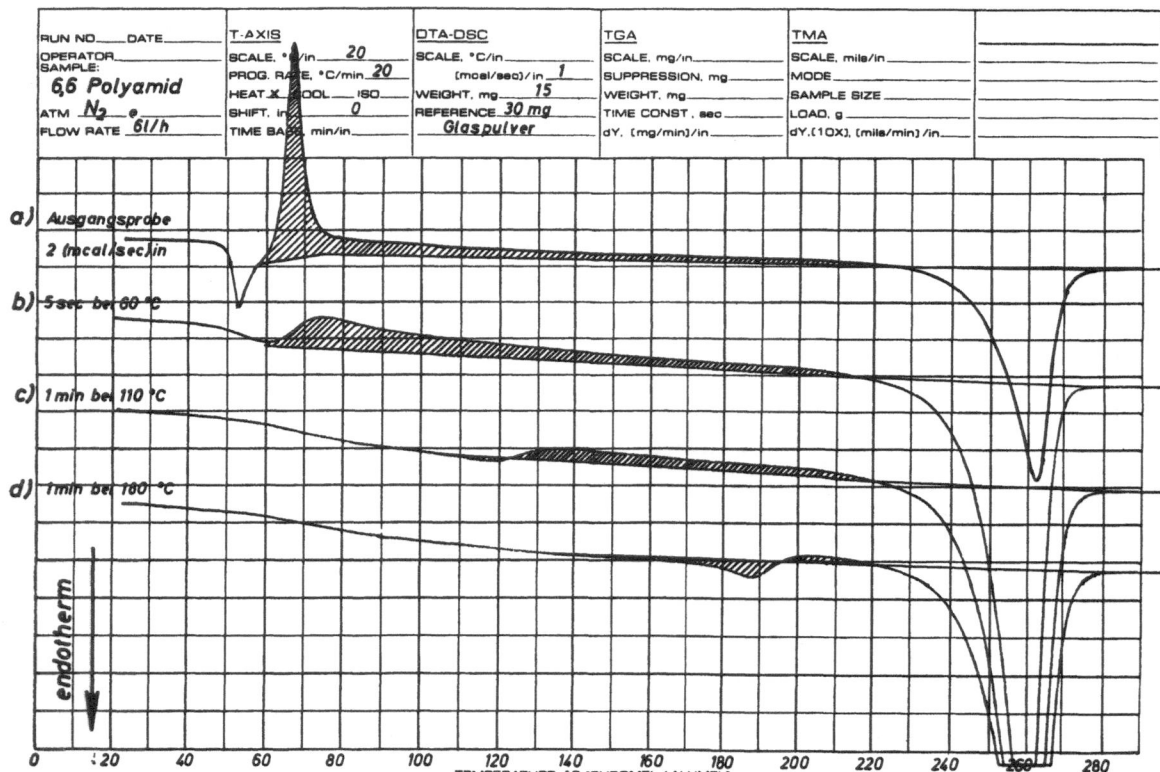

Abb. 1. Thermogramme abgeschreckter und getemperter 6.6-Polyamidproben, aufgenommen mit dem Thermal-Analyzer-990, Aufheizgeschwindigkeit 20 °C/min, Einwaage ca. 15 mg. Bei Kurve a) beträgt die Empfindlichkeit 2 mcal/sec/inch, bei b) bis d) 1 mcal/sec/inch.

a) In Eiswasser abgeschreckte Preßplatte
b) Probe a) 5 sec bei 60 °C getempert

c) Probe a) 1 min bei 110 °C getempert
d) Probe a) 1 min bei 180 °C getempert

stanzen ermittelten Eichfaktoren stimmten untereinander besser als $\pm 1\,^0/_0$ überein.

Ausgewählte Proben wurden außerdem im DuPont-Thermal-Analyzer-990 untersucht. Dieses Gerät besitzt auch bei höherer Empfindlichkeit eine lineare Basislinie und erlaubt unter Verzicht auf die Registrierung des vollständigen Hauptschmelzpeaks eine genauere Untersuchung des sich über einen breiten Temperaturbereich erstreckenden Vorschmelzens. Aus diesen Messungen erhält man Informationen über den genauen Temperaturbereich, in dem die Planimetrierung der mit dem DSC-1 B-Kalorimeter aufgenommenen Schmelzpeaks durchzuführen ist.

III. Meßergebnisse

1. Kalorische Messungen

Im folgenden wird zunächst an Hand der mit dem Thermal-Analyzer aufgenommenen Thermogramm das Schmelzverhalten der abgeschreckten und bei verschiedenen Temperaturen getemperten Preßplatten qualitativ diskutiert. Dabei gilt das besondere Augenmerk dem oben erwähnten Vorschmelzbereich, dessen vollständige Erfas-

sung beim Planimetrieren der DSC-Kurven für die Genauigkeit der Schmelzwärmebestimmung entscheidend ist.

Beim Erwärmen der abgeschreckten ungetemperten Probe (Dichte 1,098 g/cm³) tritt nach Abb. 1, Kurve a, oberhalb der Glastemperatur T_g von ca. 50 °C ein endothermes Überhitzungsmaximum auf, hervorgerufen durch die zwischen Probenherstellung und Messung stattgefunden Enthalpierelaxation (4). Bei einer unmittelbar nach dem Abschrecken durchgeführten Messung ist der Überhitzungspeak nicht vorhanden, so daß man aus Messungen mit hoher Aufheizgeschwindigkeit die Größe der c_p-Stufe bei T_g ermitteln kann. Oberhalb des Glasübergangs führt die rasch einsetzende Kaltkristallisation zum Auftreten eines scharfen exothermen Maximums zwischen 60 und 75 °C, dem sich ein breiter kontinuierlicher Nachkristallisationsbereich bis zu einer Temperatur von 220 °C anschließt. Oberhalb 220 °C überwiegt der endotherme Schmelzvorgang.

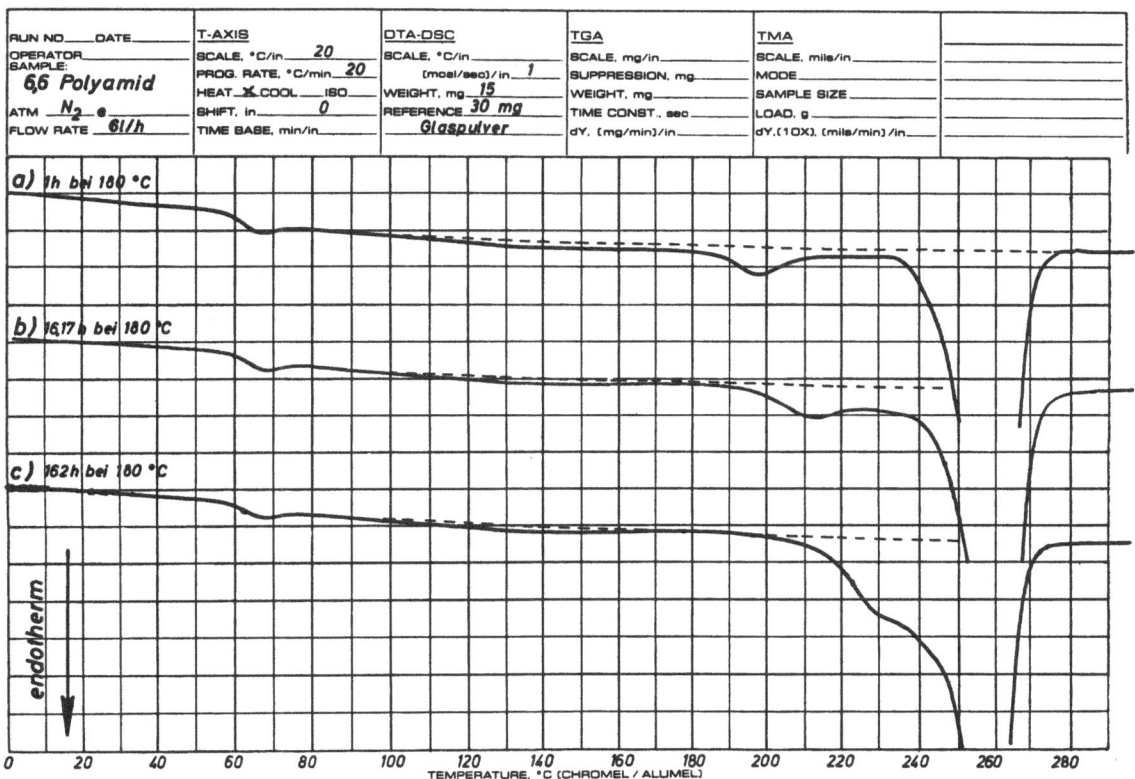

Abb. 2. Thermogramme abgeschreckter und getemperter 6.6-Polyamidproben, aufgenommen mit dem Thermal-Analyzer-990, Aufheizgeschwindigkeit 20 °C/min, Einwaage ca. 15 mg. Empfindlichkeit 1 mcal/sec/inch.
a) Abgeschreckte Preßplatte 1 Std. bei 180 °C getempert
b) abgeschreckte Preßplatte 16 Std. bei 180 °C getempert
c) abgeschreckte Preßplatte 162 Std. bei 180 °C getempert

Wenn sich wie in Abb. 1 das Schmelzen und Kristallisieren über einen Temperaturbereich von ca. 200 °C erstreckt, ist das Einzeichnen der Basislinie in das Thermogramm extrem schwierig. Zur Ermittlung der Kristallisationswärme ΔH_k (schraffierte Fläche) wurde folgendermaßen vorgegangen. Unmittelbar oberhalb T_g ist die Basislinie in einem sehr engen Temperaturbereich durch die c_p-Stufe bei T_g gegeben. Infolge der zwischen 60 und 75 °C erfolgenden Kaltkristallisation muß die Basislinie in diesem Temperaturbereich in Abb. 1 ansteigen. Die Größe des als linear angenommenen Anstiegs wurde ermittelt, indem eine abgeschreckte Probe mit 20 °C/min auf 75 °C erwärmt, dann sehr schnell auf Raumtemperatur abgekühlt und anschließend gemessen wurde. Durch diese letzte Messung erhält man den c_p-Wert bei 75 °C für die bis zu dieser Temperatur kaltkristallisierte Probe, wobei angenommen ist, daß unterhalb T_g die spezifische Wärme nicht

von der Kristallinität abhängt. Von 75 °C bis zum Schmelz-Ende bei ca. 275 °C wurde als Basislinie eine fast lineare Leermessung benutzt, bei der die Probe durch eine Glaspulvereinwaage von gleicher Wärmekapazität ersetzt war. Es läßt sich abschätzen, daß die auf diese Weise im Temperaturbereich zwischen T_g und 220 °C ermittelte Kristallisationswärme der abgeschreckten ungetemperten Probe um max. 5 J/g zu groß sein kann. Bei allen getemperten Proben ist der Fehler kleiner. Auf die Benutzung einer Leermessung als Basislinie wird weiter unten ausführlicher eingegangen.

Tempert man die abgeschreckte Probe 5 sec bei 60 °C und schreckt sie dann wieder in Eiswasser ab, so ist im anschließend aufgenommenen Thermogramm (Abb. 1, Kurve b) der scharfe Kaltkristallisationspeak verschwunden, und es tritt nur noch der kontinuierliche Nachkristallisationsbereich auf. Tempert man abgeschrecktes 6.6-Polyamid 1 min bei Temperaturen zwischen

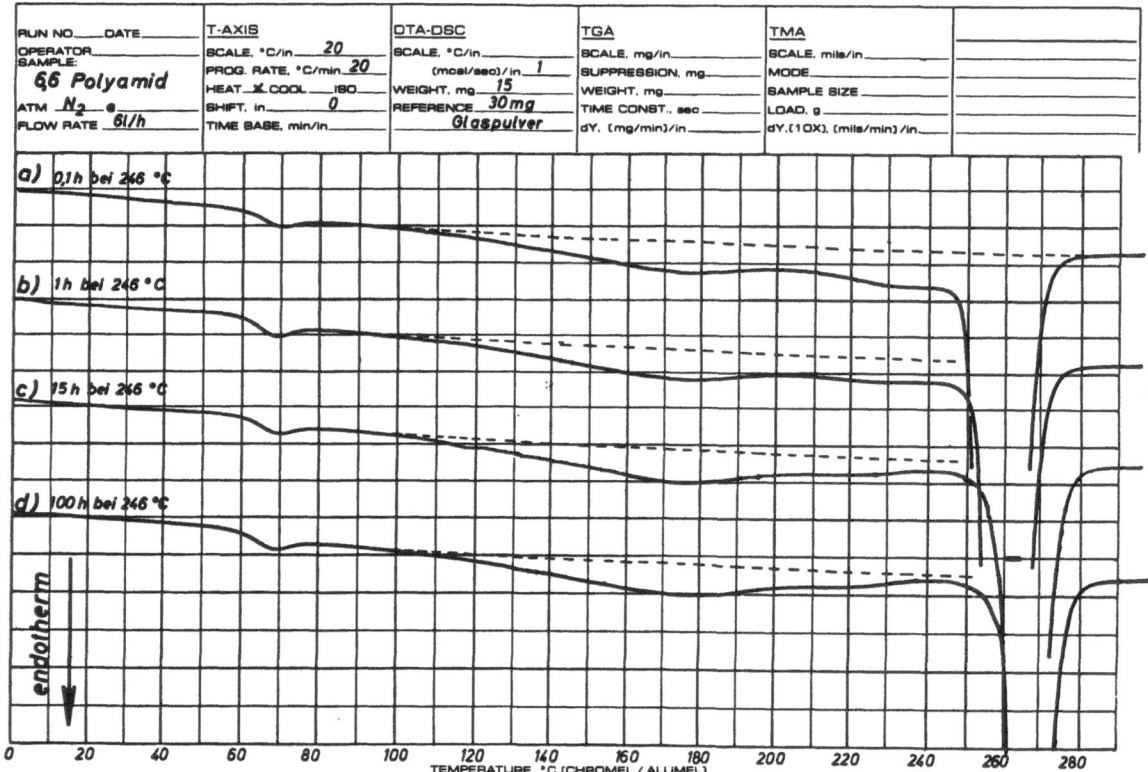

Abb. 3. Thermogramme abgeschreckter und getemperter 6.6-Polyamidproben, aufgenommen mit dem Thermal-Analyzer-990, Aufheizgeschwindigkeit 20 °C/min, Einwaage ca. 15 mg, Empfindlichkeit 1 mcal/sec/inch.
a) Abgeschreckte Preßplatte 0,1 Std. bei 246 °C getempert
b) abgeschreckte Preßplatte 1 Std. bei 246 °C getempert
c) abgeschreckte Preßplatte 15 Std. bei 246 °C getempert
d) abgeschreckte Preßplatte 100 Std. bei 246 °C getempert

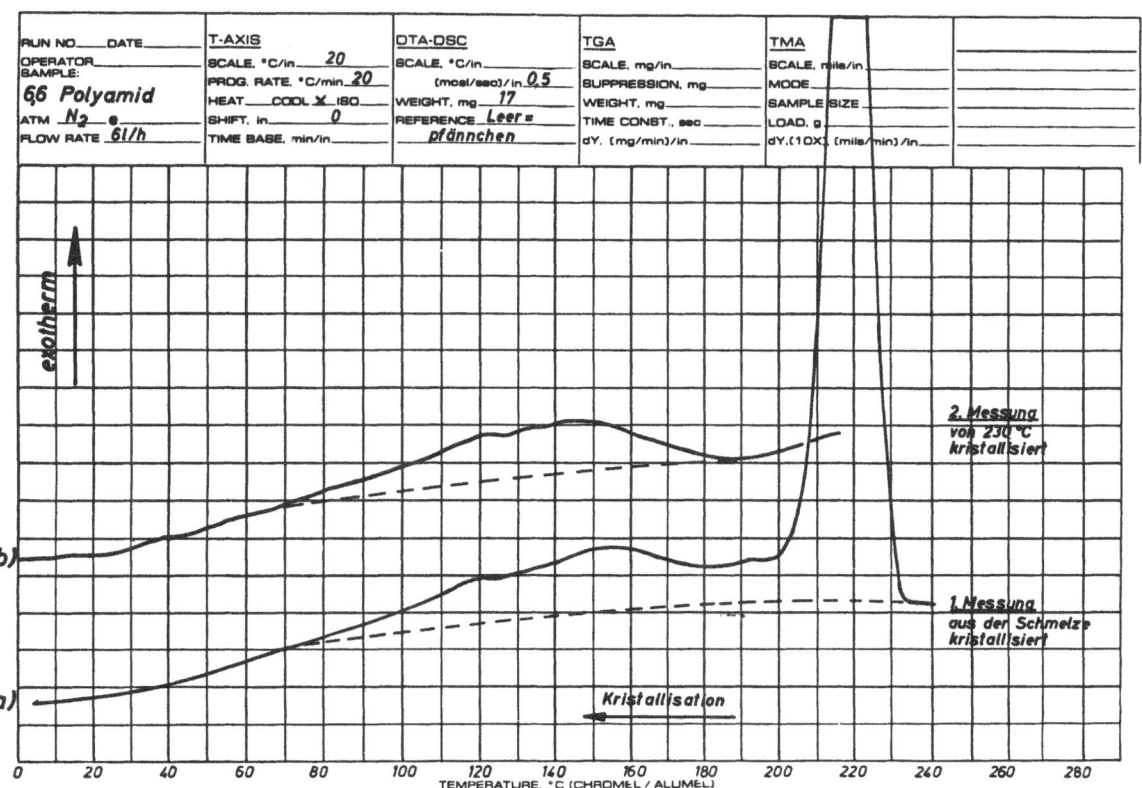

Abb. 4. Während der Abkühlung mit 20 °C/min aufgenommene Kristallisationskurven von 6.6-Polyamid. Einwaage ca. 17 mg, Empfindlichkeit 0,5 mcal/sec/inch.
a) Von 280 °C aus abgekühlt
b) im Anschluß an a) auf 230 °C erwärmt und während der erneuten Abkühlung gemessen

60 und 180 °C, so setzt die restliche Nachkristallisation oberhalb der Tempertemperatur ein (Abb. 1, Kurven c und d). Außerdem treten kurz oberhalb der Tempertemperatur kleine endotherme Schmelzmaxima auf, bei denen es sich um die bei fast allen kristallinen Polymeren wohlbekannten „*Temperpeaks*" handelt.

Das Schmelzverhalten abgeschreckter und bei 180 °C getemperter Proben ist in Abb. 2 für längere Temperzeiten dargestellt. Mit steigender Temperdauer verschiebt sich der Temperpeak wie üblich zu höherer Temperatur und wird intensiver. Ein kleiner Kristallanteil schmilzt bereits im Temperaturbereich zwischen 100 °C und dem Temperpeak auf. Eine Vernachlässigung dieses *Vorschmelzens* beim Planimetrieren der DSC-Kurven führt zu einem merklichen Fehler in der Schmelzwärme.

Wie ein Vergleich von Abb. 2 und 3 zeigt, nimmt der im Vorschmelzbereich schmelzende Kristallanteil mit steigender Tempertemperatur zu. Die im Vorschmelzbereich verbrauchte

Schmelzwärme ist unabhängig von der Temperdauer, weil sie durch das Schmelzen von Kristallen zustande kommt, die nach Beendigung der Temperung während der raschen Abkühlung der Probe entstanden sind[1]). Dies ist in Abb. 4, Kurve a, durch das während der Abkühlung aus der Schmelze aufgenommene Thermogramm nachgewiesen. Unterhalb des durch die primäre Kristallisation verursachten Hauptkristallisationsmaximums (235 bis ca. 200 °C) findet in einem breiten, sich bis etwa 80 °C erstreckenden Temperaturbereich eine sekundäre Kristallisation statt. Erwärmt man die Probe wieder auf 230 °C (bei dieser Temperatur sind die primär entstandenen Kristalle noch nicht aufgeschmolzen), so tritt bei der anschließenden Abkühlung die sekundäre Kristallisation im Thermogramm erneut auf (Kurve b).

---

[1]) Durch extrem langsame Abkühlung oder sehr lange nachträgliche Temperung im Temperaturbereich des Vorschmelzens kann man das Vorschmelzen vollständig zum Verschwinden bringen.

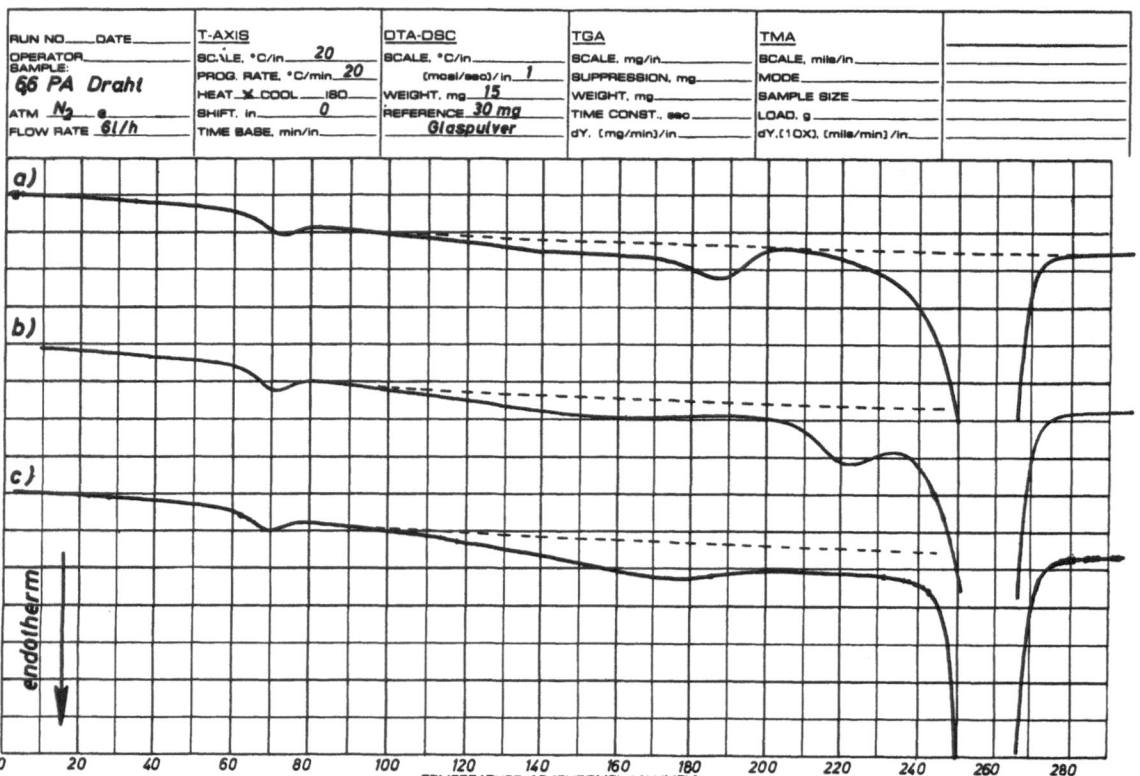

Abb. 5. Thermogramme von in CCl_4-Bad gesponnenem 6.6-Polyamiddraht ,unverstreckt. Aufheizgeschwindigkeit 20 °C/min, Einwaage ca. 15 mg, Empfindlichkeit 1 mcal/sec/inch. Temperdauer 1 Std. bei
a) 160 °C,   b) 200 °C,   c) 240 °C

Der bei der abgeschreckten und getemperten Preßplatte beobachtete sehr breite Vorschmelzbereich ist auch bei den getemperten unverstreckten Drähten vorhanden und beginnt nach Abb. 5 ebenfalls bei 100 °C. Bei heißverstreckten Drähten ist nach der Temperung kein Vorschmelzbereich zu erkennen, und es tritt nur ein relativ scharfer Hauptschmelzpeak auf.

Die Ermittlung der Schmelzwärme aus einer DSC-Kurve setzt strenggenommen die Kenntnis der Temperaturabhängigkeit der spezifischen Wärme der Schmelze und des Kristalls sowie der Kristallinität im untersuchten Temperaturbereich voraus. Diese Funktionen sind jedoch im allgemeinen nicht bekannt. Ein in (5) entwickeltes Iterationsverfahren ist nicht anwendbar, wenn im Temperaturbereich unterhalb des Hauptschmelzpeaks ein breiter Vorschmelzbereich vorhanden ist. In der vorliegenden Arbeit wurde die Schmelzwärme aus der Fläche zwischen der DSC-Kurve und einer linearen Basislinie ermittelt, die den Schmelzbeginn und das Schmelz-Ende verbindet. Dabei muß natür-

lich sichergestellt sein, daß die Basislinie in diesem Temperaturbereich nicht aus apparativen Gründen gekrümmt ist. Für alle unsere Untersuchungen mit dem DSC-1 B-Kalorimeter wurde als Basislinie eine Leermessung benutzt, die in den Temperaturbereichen oberhalb T_g und oberhalb des Schmelz-Endes, in denen kein Schmelzen stattfindet, mit der DSC-Kurve der Probe zur Deckung gebracht wurde. Die Messungen mit und ohne Polyamidprobe wurden mit der gleichen Aufheizgeschwindigkeit und Kalorimeterempfindlichkeit durchgeführt. Die Leermessung ergab stets eine zwischen 50 und 350 °C lineare oder höchstens sehr schwach gekrümmte Basislinie, die täglich neu aufgenommen und auch zwischen den einzelnen Messungen häufig kontrolliert wurde.

Zur Abschätzung des Fehlers, den man bei Verwendung einer Leermessung als Basislinie macht, wurde wie folgt vorgegangen. An einer gut kristallisierten Probe (Dichte 1,147 g/cm³) wurden absolute c_p-Messungen im Temperaturbereich zwischen —50 und 320 °C durchgeführt.

Abb. 6. Zusammenhang zwischen Schmelzwärme und Temperdauer von abgeschreckten
und anschließend bei den angegebenen Temperaturen getemperten 6.6-Polyamid-Preß-
platten. Messungen mit dem DSC-1 B-Kalorimeter

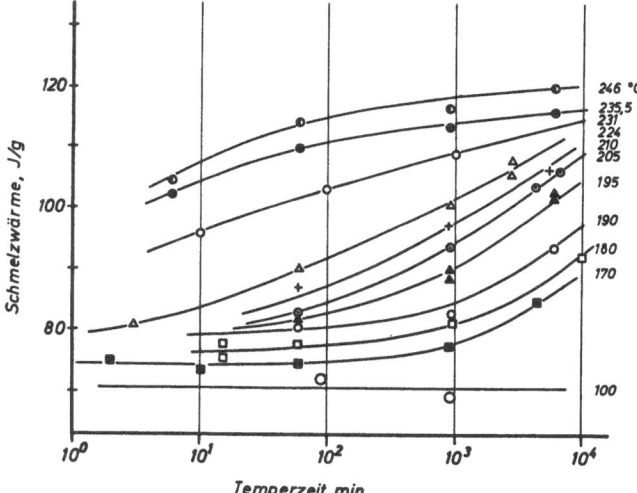

Zwischen —50 und 30 °C und oberhalb des
Schmelz-Endes nimmt c_p linear mit steigender
Temperatur zu, mit dem größeren Anstieg bei
tiefen Temperaturen. Aus dem Temperaturbereich unterhalb T_g wird c_p linear in den
Schmelzbereich extrapoliert als Näherung für
die spezifische Wärme des Kristalls. Wir nehmen
an, daß eine 100% kristalline Probe im Temperaturbereich des Hauptschmelzpeaks zwischen
220 und 280 °C schmilzt, und wählen als Näherung für die Basislinie die lineare Verbindung
zwischen dem Schmelzbeginn auf der (extra-
polierten) c_p-Kurve des Kristalls und dem
Schmelz-Ende auf der c_p-Kurve der Schmelze.
Weiter betrachten wir die Näherung von *Adam*
und *Müller* (6), nach der die Basislinie bis zur
Temperatur des Schmelzpeaks auf der c_p-Kurve
des Kristalls verläuft und dort auf die zu tieferer
Temperatur extrapolierte c_p-Kurve der Schmelze
springt. Bei kristallinem 6.6-Polyamid beträgt
die Differenz zwischen der linearen Verbindung
von Schmelzbeginn und -ende und der *Adam-
Müller*schen Näherung 6 J/g. Die wirkliche
Basislinie verläuft nach (5) im Schmelzbereich
zwischen diesen beiden Näherungen. Es läßt
sich leicht abschätzen, daß bei Verwendung
einer linearen Basislinie eine um ca. 4 J/g zu
kleine Schmelzwärme für das 100% kristalline
6.6-Polyamid gefunden wird. Dieser Fehler verringert sich stark mit abnehmender Kristallinität und dürfte bei den von uns untersuchten
Proben innerhalb der durch Planimetrieren und
Wägen gegebenen Meßgenauigkeit liegen.
Die experimentelle Schmelzwärme abgeschreckter und anschließend getemperter Preß-

platten ist in Abb. 6 als Funktion der Temperdauer aufgetragen. Vor der Temperung besitzt
die frisch abgeschreckte Probe eine Schmelzwärme von 10—15 J/g[2]). Die Zunahme der
Schmelzwärme beim Tempern erfolgt in zwei
Stufen. Beim Eintauchen der Probe in das auf
die Tempertemperatur erwärmte *Wood*-Metall
läuft zunächst die dem scharfen exothermen
Kaltkristallisationspeak in Abb. 1 entsprechende
Zunahme der Schmelzwärme auf ca. 70 J/g ab.
Dieser schnelle Vorgang ist oberhalb 70 °C in
Zeiten < 1 min beendet und geht daher den
in Abb. 6 dargestellten Schmelzwärmeänderun-
gen voraus. Bei Temperaturen bis zu 150 °C
bleibt die Schmelzwärme im Anschluß an die
schnelle Kaltkristallisation für Temperzeiten
bis zu 10^4 min konstant. Oberhalb 150 °C erfolgt
ein erneuter Schmelzwärmeanstieg, der mit steigender Tempertemperatur zunehmend früher
einsetzt.

Die Abhängigkeit der Schmelzwärme abgeschreckter Proben von der Tempertemperatur
für verschiedene konstante Temperzeiten ist in
Abb. 7 dargestellt. Bei kurzen Temperzeiten ist
ein sehr starker Anstieg der Schmelzwärme im
Temperaturbereich um 230 °C besonders augenfällig. Bei längerer Temperdauer verschwindet
diese Erscheinung.

Für die übrigen, 1 Std. getemperten Proben
ist die Abhängigkeit der Schmelzwärme von der
Tempertemperatur in Abb. 8 wiedergegeben.

---

[2]) Bei Proben mit Nachkristallisation (vgl. Abb. 1)
ergibt sich die Schmelzwärme als Differenz der endo-
thermen und exothermen Effekte.

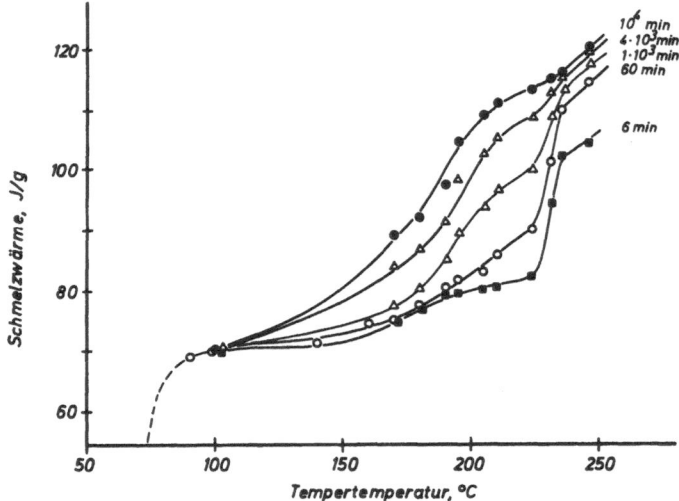

Abb. 7. Zusammenhang zwischen Schmelzwärme und Tempertemperatur für konstante Temperzeiten. Abgeschreckte und
anschließend getemperte 6.6-Polyamid-
Preßplatten (DSC-1 B-Kalorimeter)

Bei dem in CCl_4 abgeschreckten unverstreckten Draht und weniger deutlich bei der in der Presse schnell abgekühlten Preßplatte tritt ebenfalls eine stärkere Erhöhung der Schmelzwärme durch Temperung oberhalb 220 °C ein. Obwohl die in Abb. 8 untersuchten Proben vor der Temperung sehr unterschiedliche Schmelzwärmen besitzen (10—110 J/g), erhält man nach einer Temperung oberhalb ca. 230 °C für alle Proben innerhalb der Fehlergrenze den gleichen Wert.

Die von *Hinrichsen* an unverstrecktem Draht gemessene Schmelzwärme ist ebenfalls in Abb. 8 eingetragen. Die genauen Herstellungsbedingungen und die Temperdauer wurden in (3) nicht angegeben. Beim Vergleich mit unseren Messungen fällt auf, daß oberhalb einer Tempertemperatur von 200 °C die von *Hinrichsen* angegebenen Schmelzwärmen kleiner als die eigenen Werte sind.

2. Dichtemessungen

An den abgeschreckten Preßplatten wurde die Dichte als Funktion der Temperdauer bei Tempertemperaturen zwischen 100 und 246 °C gemessen. Die Ergebnisse sind in Abb. 9 wiedergegeben. Beim Vergleich mit Abb. 6 fällt auf, daß bei Tempertemperaturen bis zu 180 °C und Temperzeiten $< 10^4$ min die Dichte mit zunehmender Temperdauer linear ansteigt, während sich die Schmelzwärme innerhalb der Meßgenauigkeit nicht ändert.

Die aus Abb. 9 bei verschiedenen konstanten Zeiten entnommenen Dichten sind in Abb. 10 als Funktion der Tempertemperatur dargestellt. Die Zunahme der Dichte erfolgt wie bei der Schmelzwärme in 2 Stufen.

Für die übrigen in Abschn. II.1 beschriebenen Proben wurde die Dichte nur nach einstündiger

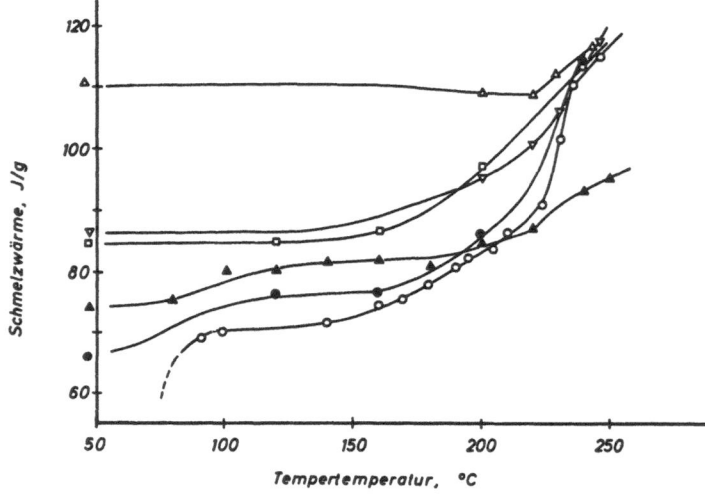

Abb. 8. Zusammenhang zwischen Schmelzwärme und Tempertemperatur für 1 Std. getemperte 6.6-Polyamidproben. DSC-1 B-Kalorimeter.

○ abgeschreckte Preßplatte
▽ schnell in der Presse abgekühlte Platte
△ langsam in der Presse abgekühlte Platte
● unverstreckter Draht, in CCl_4-Bad abgeschreckt
□ unverstreckter Draht, in Wasserbad abgeschreckt
▲ unverstreckter Draht nach *Hinrichsen* (3)

Abb. 9. Zusammenhang zwischen Dichte (20 °C) und Temperdauer von abgeschreckten und anschließend bei den angegebenen Temperaturen getemperten 6.6-Polyamid-Preßplatten

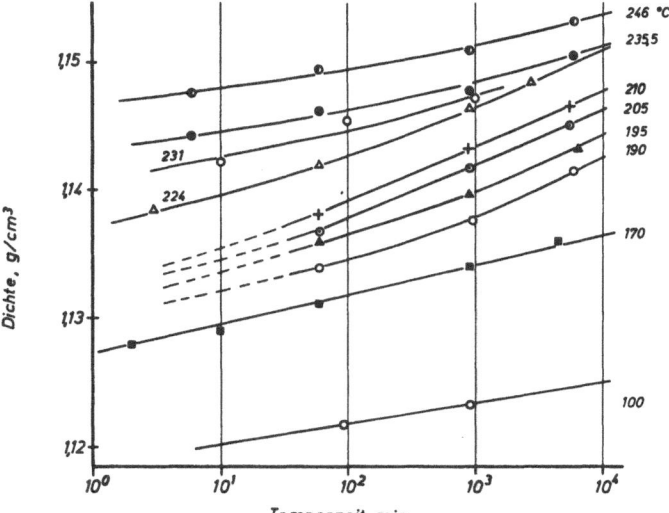

Temperung gemessen. Die Ergebnisse sind in Abb. 11 als Funktion der Tempertemperatur aufgetragen. Die Werte für die abgeschreckte Preßplatte wurden aus Abb. 10 übernommen. Trotz großer Dichteunterschiede der Ausgangsproben laufen die Kurven in Abb. 11 bei hohen Tempertemperaturen zusammen. Nach einer einstündigen Temperung bei 245 °C besitzen Proben mit Ausgangsdichten zwischen 1,098 und 1,147 g/cm³ die gleiche Dichte von 1,150 ± 0,002 g/cm³.

Die im Vergleich zu den übrigen Proben größere Dichte des feucht gelagerten Drahtes ist auf die erhöhte Effektivität der Temperung bei Anwesenheit von Wasser im Anfangsstadium

der Temperung zurückzuführen. Im Verlauf der einstündigen Temperung bei 160, 200 bzw. 240 °C verlieren die Proben schließlich das vorher aufgenommene Wasser, so daß die Dichtemessung im trockenen Zustand erfolgte. Der 1 Std. bei 120 °C getemperte feuchte Draht enthält wahrscheinlich noch eine geringe Restfeuchtigkeit.

Zum Vergleich ist in Abb. 11 ferner die von *Hinrichsen* (3) an unverstrecktem Draht gemessene Dichtekurve eingezeichnet. Sie läuft im Temperaturbereich zwischen 150 und 250 °C wesentlich oberhalb unserer Messungen. Die dieser Differenz zugrunde liegende Ursache wird weiter unten diskutiert.

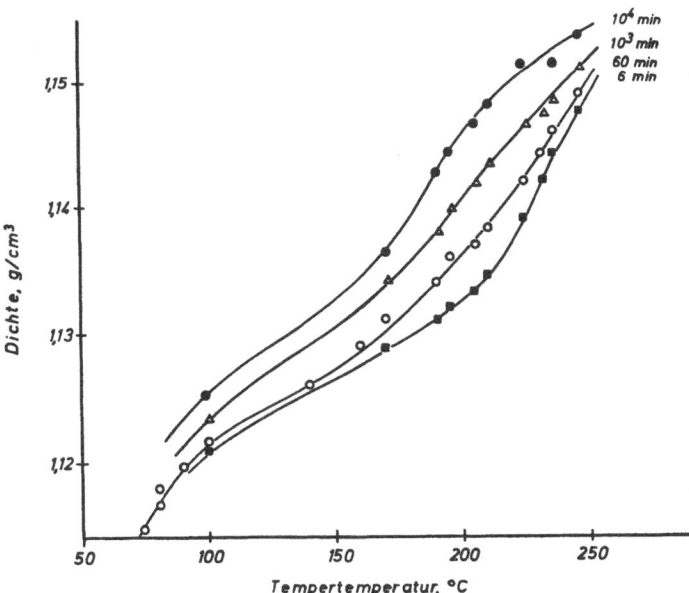

Abb. 10. Zusammenhang zwischen Dichte (20 °C) und Tempertemperatur für konstante Temperzeiten bei abgeschreckten 6.6-Polyamid-Preßplatten

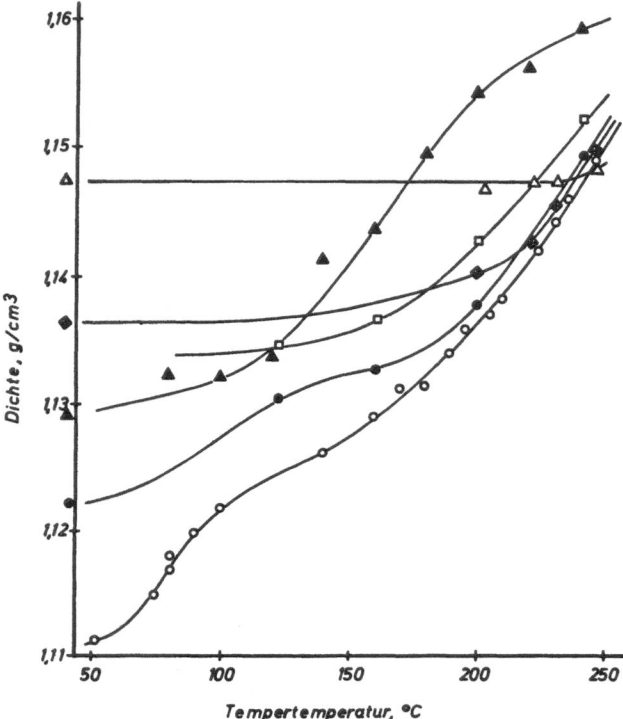

Abb. 11. Zusammenhang zwischen Dichte (20 °C) und Tempertemperatur für 1 Std. getemperte 6.6-Polyamidproben.

○ Abgeschreckte Preßplatte
|⊕| schnell in der Presse abgekühlte Platte
△ langsam in der Presse abgekühlte Platte
● unverstreckter Draht, in CCl₄-Bad abgeschreckt
□ unverstreckter Draht, in Wasserbad abgeschreckt
△ unverstreckter Draht nach *Hinrichsen* (3)

IV. Diskussion

1. Einfluß der Probenfeuchtigkeit auf Dichte und Schmelzwärme

Die von *Hinrichsen* untersuchten Proben wurden vor und nach der Temperung unter Laborbedingungen gelagert. Dabei können die Proben Luftfeuchtigkeit aufnehmen. Von *Hinrichsen* (3) wird jedoch ausdrücklich betont, daß die Unterschiede zwischen seinen und unseren Ergebnissen „*nicht dadurch wegzudeuten sind, daß ungetrocknete Proben verwendet wurden, da durchgeführte Vergleichsmessungen an über P₂O₅ getrockneten Proben keine nennenswerten Unterschiede der Meßwerte erbrachten*". Diese Behauptung bedarf einer näheren Untersuchung.

Aus den Untersuchungen von *Starkweather* (7) ist seit langem bekannt, daß die Dichte von 6.6-Polyamid mit steigendem Wassergehalt zu-

nimmt. Wie aus Tabelle 1 hervorgeht, gilt dies auch für die in dieser Arbeit untersuchten Proben. Durch eine 10tägige Lagerung in der feuchten Laboratmosphäre tritt eine Wasseraufnahme von 1,5—1,8% ein. Dies verursacht eine Dichtezunahme um 0,005—0,007 g/cm³. Durch eine 24stündige Vakuumtrocknung bei 100 °C läßt sich das bei der Lagerung aufgenommene Wasser fast vollständig wieder entfernen, und die Dichte geht wieder ungefähr auf den Wert der trockenen Ausgangsprobe zurück. Die höhere Dichte der feuchten Proben ist also nicht auf eine Nachkristallisation unter dem Einfluß der Feuchtigkeit zurückzuführen.

Da nach den Angaben in (3) eine längere Lagerung der Proben über P₂O₅ keine nennenswerte Änderung der Dichte erbrachte, bestand der Verdacht, daß durch eine derartige Trocknung bei Raumtemperatur sich das Wasser aus den Proben nicht oder nur unvollständig entfernen läßt. Wir haben dies an verstreckten und unverstreckten Drähten überprüft. Das Ergebnis ist in Tabelle 2 für Proben wiedergegeben, die nach der Herstellung ca. ¹/₂ Jahr in feuchter Atmosphäre gelagert waren. Obwohl in (3) keine Angaben über die Trocknungsdauer gemacht wurden, muß man nach Tabelle 2 annehmen, daß *Hinrichsen* für seine Vergleichsmessungen im wesentlichen trockene Proben benutzt hat. Angesichts der in Tabelle 1 und 2 aufgeführten Ergebnisse ist es jedoch nicht verständlich, daß eine P₂O₅-Trocknung „*keine nennenswerten Unterschiede der Meßwerte*" erbracht haben soll.

Im Verlauf der kalorischen Messung wird das in feuchtem Polyamid gelöste Wasser wieder abgegeben. Diese Desorption läßt sich z. B. mit Hilfe des im DSC-Kalorimeter eingebauten Wärmeleitfähigkeitsdetektors leicht nachweisen. In Abb. 12 sind die Thermogramme einer 16 Std. bei 180 °C getemperten Probe im trockenen Zustand (Kurve a) und nach 14tägiger Lagerung in der feuchten Laboratmosphäre (Kurve b) wiedergegeben. Das Thermogramm der wasserhaltigen Probe weist zwischen 70 und 200 °C ein breites endothermes Maximum auf, das im trockenen Zustand nicht vorhanden ist und durch die Desorptionswärme des Wassers verursacht wird. Für die Verschmierung der Desorptionswärme über einen Temperaturbereich von mehr als 100 °C sind die relativ große Probendicke und die hohe Aufheizgeschwindigkeit verantwortlich. Bei der Untersuchung dünner 6.6-Polyamidfasern im luftfeuchten Zustand tritt ein wesentlich schärferer endothermer Peak bei ca. 70 °C auf (Kurve c).

Das während der kalorischen Messung aus feuchten Proben entweichende Wasser kann also einen Vorschmelzbereich vortäuschen bzw. den bei höherer Tempertemperatur oder Kristallisation aus der Schmelze vorhandenen Vorschmelzbereich scheinbar verstärken. Man erhält für die experimentelle Schmelzwärme zu *große* Werte, wenn man bei feuchten Proben den Desorptionspeak mitplanimetriert.

Tabelle 1. Änderung der Dichte durch Feuchtigkeitsaufnahme

| | Temperung | Dichte bei 20 °C, g/cm | | |
		Im trockenen Zustand direkt nach der Temperung	Nach 10tägiger Lagerung in der Laboratmosphäre	Nach 24stündiger Vakuumtrocknung bei 100 °C
abgeschreckte Preßplatte	1 Std. 170 °C	1,1313	1,1366	1,1316
abgeschreckte Preßplatte	1 Std. 205 °C	1,1368	1,1444	1,1375
Draht, unverstreckt in CCl₄ gesponnen	1 Std. 160 °C	1,1322	1,1369	
Draht unverstreckt in CCl₄ gesponnen	1 Std. 200 °C	1,1376	1,1427	
Draht, unverstreckt in CCl₄ gesponnen	1 Std. 240 °C	1,1493	1,1551	1,1500
Draht, unverstreckt in H₂O gesponnen (feucht gelagert)	1 Std. 240 °C	1,1522	1,1592	1,1523

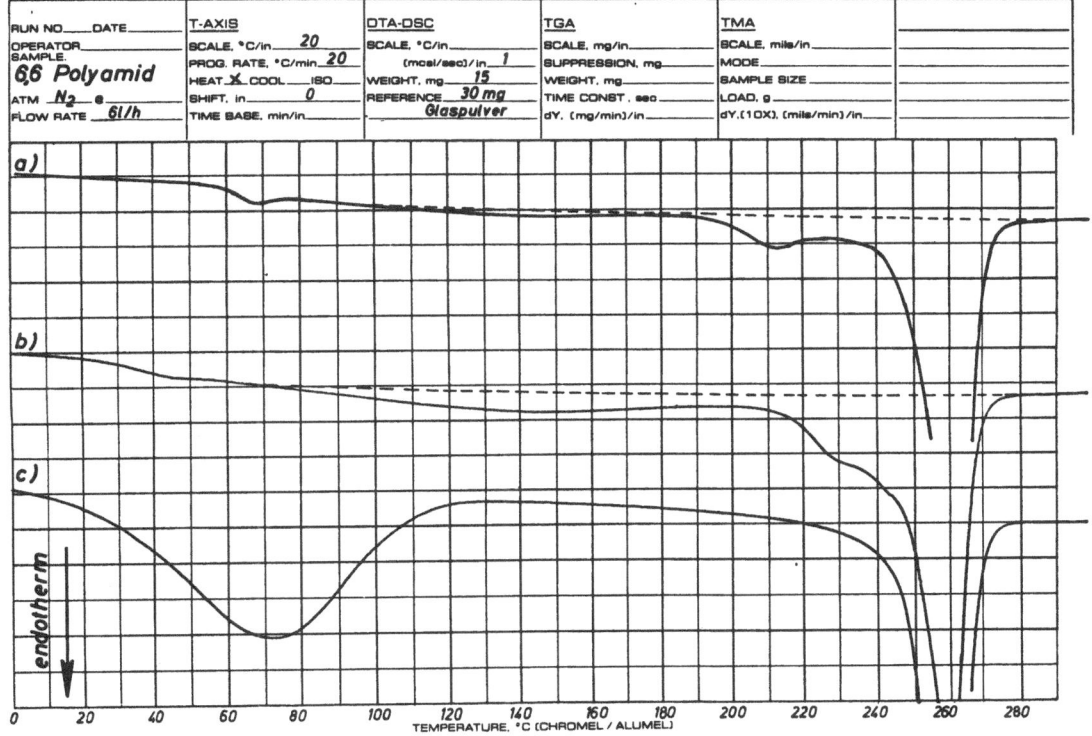

Abb. 12. Einfluß des Wassergehaltes auf die Thermogramme von 6.6-Polyamid.
a) 16 Std. bei 180 °C getemperte abgeschreckte Preßplatte (trocken)
b) wie a), jedoch 14 Tage in der Laboratmosphäre gelagert (feucht)
c) *lange in der* Laboratmosphäre gelagerte 6.6-Polyamid-Faser, Titer 70/18 (feucht)

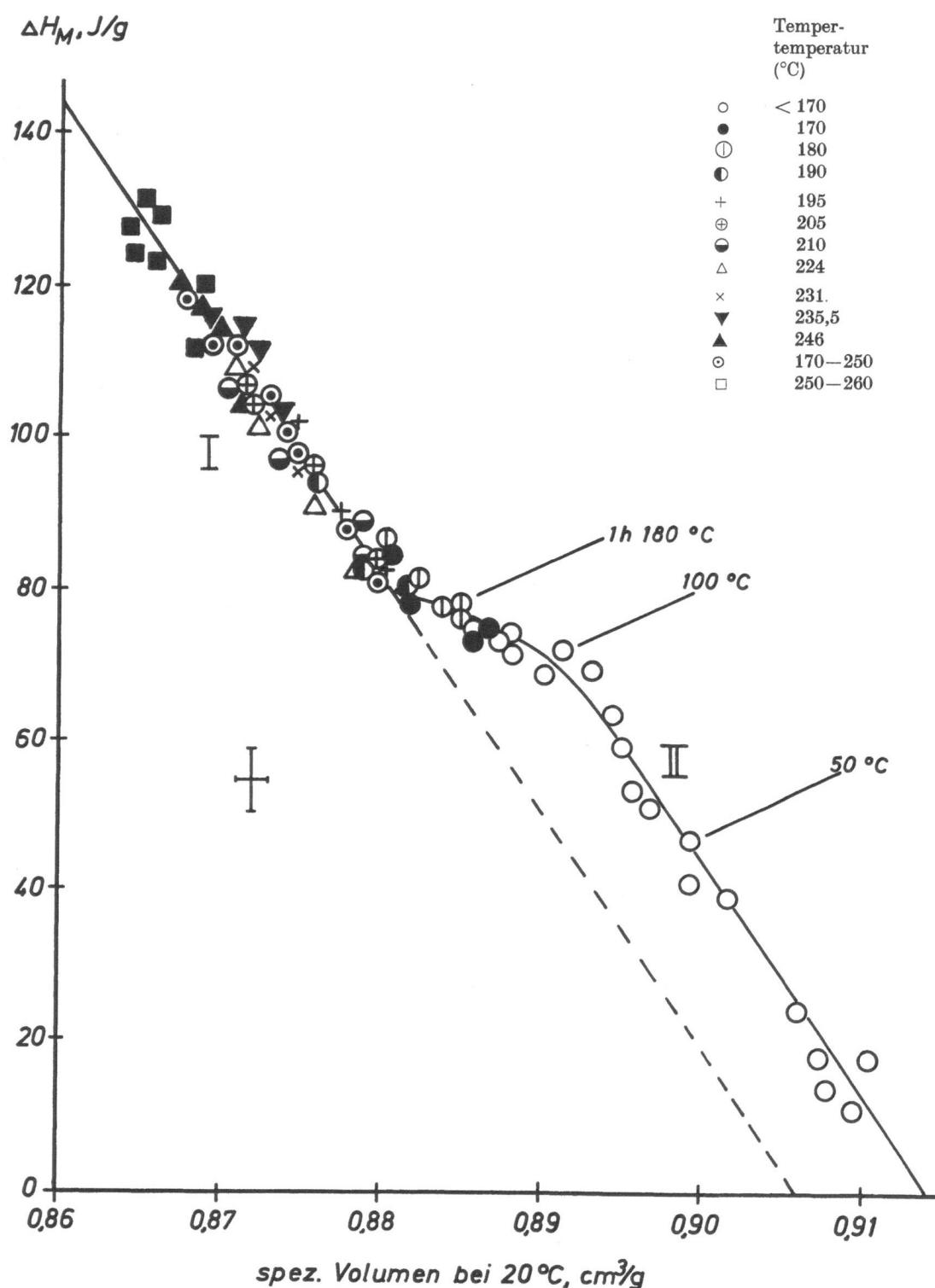

Abb. 13. Zusammenhang zwischen Schmelzwärme und spez. Volumen von abgeschreckten und anschließend getemperten Preßplatten aus 6.6-Polyamid (nicht sphärolithisch). Temperdauer: 0,03 bis max. 162 Std

Tabelle 2. Trocknung von 6.6-Polyamid durch Lagerung über P_2O_5

	Nach ¹/₂jähriger Lagerung in feuchter Atmosphäre		17 Tage über P_2O_5 getrocknet	
	Wassergehalt (⁰/₀)	Dichte (g/cm³)	Dichte (g/cm³)	Gewichtsverlust (⁰/₀)
Unverstreckter Draht, in H_2O gesponnen, 3 Std. bei 130 °C im Vakuum getempert	2,20	1,1271	1,1196	2,14
heißverstreckter Draht, 1 : 4,3	1,74	1,1410	1,1340	1,65

2. Zusammenhang zwischen Schmelzwärme und spezifischem Volumen

2.1. Abgeschreckte und getemperte Proben

In Abb. 13 ist die experimentelle Schmelzwärme aller nach scharfem Abschrecken getemperten Proben (Preßplatte) als Funktion des bei 20 °C gemessenen spez. Volumens aufgetragen. In dieser Darstellung sind die Meßwerte auf *zwei* Geraden angeordnet. Auf der Geraden II liegen alle Proben mit „pseudohexagonaler" Struktur (8, 9), die bei Temperaturen bis zu 100 °C getempert wurden. Auf der Geraden I befinden sich alle oberhalb 180 °C getemperten Proben mit gut ausgebildeter trikliner Struktur (8, 9). Auf Grund neuer Messungen und der dabei gewonnenen Erfahrungen bezüglich des breiten Vorschmelzbereiches ergibt sich für die Geraden I und II in Abb. 13 ein etwas anderer Verlauf als in der ursprünglichen Mitteilung (2).

Der Übergang zwischen den Geraden I und II wird von den zwischen 100 und 180 °C getemperten Proben gebildet. Bei diesen Proben hängt die Schmelzwärme nach Abb. 6 und 7 nur wenig von Tempertemperatur und Temperdauer ab, während ihre Dichte (Abb. 9 und 10) stärker ansteigt. Eine Strukturänderung ruft also auch bei 6.6-Polyamid eine Verschiebung der ΔH_M-v-Geraden hervor. Der Betrag dieser Verschiebung ist jedoch geringer, und der Übergang zwischen den beiden Geraden erfolgt allmählicher als bei 6-Polyamid (1).

2.2. Sphärolithisch kristallisierte Proben

Wie bei 6-Polyamid führt die beim Tempern gut abgeschreckter Proben stattfindende Kaltkristallisation auch bei 6.6-Polyamid nicht zu einer sphärolithischen Morphologie. Um zu

überprüfen, ob die sphärolithische Überstruktur einen Einfluß auf den Zusammenhang zwischen Schmelzwärme und spez. Volumen besitzt, wurden isotherm bei 240 °C aus der Schmelze kristallisierte Proben und langsam abgekühlte Proben nach unterschiedlicher Temperung untersucht. Die in Abb. 14 wiedergegebenen Messungen liegen sämtlich auf der aus Abb. 13 übernommenen Geraden I. Eine Beziehung zwischen der Überstruktur und dem Zusammenhang zwischen Schmelzwärme und spez. Volumen besteht demnach nicht.

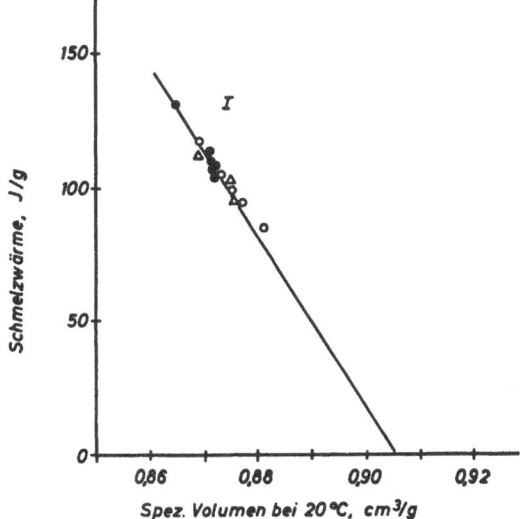

Abb. 14. Zusammenhang zwischen Schmelzwärme und spez. Volumen (20 °C) bei aus der Schmelze sphärolithisch kristallisierten 6.6-Polyamid-Preßplatten.

△ Isotherm bei 240 °C aus der Schmelze kristallisiert

○ in der Presse schnell abgekühlt, anschließend getempert

● in der Presse langsam abgekühlt, anschließend getempert

3. Einfluß von Oxidation, Nachkondensation und Zersetzungserscheinungen beim Tempern auf Schmelzwärme und spezifisches Volumen

Von *Hinrichsen* (3) wird darauf hingewiesen, daß bekanntermaßen bei Polyamiden bei erhöhter Temperatur Nachkondensations- und Abbauvorgänge sowie oxidative Schädigung eintreten können. Wir haben daher zu prüfen, ob unsere Meßergebnisse durch derartige Effekte beeinflußt sind. Eine Oxidation findet bei dem von uns gewählten Temperungsverfahren zwischen fest anhaftenden Metallfolien im *Wood*-Metallbad nicht statt, wie durch IR-Messungen überprüft wurde.

An abgeschreckten Proben, die 1 bzw. 16 Std. bei 200—250 °C getempert worden waren, wurde in 1% Schwefelsäurelösung die relative Viskosität ermittelt und eine Endgruppenbestimmung durchgeführt. Die Meßergebnisse sind in Tabelle 3 zusammengestellt. Die Viskositätswerte zeigen an, daß mit steigender Tempertemperatur und -dauer das mittlere Molekulargewicht durch Weiterkondensation zunimmt. Hierdurch kann allenfalls der Verlauf der Dichte- *und* Schmelzwärmekurven in Abb. 6—11 beeinflußt werden. Es ist jedoch nicht zu erwarten, daß der in Abb. 13 dargestellte Zusammenhang zwischen Schmelzwärme und spezifischem Volumen durch die Nachkondensation beeinflußt ist. Geht man nämlich von einem 6.6-Polyamid aus, das vor der Temperung bereits ein höheres Molgewicht besitzt, so stimmen die Messungen an abgeschreckten und dann getemperten Proben mit den in Abb. 13 wiedergegebenen völlig überein.

Falls bei der Temperung ausschließlich eine Nachkondensation stattfindet, bleibt die Differenz der COOH- und NH₂-Endgruppen kon-

stant. Bei den 1 Std bei Temperaturen bis zu 250 °C getemperten Proben ist dies nach Tabelle 3 (letzte Spalte) der Fall. Bei längerer Temperung oberhalb 220 °C treten jedoch Zersetzungsvorgänge ein, wodurch COOH-Gruppen verschwinden und NH₂-Gruppen gebildet werden, so daß die Endgruppendifferenz in der letzten Spalte von Tabelle 3 abnimmt. Diese Zersetzungserscheinungen haben auf den in Abb. 13 dargestellten Zusammenhang zwischen Schmelzwärme und spez. Volumen jedoch offensichtlich keinen Einfluß, da sowohl die nicht zersetzten, 1 Std. getemperten Proben als auch die chemisch veränderten, länger getemperten Proben auf der Geraden I liegen. Eine bei 260 °C 90 Std. getemperte Probe wurde bei 300 °C aufgeschmolzen, in Eiswasser abgeschreckt und dann erneut bei 50—160 °C getempert. Die an diesen Proben gemessenen Schmelzwärmen und spez. Volumina liegen wieder auf der Geraden II in Abb. 13. Damit ist bewiesen, daß die aus der Endgruppenbestimmung erkennbare chemische Veränderung von lange bei hoher Temperatur getempertem 6.6-Polyamid keinen Einfluß auf den Zusammenhang zwischen Schmelzwärme und spez. Volumen hat.

4. Vergleich mit den Messungen von Hinrichsen

Vergleicht man in Abb. 15 unsere Messungen an unverstreckten und heißverstreckten Drähten mit denen von *Hinrichsen* (3), so stellt man fest, daß unsere Messungen an sämtlichen Drahtproben wieder auf der aus Abb. 13 übertragenen Geraden I angeordnet sind, während diejenigen von *Hinrichsen* stark von dieser abweichen.

Wenn wir bei unseren unverstreckten Proben jedoch fälschlicherweise die Schmelzwärmen nur

Tabelle 3. Änderung der relativen Viskosität und der Endgruppenkonzentration beim Tempern von 6.6-Polyamid

Tempertemperatur (°C)	Temperdauer (Std.)	Relative Viskosität	Endgruppen (m-Äquivalent/kg)		
			[COOH]	[NH₂]	[COOH]−[NH₂]
ungetempert	—	2,8	67	51	16
230	1	3,2	52	37	15
240	1	3,2	51	35	16
250	1	3,5	50	35	15
200	16	4,0	43	27	16
220	16	4,7	34	20	14
230	16	5,1	32	20	12
240	16	5,2	28	20	8
250	16	5,1	23	24	−1

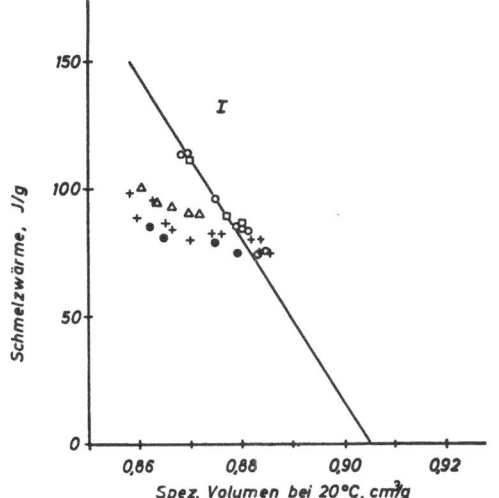

Abb. 15. Zusammenhang zwischen Schmelzwärme und spez. Volumen (20 °C) bei 6.6-Polyamid-Drähten.

○ In CCl₄- bzw. Wasserbad gesponnen, unverstreckt
□ verstreckt
+ unverstreckter Draht nach *Hinrichsen* (3)
△ heißverstreckter Draht nach *Hinrichsen* (3)
● unverstreckter Draht (eigene Messung). Die Schmelzwärme wurde nur aus dem Hauptschmelzpeak ermittelt und das spez. Volumen der luftfeuchten Probe benutzt

aus dem oberhalb 200 °C liegenden Hauptschmelzpeak ermitteln und außerdem das spez. Volumen der luftfeuchten Proben verwenden, ergibt sich in Abb. 15 eine gute Übereinstimmung mit den entsprechenden Werten von *Hinrichsen*. Diese Übereinstimmung bleibt in etwa erhalten, wenn wir bei unseren Messungen das spez. Volumen der trockenen Proben benutzen.

In (3) wurden keine DSC-Kurven wiedergegeben und keine Angaben über deren Auswertung gemacht. Aus Abb. 15 ergibt sich jedoch der zwingende Verdacht, daß der Vorschmelzbereich in (3) beim Planimetrieren vernachlässigt worden ist[1]). Dies führt nach Abb. 3, 5 und 8 zu einem mit steigender Tempertemperatur stark anwachsenden Fehler in der Schmelzwärme. Durch eine evtl. Benutzung des spez. Volumens luftfeuchter Proben kann die Abweichung der Meßpunkte von unserer Geraden I noch zusätzlich vergrößert werden.

Aus der wahrscheinlich auf diese Weise zustande kommenden nur geringfügigen Zunahme

---

[1]) Nach einer nachträglichen persönlichen Mitteilung von Hinrichsen wurden die DSC-Kurven in (3) ab *180–200 °C* integriert.

der Schmelzwärme zieht *Hinrichsen* die Schlußfolgerung, daß sich die Kristallinität des 6.6-Polyamids beim Tempern kaum ändert. Dies trifft nach unseren Untersuchungen bei Temperung oberhalb 180 °C nicht zu. Aus der Diskussion der aus Röntgenweitwinkelmessungen ermittelten Winkeldifferenz $\Delta \Theta$ der Äquatorhauptreflexe (100) und (010) wird die beim Tempern eintretende Dichtezunahme hauptsächlich auf eine Zunahme der „*Kristalldichte*" von 1,14 auf 1,25 g/cm³ zurückgeführt. Wir kommen auf diesen Punkt in einer nachfolgenden Arbeit zurück (9), in der Schmelzpunkt, Schmelzwärme und spez. Volumen im Zusammenhang mit Röntgenweitwinkel- und -kleinwinkeluntersuchungen besprochen werden

Ich danke Herrn Dr. *Mertens* für die Viskositäts- und Endgruppenbestimmungen und Herrn Dr. *Haberkorn* für zahlreiche Diskussionen.

Zusammenfassung

An scharf abgeschrecktem und an langsam kristallisiertem 6.6-Polyamid in Form von Preßplatten und unverstreckten bzw. verstreckten Drähten wird die Abhängigkeit der Dichte (20 °C) und der Schmelzwärme von Tempertemperatur (50—260 °C) und Temperdauer (2—10⁴ min) untersucht. Bei der Ermittlung der Schmelzwärme aus den DSC-Kurven muß ein zwischen 100 und 200 °C auftretender Vorschmelzbereich mitberücksichtigt werden, dessen Anteil bei Temperung oberhalb 200 °C stark zunimmt. Bei trockenen Proben erhält man für den Zusammenhang zwischen Schmelzwärme und spez. Volumen wie bei 6-Polyamid zwei Geraden. Auf der Geraden II liegen die abgeschreckten und unterhalb 100 °C getemperten Proben mit „pseudohexagonaler" Struktur. Auf der Geraden I befinden sich alle Proben mit gut ausgebildeter trikliner Struktur (abgeschreckte und oberhalb 180 °C getemperte Platte, alle langsam kristallisierten sphärolithischen Proben, verstreckte und unverstreckte Drähte). Der Übergang zwischen den Geraden I und II wird durch die zwischen 100 und 180 °C getemperten abgeschreckten Proben gebildet. Die bei höheren Tempertemperaturen und längerer Temperdauer auftretenden Nachkondensations- und Zersetzungserscheinungen haben keinen Einfluß auf die Abhängigkeit der Schmelzwärme vom spez. Volumen.

Entgegen der von *Hinrichsen* vertretenen Meinung ist die Dichte luftfeuchter Proben um 0,005–0,007 g/cm³ größer als im trockenen Zustand. Die beim Erwärmen feuchter Proben im Temperaturbereich des Vorschmelzens gleichzeitig auftretende Desorptionswärme des Wassers verursacht eine scheinbare Erhöhung der Schmelzwärme. Die von *Hinrichsen* gefundene ungefähre Konstanz der Schmelzwärme beim Tempern bei gleichzeitiger starker Zunahme der Dichte kommt sehr wahrscheinlich außer durch Benutzung feuchter Proben vor allem durch Nichtberücksichtigung des Vorschmelzbereiches zustande.

Summary

The dependence of density (20 °C) and heat of fusion on annealing temperature (50—260 °C) and time (2 to 10^4 min) is investigated for quenched and slowly cooled 6.6-polyamide. The starting materials are pressure-molded films as well as drawn and undrawn wires.

The correct determination of the heat of fusion from the area of the DSC-curve must include a broad premelting process between 100 and 200 °C, the fraction of which strongly increases with increasing annealing temperature above 200 °C. For dry samples the relation between heat of fusion and specific volume consists of two straight lines, similar as for 6-polyamide. The quenched and below 100 °C annealed samples with "pseudohexagonal" structure form line II. All samples with well developed triclinic structure (annealed above 180 °C after quenching, all slowly crystallized spherulithic samples, drawn and undrawn wires) are on line I. A transition between line I and II is formed by quenched samples after annealing between 100 and 180 °C. After-condensation and chemical changes occurring at high annealing temperatures and long annealing periods have no influence upon the relation between heat of fusion and specific volume.

In contrast to *Hinrichsen* the density of moisture-containing samples is found to be 0.005—0.007 g/cm³ higher compared with the dry state. On heating moisture containing samples the heat of water-desorption adds to the melting endotherm causing an apparent increase of the heat of fusion. *Hinrichsen* reported a nearly constant heat of fusion on annealing and at the same time a strong increase of density. Most probably this effect is caused by the use of water-containing samples and above all by nonconsideration of the premelting.

Literatur

1) *Illers, K. H.* und *H. Haberkorn*, Makromol. Chem. **142**, 31 (1971).
2) *Illers, K. H.* und *H. Haberkorn*, Makromol. Chem. **146**, 267 (1971).
3) *Hinrichsen, G.*, Colloid & Polymer Sci. **250**, 1162 (1972).
4) *Illers, K. H.*, Makromol. Chem. **127**, 1 (1969).
5) *Hinrichsen, G.*, Faserforschg. Textiltechnik **20**, 529 (1969).
6) *Adam, G.* und *F. H. Müller*, Colloid & Polymer Sci. **192**, 29 (1963).
7) *Starkweather, H. W.*, J. Appl. Polymer Sci. **2**, 129 (1959).
8) *Starkweather, H. W., J. F. Whitney* und *D. R. Johnson*, J. Polymer Sci. A, **1**, 715 (1963).
9) *Haberkorn, H.* und *K. H. Illers* (in Vorbereitung).

Adresse des Autors:

Dr. K. H. Illers
Meß- und Prüflaboratorium
der BASF Aktiengesellschaft
D-67 Ludwigshafen/Rhein

Progr. Colloid & Polymer Sci. **58**, 77—80 (1975)

Hoechst Aktiengesellschaft 6230 Frankfurt (M) 80

Das Verhalten von Polypropylen bei der Einwirkung ionisierender Strahlen in Luft

H. Wilski

Mit 6 Abbildungen und 2 Tabellen

(Eingegangen am 17. Mai 1974)

Polypropylen wird durch die Einwirkung ionisierender Strahlen bei strengstem Sauerstoffausschluß vernetzt. Führt man die Bestrahlung in Gegenwart von Sauerstoff durch, so wird das Polypropylen abgebaut. Der Reaktionsmechanismus für den Abbau konnte von *Fischer, Hellwege, Johnsen* und *Neudörfl* (1, 2) aufgeklärt werden. Danach reagieren die durch die Bestrahlung gebildeten Radikale, die bei Sauerstoffabwesenheit zur Vernetzung geführt hätten, in Gegenwart von Sauerstoff zu labilen Peroxiden, die unter Spaltung des Moleküls in zwei sauerstoffhaltige Bruchstücke zerfallen. Mit diesem Abbau geht eine Verschlechterung der mechanischen Eigenschaften einher (3, 4). Im folgenden soll gezeigt werden, wie sich die mechanischen Eigenschaften des Polypropylens bei Bestrahlung in Luft in Abhängigkeit von der Dosis und der Dosisleistung (d.h. der Zeit, die für die Diffusion des Sauerstoffs in die Probe zur Verfügung steht) ändern. Die Ergebnisse sind ähnlich wie bei dem früher untersuchten Niederdruck-Polyäthylen (5) und zeigen, daß die dort gefundenen Gesetzmäßigkeiten allgemeinere Gültigkeit beanspruchen können. Wegen Einzelheiten sei auf (5) verwiesen.

Experimentelle Angaben

Material. Das verwendete isotaktische Polypropylen war ein Produkt der *Hoechst Aktiengesellschaft* (Hostalen®PP). Das Material wurde bei 260 °C versponnen. Durch verschiedene Streckprozesse wurden zwei Drahtsorten von je 0,4 mm Durchmesser, aber verschiedener Festigkeit hergestellt. Beide Sorten wurden bei 140 °C fixiert. Die wichtigsten Eigenschaften der Drähte sind in Tabelle 1 zusammengefaßt. Infolge unterschiedlichen Abbaues während der Verarbeitung sind die reduzierten *Viskositäten und die daraus berechneten Molekular-*

gewichte geringfügig verschieden. Der Unterschied ist jedoch so gering, daß er bei der Auswertung der Versuche vernachlässigt werden konnte. In den Abbildungen sind die Proben mit der geringeren Ausgangsfestigkeit stets durch hohle Symbole, die mit der höheren durch fette Symbole gekennzeichnet. Die Drähte waren in der üblichen Weise stabilisiert.

Bestrahlung. Die Proben wurden sämtlich im Langzeitversuch in Gegenwart von Luft mit den Gamma-Strahlen einer Kobalt-60-Quelle bestrahlt[1]). Die experimentellen Bedingungen sind exakt die gleichen wie früher (5).

Viskositätsmessung und Molekulargewichte. Die Viskositäten wurden bei einer Konzentration von 0,1 g/dl bei 135 °C in Dekahydronaphthalin, stabilisiert mit 0,5% Phenyl-β-napththylamin, gemessen. Aus den η_{spez}/c-Werten („reduzierte Viskositäten") wurden in der gleichen Weise wie früher für Polyäthylen (5) die Grenzviskositäten berechnet und daraus nach der Beziehung von *Kinsinger* und *Hughes* (6)

$$[\eta] = 1{,}10 \cdot 10^{-4} \cdot M^{0{,}80}$$

die Molekulargewichte.

Tabelle 1. Eigenschaften des Ausgangsmaterials. Beide Polypropylendrähte bei 260 °C gesponnen, aber in verschiedener Weise verstreckt

		1	2
Reduzierte Viskosität	dl/g	2,75	2,82
Molekulargewicht	g/Mol	285 000	292 000
Dichte bei 20 °C	g/cm³	0,900	0,899
Reißfestigkeit	N/mm²	397	507
Reißdehnung	%	23	21
Kochschrumpf	%	6,7	7,0

---

[1]) Herrn Dr. *S. Rösinger*, Frankfurt, sei auch an dieser Stelle für die Durchführung der Bestrahlungen gedankt.

Mechanische Prüfung. Die Kraftverlängerungsdiagramme wurden nach DIN 53816 bei 20 °C, 65% relativer Luftfeuchtigkeit und mit einem Vorschub von 100 mm/min aufgenommen. Sämtliche angegebenen Werte sind Mittelwerte aus 5 Einzelmessungen. Die Proben lagerten vor der Messung stets 4 Wochen im klimatisierten Raum.

Kochschrumpf. Der Kochschrumpf wurde in der gleichen Weise wie früher (5) bestimmt. Die Werte für das Ausgangsmaterial sind in Tabelle 1 angegeben.

Ergebnisse

Reißfestigkeit und Reißdehnung sind in Abb. 1 in Abhängigkeit von der Bestrahlungsdosis dargestellt. Wie man sieht, werden die mechanischen Eigenschaften schon bei sehr kleinen Dosen erheblich herabgesetzt, und zwar

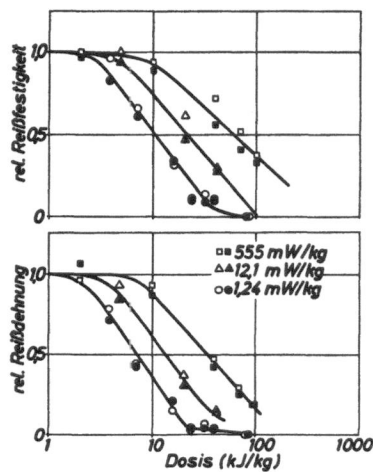

Abb. 1. Relative Reißfestigkeit und Reißdehnung von 0,4-mm-Polypropylendrähten, offen an der Luft bestrahlt. Die fetten Symbole bezeichnen den Draht mit 507 N/mm², die hohlen Symbole den Draht mit 397 N/mm² Ausgangsfestigkeit

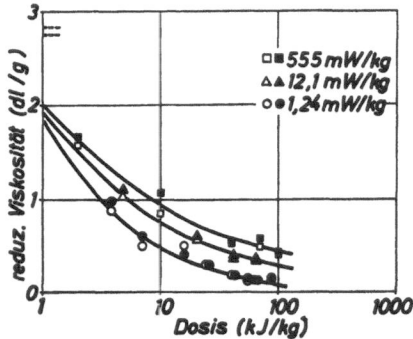

Abb. 2. Relativer Kochschrumpf von 0,4-mm-Polypropylendrähten, offen an der Luft bestrahlt

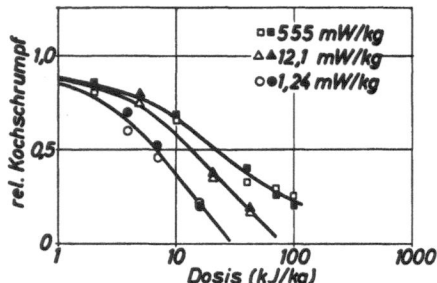

Abb. 3. Reduzierte Viskosität von 0,4-mm-Polypropylendrähten, offen an der Luft bestrahlt

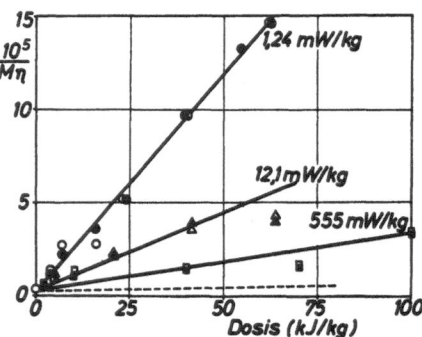

Abb. 4. Strahlenabbau des Polypropylens. Ausgezogene Geraden: Bestrahlung in Luft; gestrichelte Gerade: Bestrahlung im Vakuum nach (7)

tritt, ähnlich wie beim Polyäthylen, die Schädigung bei um so kleineren Dosen ein, je kleiner die Dosisleistung ist. Ganz Entsprechendes gilt für den Kochschrumpf, wie Abb. 2 zeigt. Die Ursache für die Verkleinerung der Werte ist die Verringerung des Molekulargewichts durch Strahlenoxydation, die um so wirkungsvoller verläuft, je mehr Sauerstoff in den Proben vorhanden ist. Abb. 3 zeigt die entsprechenden Abnahmen der reduzierten Viskositäten, während in Abb. 4 die hieraus berechneten Reziprokwerte der Molekulargewichte in Abhängigkeit von der Dosis dargestellt sind. Es ergibt sich der erwartete lineare Zusammenhang, der zeigt, daß die Kettenbrüche statistisch erfolgen. Nach Untersuchungen von *Keyser, Clegg* und *Dole* (7) sinkt die Grenzviskosität des Polypropylens in dem hier interessierenden Dosisbereich (d.h. unterhalb der Geldosis) auch bei Bestrahlung unter strengstem Sauerstoffausschluß. Dies hat jedoch, wie *Salovey* und *Dammont* (8) zeigen konnten, in diesem Fall nichts mit einer Abnahme des Molekulargewichts zu tun, sondern ist eine Folge der (als Vorstufe der Vernetzung) gebildeten Langkettenverzweigungen. In Abb. 4 sind die aus den Grenzviskositäten von *Keyser* et al. berechneten scheinbaren Molekulargewichte für isotaktisches Polypropylen gestrichelt eingezeichnet. Die gestrichelte Gerade entspricht einem scheinbaren Abbaukoeffizienten $p_0 < 1 \cdot 10^{-6}$. Im Falle des Abbaues durch Strahlenoxydation wird die Bildung von Langkettenverzweigungen entsprechend zurückgedrängt, so daß auch bei der höchsten angewandten Dosisleistung der Fehler in den Molekulargewichten vernachlässigbar sein dürfte.

Die aus den Daten der Abb. 4 berechneten Abbaukoeffizienten p_0 sind in Tabelle 2 zusammen mit den aus den vorhergehenden Abbildungen entnommenen Halbwertsdosen wiedergegeben. Ein Vergleich der Zahlen mit (5) zeigt, daß Polypropylen rund dreimal empfindlicher gegen Strahlenoxydation ist als Niederdruck-Polyäthylen. Halbwertsdosen für höhere Dosisleistungen und andere Bestrahlungsbedingungen lassen sich aus (3), (4) und (9) entnehmen.

Tabelle 2. Halbwertsdosis (kJ/kg) von Polypropylen, gemessen an verstreckten Drähten von 0,4 mm Durchmesser. Bestrahlung in Luft

Dosisleistung (mW/kg)	555	12,1	1,24
Reißfestigkeit	60	21	10
Reißdehnung	32	13	7
Kochschrumpf	22	13	6,5
Abbaukoeffizient	$6,4 \cdot 10^{-6}$	$18 \cdot 10^{-6}$	$49 \cdot 10^{-6}$

Der Zusammenhang zwischen Festigkeit (bzw. Dehnung) und Molekulargewicht wird besonders deutlich, wenn man diese Größen in Abhängigkeit vom Molekulargewicht aufträgt (Abb. 5 und 6). Sämtliche Meßpunkte für die Festigkeit ordnen sich jetzt auf einer gemeinsamen Kurve an, aus der hervorgeht, daß die Festigkeit eine eindeutige Funktion des Molekulargewichts ist. Es spielt ganz offensichtlich keine Rolle, ob ein bestimmtes Molekulargewicht durch Bestrahlen mit hoher Dosisleistung und hoher Dosis oder mit kleiner Dosisleistung und kleiner Dosis erzeugt wurde. Die Abb. 5 und 6 zeigen darüber hinaus den bekannten Zusammenhang, nach dem bei hohen Molekulargewichten Festigkeit und Dehnung einen (von der mechanisch-thermischen Vorbehandlung abhängigen) Grenzwert erreichen, *der durch weitere Erhöhung des Molekular-*

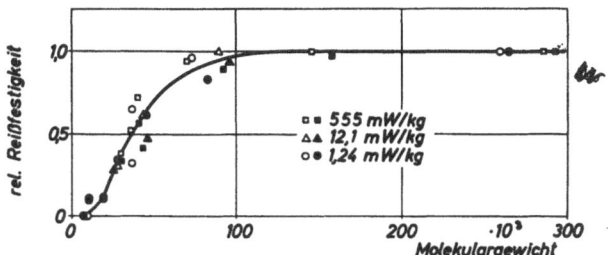

0,4 mm - Polypropylen - Drähte, verstreckt, in Luft bestrahlt mit Co-60

Abb. 5. Relative Reißfestigkeit von abgebautem Polypropylen

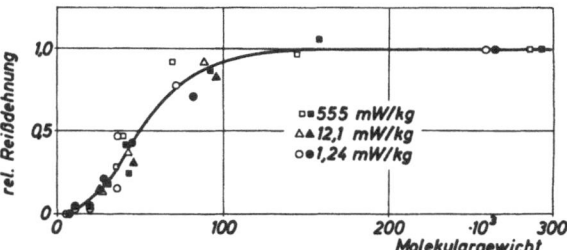

Abb. 6. Relative Reißdehnung von abgebautem Polypropylen

gewichtes nicht zu überschreiten ist. Unterhalb von einem Molekulargewicht von $M_\eta = 100000$ nehmen, ähnlich wie beim Niederdruck-Polyäthylen, die Werte für Festigkeit und Dehnung ab, um bei Werten um $M_\eta \approx 1000$ praktisch Null zu werden.

Zusammenfassung

Polypropylen, das bei Bestrahlen mit ionisierenden Strahlen bei Ausschluß von Sauerstoff vernetzt wird, wird beim Bestrahlen in Gegenwart von Sauerstoff abgebaut. Die Strahlenschädigung ist von der Dosis und der Dosisleistung abhängig. Die Dosis, die erforderlich ist, um die Reißdehnung des Materials auf die Hälfte des Ausgangswertes zu verkleinern (Halbwertsdosis) liegt bei Langzeitversuchen (1,24 mW/kg in Luft) bei nur 7 kJ/kg. Die Verminderung der Reißfestigkeit und Reißdehnung läßt sich auf die durch die Bestrahlung bewirkte Abnahme des Molekulargewichtes zurückführen.

Summary

Poly(propylene), which is crosslinked by irradiation by exclusion of oxygen, will be degraded by irradiation in the presence of oxygene. The degree of degradation depends on the dose and the dose rate. The dose, which is necessary to reduce the ultimate elongation of the material under investigation to 50% of its initial value, lies under the conditions of long time irradiation

(1,24 mW/kg in air) by only 7 kJ/kg. It is shown that the reduction of the ultimate tensile strength and the ultimate elongation, resp., is a consequence of the radiation induced reduction of the molecular weight.

Literatur

1) *Fischer, H., K.-H. Hellwege* und *P. Neudörfl,* J. Polym. Sci. A 1, 2109 (1963).
2) *Fischer, H., K.-H. Hellwege, U. Johnsen* und *P. Neudörfl,* Kolloid-Z. u. Z. Polymere **195,** 129 (1964).
3) *Wilski, H.,* Kunststoffe **58,** 18 (1968).
4) *Fischer, H., K.-H. Hellwege* und *W. Langbein,* Kunststoffe **58,** 625 (1968).
5) *Wilski, H.,* Kolloid-Z. u. Z. Polymere **251,** 703 (1973).
6) *Kinsinger, J. B.,* und *R. E. Hughes,* J. Phys. Chem. **63,** 2002 (1959).
7) *Keyser, R. W., B. Clegg* und *M. Dole,* J. Phys. Chem. **67,** 300 (1963).
8) *Salovey, R.* und *F. R. Dammont,* J. Polym. Sci. A1. 2155 (1963).
9) *Benderly, A. A.* und *B. S. Bernstein,* J. Appl. Polym. Sci. **13,** 505 (1969).

Adresse des Autors:

H. Wilski
Hoechst Aktiengesellschaft
6230 Frankfurt (M) 80

Progr. Colloid & Polymer Sci. **58**, 81—89 (1975)

Erich-Schmid-Institut für Festkörperphysik der Österreichischen Akademie der Wissenschaften (Leoben, Austria)

An analysis of thermally activated macro-deformation in polymers

E. Pink

With 11 figures

(Received January 1, 1974)

The kinetics of the viscoelastic behaviour of polymers have been extensively investigated during the past decade, but it is only in recent years that the interest in irreversible plastic deformation, and especially in the initial stages of yielding, has been intensified. Background for such work is the site-model theory which is based on the probability

$$\omega = \nu \exp[-\Delta G/kT] \qquad [1]$$

known from statistical mechanics. The goal of these investigations is the determination of the activation enthalpy ΔG which is necessary to overcome potentials impeding the deformation process. The knowledge of ΔG (and of other terms to be introduced later) enables the choice of a relevant deformation mechanism. In this report we shall discuss a method which has been applied often to study the thermally activated deformation of metals (1—3), and we shall extend it to fit the special needs of polymers which are strained beyond yielding.

The Theory of Thermally Activated Deformation

Eq. [1] is an expression for the probability that a forward jump of a structural element over the deformation barrier occurs, caused by thermal vibration with a frequency ν. The possibility of jumps back to the starting position, which would reduce the probability ω, can be neglected if the applied external stress is reasonably large (4), as can be expected in low-temperature tests.

In the theory the type of barrier is not specified: in the case of metals such obstacles are relevant which affect the movement of lattice dislocations, in amorphous or partially

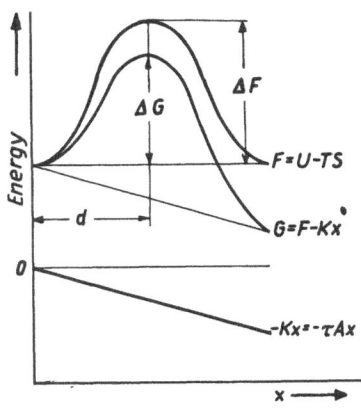

Fig. 1. Schematic potential energy around a barrier

crystalline polymers the mobility of chains will be of importance.

During the translation of a structural element along a direction x, the *Helmholtz* energy

$$F = U - TS \qquad [2]$$

will be modified, as demonstrated in Fig. 1. The obstructing interatomic or intermolecular forces cause a localized increase of F. Starting with an equation of state

$$f(p, v, T)_{\dot{\varepsilon}, \tau} = \text{const.} \qquad [3]$$

and following classical thermodynamics, the *Gibbs* free energy is

$$G = F + p v. \qquad [4]$$

In polymers this condition is certainly of importance, since it is known that their strength properties are strongly influenced by the pressure p. It is, on the other hand, known from metals that another equation of state

$$f(\tau, \dot{\varepsilon}, T)_{p, v} = \text{const.} \qquad [5]$$

(τ shear stress, $\dot{\varepsilon}$ strain rate) describes the mechanical performance. Since the strength values of polymers are temperature dependent, too, we shall at first maintain this formalism although it neglects the pressure dependence.

The shear stress τ is caused by a force K acting on an area A. While the obstacle is overcome, a work Kx is done which now decreases the *Helmholtz* energy, i.e. supports the thermal activation (Fig. 1). The critical energy necessary to overcome the obstacle is then

$$\Delta G = \Delta F - Kd = \Delta F - \tau A d \qquad [6]$$
$$= \Delta F - \tau v'_\tau .$$

A is the "activation area", the product $A d$ is, due to its dimension, denoted as "stress-activation volume" v'_τ (it does not, however, characterize a volume change). A further thermodynamic value is the activation enthalpy ΔH. Taking into account equation of state [5] it is

$$[\Delta H]_p = \Delta U - \tau v_\tau . \qquad [7]$$

ΔU is the difference in potential energy at absolute zero in terms of the internal energy. It is often denoted as ΔH_0.

The application of the rate theory in macro deformation relates the *Gibbs* free energy to the deformation (or strain) rate:

$$\dot{\varepsilon} = \dot{\varepsilon}'_0 \exp[-\Delta G/kT] . \qquad [8]$$

Using the activation enthalpy

$$\Delta H = \Delta G + T \Delta S \qquad [9]$$

gives

$$\dot{\varepsilon} = \dot{\varepsilon}'_0 \exp[-\Delta H/kT] \exp[\Delta S/k] . \qquad [10]$$

When the entropy term is combined with $\dot{\varepsilon}'_0$ to form a new frequency factor

$$\dot{\varepsilon}_0 = \dot{\varepsilon}'_0 \exp[\Delta S/k] , \qquad [11]$$

the final rate equation is

$$\dot{\varepsilon} = \dot{\varepsilon}_0 \exp[-\Delta H/kT] . \qquad [12]$$

This form is directly related to experimental results. The stress-activation volume, defined as

$$[\partial \Delta G/\partial \tau]_{T, p, v} = - v'_\tau \qquad [13]$$

can be obtained from eq. [8] as

$$v'_\tau = kT/\lambda'_\tau , \qquad [14]$$

where λ'_τ is the strain-rate sensitivity:

$$\lambda'_\tau = [\partial \tau/\partial \ln (\dot{\varepsilon}/\dot{\varepsilon}'_0)]_{T, p, v} . \qquad [15]$$

The eqs. [13]—[15] are derived from the *Gibbs* free energy, values for which are not necessarily accessible through the experiment (23). An apparent activation volume v_τ is therefore introduced by substituting ΔH, λ_τ and $\dot{\varepsilon}_0$ in the previous equations:

$$[\partial \Delta H/\partial \tau]_{T, p, v} = - v_\tau , \qquad [16]$$

$$v_\tau = kT/\lambda_\tau , \qquad [17]$$

$$\lambda_\tau = [\partial \tau/\partial \ln (\dot{\varepsilon}/\dot{\varepsilon}_0)]_{T, p, v} . \qquad [18]$$

By differentiating the eqs. [8] and [12] with respect to τ, one can show that the relation between true and apparent activation volumina is given by

$$v_\tau = v'_\tau - T [\partial \Delta S/\partial \tau]_T . \qquad [19]$$

When the derivative of eq. [12] with respect to $1/T$ is taken, an expression for the activation enthalpy is obtained:

$$\Delta H = - k[\partial \ln (\dot{\varepsilon}/\dot{\varepsilon}_0)/\partial (1/T)]_{\tau, p, v} . \qquad [20]$$

Differentiated with respect to T, eq. [12] yields

$$\Delta H = kT^2[\partial \ln (\dot{\varepsilon}/\dot{\varepsilon}_0)/\partial T]_{\tau, p, v} . \qquad [21]$$

With the well-known condition

$$[\partial \tau/\partial T]_{\dot{\varepsilon}, p, v} [\partial \ln (\dot{\varepsilon}/\dot{\varepsilon}_0)/\partial \tau]_{T, p, v}$$
$$\times [\partial T/\partial \ln (\dot{\varepsilon}/\dot{\varepsilon}_0)]_{\tau, p, v} = -1 , \qquad [22]$$

which is derived from the equilibrium conditions of [5], eq. [21] becomes

$$\Delta H = - kT^2[\partial \ln (\dot{\varepsilon}/\dot{\varepsilon}_0)/\partial \tau]_{T, p, v} [\partial \tau/\partial T]_{\dot{\varepsilon}, p, v}$$
$$= - v_\tau T [\partial \tau/\partial T]_{\dot{\varepsilon}, p, v} . \qquad [23]$$

All the terms in eqs. [18], [20], [21] and [23] necessary to evaluate v_τ and ΔH are available by means of simple tests. The maximum shear stress τ is taken as $\sigma/2$. $[\partial \tau/\partial \ln (\dot{\varepsilon}/\dot{\varepsilon}_0)]_{T, p, v}$ can be obtained from a strain-rate change executed

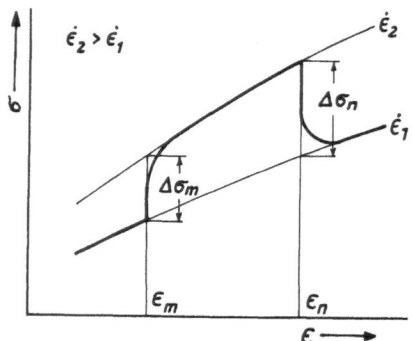

Fig. 2. The determination of $\Delta \sigma (= 2 \Delta \tau)$ by means of strain-rate changes

during a tensile or compression test ($\dot{\varepsilon}_1 \to \dot{\varepsilon}_2 \to \dot{\varepsilon}_1$, cf. Fig. 2)[1]). Assuming that the pre-exponential factors $\dot{\varepsilon}_0'$ or $\dot{\varepsilon}_0$ are not affected during the strain-rate change, eq. [18] can be simplified:

$$\lambda_\tau = [\Delta\tau/\ln(\dot{\varepsilon}_2/\dot{\varepsilon}_1)]_{\tau, p, v}. \tag{24}$$

Eq. [20] is experimentally verified by measuring the increase of temperature which, while a higher strain rate $\dot{\varepsilon}_2$ is applied, brings about the same stress τ. Data for eq. [21] could be measured in creep tests, but the modification leading to eq. [23] allows the further use of λ_τ. The additional term $[\partial\tau/\partial T]_{\dot{\varepsilon}, p, v}$ which is the temperature dependence of the shear stress, can easily be taken from stress-temperature relations.

So far the entire derivation has been based on the equation of state [5], i.e. on the assumption of constant pressure. Especially when applied to polymers, this would limit the validity of the theory due to the pressure dependence of the strength characteristics. In analogy to eq. [7], we have therefore to write for constant stress

$$[\Delta H]_\tau = \Delta H_0 + p\,v_p. \tag{25}$$

v_p is the pressure-activation volume, and it characterizes the actual change of the glide element during the thermal event under the influence of an external pressure p. The combined influence of external force and pressure yields (9, 12)

$$\Delta H = \Delta H_0 - \tau\,v_\tau + p\,v_p. \tag{26}$$

Fig. 3 represents the dependence of the activation enthalpy ΔH on temperature, and its modification by stress and pressure.

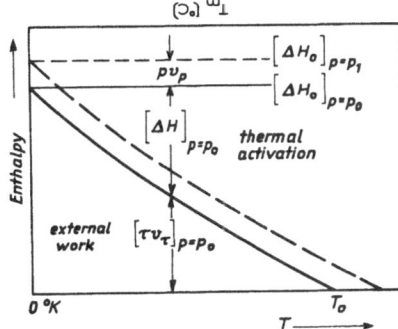

Fig. 3. The contributions of external work and thermal activation on the total enthalpy of deformation between 0 K and T_0

To obtain a formula for the pressure-activation volume v_p, the enthalpy ΔH is limited to constant stress conditions (eq. [21]). By an analogous procedure as for v_τ, we obtain

$$v_p = -kT[\partial\ln(\dot{\varepsilon}/\dot{\varepsilon}_0)/\partial p]_{T, \tau, v}. \tag{27}$$

The eqs. [17] and [18] for the stress-activation volume appear similar except for $[\partial p]_\tau$. Nevertheless v_p cannot be evaluated by a strain-rate change as v_τ, but only from the change of creep rate $\dot{\varepsilon}$ when the pressure is varied and the stress is kept constant.

A transformation of eq. [27] assuming another equation of state

$$f(\tau, p, \dot{\varepsilon})_{v, T} = \text{const.} \tag{28}$$

gives

$$\begin{aligned}v_p &= kT[\partial\ln(\dot{\varepsilon}/\dot{\varepsilon}_0)/\partial\tau]_{p, T, v}[\partial\tau/\partial p]_{\dot{\varepsilon}, T, v}\\ &= v_\tau[\partial\tau/\partial p]_{\dot{\varepsilon}, T, v}.\end{aligned} \tag{29}$$

Still another form for the pressure-activation volume comes from equation of state

$$f(p, \dot{\varepsilon}, T)_{\tau, v} = \text{const.} \tag{30}$$

and is

$$v_p = kT[\partial\ln(\dot{\varepsilon}/\dot{\varepsilon}_0)/\partial T]_{p, \tau, v}[\partial T/\partial p]_{\dot{\varepsilon}, \tau, v}. \tag{31}$$

A way to calculate the activation enthalpy at constant stress was already given through eq. [21]. Combining with [31] leads to

$$\Delta H = v_p\,T\,[\partial p/\partial T]_{\dot{\varepsilon}, \tau, v}. \tag{32}$$

As mentioned before, $[\partial\ln(\dot{\varepsilon}/\dot{\varepsilon}_0)/\partial T]_{p, \tau, v}$ can be measured in creep tests. The second differential in [31] and [32] is known from thermodynamics as

$$[\partial p/\partial T]_v = \alpha/\varkappa, \tag{33}$$

(α is the coefficient of volumetric expansion, $\varkappa$ the isothermal compressibility).

With v_p all contributions to the total activation enthalpy of eq. [26] are known.

Complications

(a) The activation analysis is only valid when one single mechanism is rate controlling. Data which are determined in transition regions (where mechanism I is gradually replaced by mechanism II) have no physical meaning.

(b) One of the reasons listed under (a) is the possible change of $\dot{\varepsilon}_0$ during strain-rate changes. The formulae for evaluating v_τ or ΔH become unreliable, since the frequency factor may not be neglected anymore in $\partial \ln (\dot{\varepsilon}/\dot{\varepsilon}_0)$. Variations of $\dot{\varepsilon}_0$ due to structural modifications are perhaps of minor importance, when the ratio $\dot{\varepsilon}_2/\dot{\varepsilon}_1$ can be kept small. Errors from an entropy contribution may also enter through $\dot{\varepsilon}_0$ in temperature-change experiments.

(c) Among others, a deformation mechanism is characterized by ΔG. [20], [21] and [23] allow, however, only the experimental determination of the enthalpy ΔH. The entropy term $T \Delta S$ which determines the difference, could be derived taking, for instance, into account the temperature dependence of the elastic modulus (24). This, however, has not yet been done.

(d) The previous discussion of the influence of stress and pressure may have given the impression that a strict partition between those two effects is always possible. In practice, the tensile test is often favoured over a torsion test, and it subjects the material, and in consequence the glide element, not only to a shear stress $\tau = \sigma/2$, but also to a hydrostatic stress component $p_h = -\sigma/3$. Any variation of the tensile stress σ, arising for instance from temperature changes, makes it necessary to consider simultaneously the negative pressure effect of a normal stress component $\bar{\tau}$, whose value is that of the hydrostatic stress component p_h. The activation enthalpy ΔH $(p = p_0)$, determined by means of eq. [23] in such a uniaxial tensile test, is therefore always misrepresented, because the condition of constant pressure cannot be maintained. The total enthalpy in this "constant-pressure" test contains still the additional pressure term as in [26]:

$$\Delta H = \Delta H_0 - \tau v_\tau + \bar{\tau} v_{ph} . \qquad [34]$$

v_{ph} may be called a pressure-activation volume arising from a normal stress $\bar{\tau}$, i.e. from the hydrostatic stress component p_h. v_{ph} could be measured in three different ways: (i) from the difference between the "apparent stress enthalpy" from the conventional method using [23] in a tensile or compression test, and the "true stress enthalpy" in a torsion test; (ii) from the medium value of the "apparent stress enthalpies" obtained in tensile and compression tests; (iii) from the difference between the "apparent stress enthalpy" using eq. [23], and the true enthalpies derived from the eqs. [20] and [21] which do not incorporate stress variations during the temperature change.

If the model were correct, v_{ph} $(p_h = \bar{\tau} = p_0)$ and v_p $(p = p_0)$ had to be identical, i.e. of the same numerical value at comparable pressure (or stress) p_0. As already implied in (ii), eq. [34] should be consistent with the wellknown difference between yield stresses of polymers as measured in tensile and compression tests.

(e) The essence of section (d), the idea that the internal pressure cannot be kept constant when the applied stress is varied, has also consequences on the pressure-activation volume, when it is determined according to eq. [29] by measuring v_τ. The eqs. [27] and [31] provide ways to avoid this difficulty.

Some Examples of Application

So far the methods described above have only been used in a few publications which deal either with the influence of temperature (5—8) or of pressure (9—10) on the activation parameters. Combined temperature and pressure tests have only been reported once (8). No work has considered the implications of eq. [34] about the influence of the normal stress in constant-pressure tests.

In the following we shall briefly review the available literature and examine whether the various experimental results are consistent with the theoretical parameters.

Low-Temperature Tests

The following data are characteristic for epoxy (EP) (6, 7). Often they are valid also for polyethylene (PE) (8). Basis for the use of the various formulae is a constant $\dot{\varepsilon}_0$ during strain-rate changes. To check this, v_τ has been measured at various ratios $\dot{\varepsilon}_2/\dot{\varepsilon}_1$. The results proved to be unaffected (6), so that the simplified form [24] of eq. [18] can be considered as a good approximation.

One important result was the division into regions with different deformation mechanisms. The limits for PE were 175 and 265 K, the activation enthalpy of the medium region $\Delta H_0^{II} = 2$ eV[2]. An increase by 0.5 eV characterizes the transition to the mechanism above 265 K. For EP, it was found: region I ($<$ 180 K) with $\Delta H_0^{I} < 1$ eV, region II (180—320 K) with $\Delta H_0^{II} = 2$—3 eV, and region III ($>$ 370 K) with $\Delta H_0^{III} > 4$ eV.

---

[2]) Anmerkung: 1 eV $=$ 23 kcal/mol.

The following detailed results were found to be valid in the deformation range II of EP (6, 7):

(a) Fig. 4 shows that λ_τ increases proportionally in the course of deformation, i.e. v_τ decreases with ε.

(b) Another interpretation of Fig. 4 is also possible: v_τ decreases with τ. In Fig. 5 v_τ-values at various elongations and from various temperatures are plotted in dependence of τ: all data points are concentrated in a single band. v_τ does of course change with ε and T, but only to such an extent as to yield a unique dependence on τ. If the present findings at low strain rates ($\dot\varepsilon = 10^{-4}$ sec^{-1}) can be generalized, v_τ is not temperature or strain dependent, but only a function of stress. The general $v_\tau - \tau$-relation of PE is comparable but it differs slightly in a quantitative respect (cf. Fig. 5). Similar tendencies are manifest in $\sigma - \log\dot\varepsilon - T$-relations of polymethylmethacrylate (PMMA) (11—13).

(c) The activation enthalpy of deformation is stress dependent (Fig. 6). The superimposed small strain dependence found in PE was not confirmed for EP.

(d) With known activation enthalpies the pre-exponential factor $\dot\varepsilon_0$ can be calculated using eq. [12]. Fig. 7 implies that $\dot\varepsilon_0$ is independent of temperature and stress, but dependent on strain, contrary to the behaviour of activation volumes and enthalpies.

It was not possible to assert as much for the temperature region I and III, although it is possible that the statements of (a), (b) and (c) at least are valid as well (6). It is, however, confirmed that the stress-activation volumes at comparable τ are higher in region I, and lower

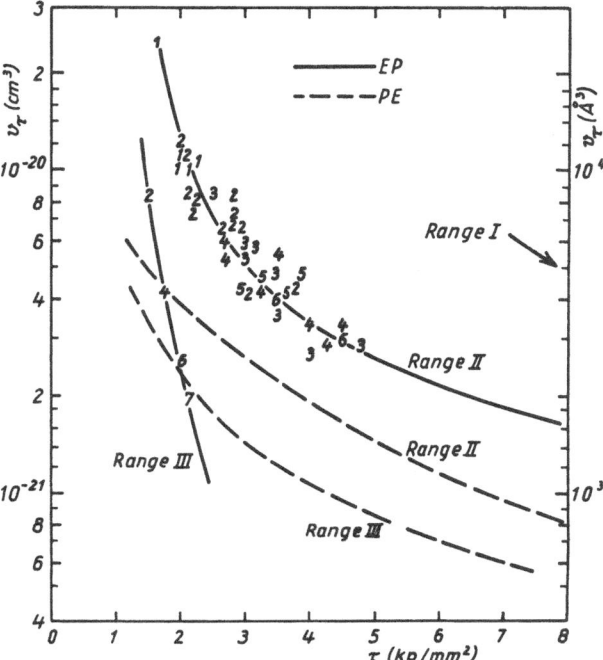

Fig. 5. The stress dependence of stress-activation volumina of epoxy (6, 7) and polyethylene (8) in various temperature regions. The numbers indicate individual data points, their values give the strain (%) at which v_τ has been measured

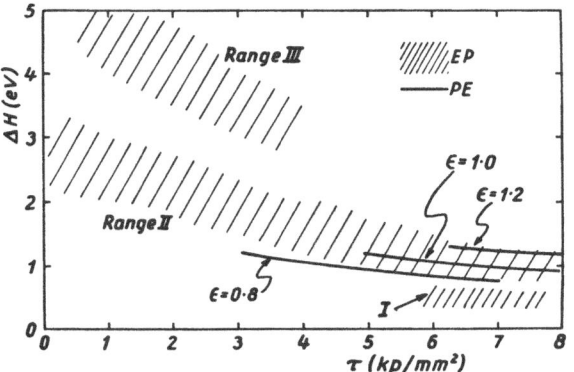

Fig. 6. The stress dependence of the activation enthalpy of epoxy (6, 7) and polyethylene (8) in various temperature regions

in region III (cf. Fig. 5). These differences lead to three distinct activation enthalpies for the three regions as quoted earlier.

High-pressure Tests

PE (8, 9) and polytetrafluorethylene (PTFE) (10) have been investigated.

(a) The pressure-activation volume v_p varies moderately at low pressure and strains (Fig. 8).

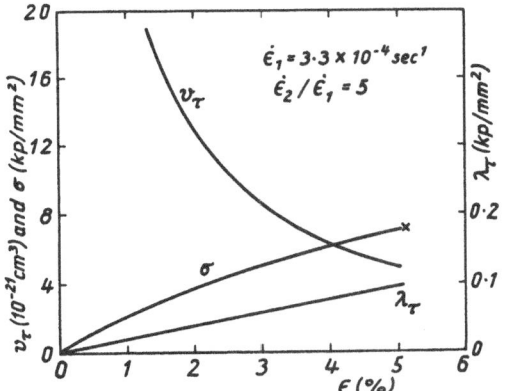

Fig. 4. Stress-strain curves and the variation of λ_τ and v_τ with ε for epoxy at room temperature (6)

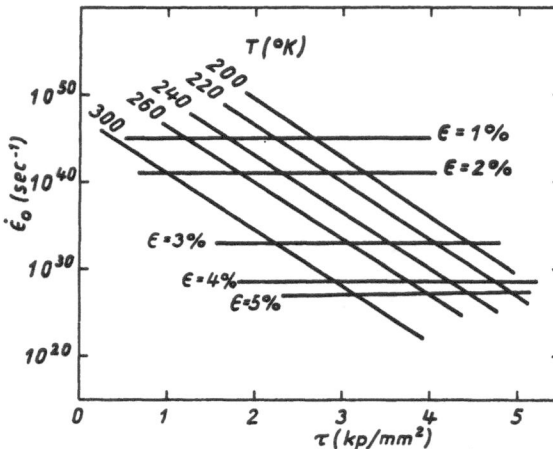

Fig. 7. Idealized curves showing the stress dependence of the frequency factor of epoxy in the temperature region II (6, 7)

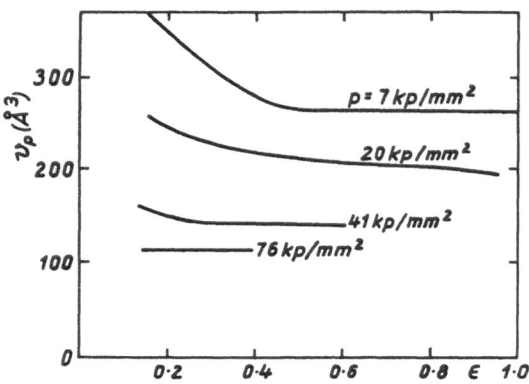

Fig. 8. The strain and pressure dependence of the pressure-activation volume of polytetrafluorethylene at room temperature (10)

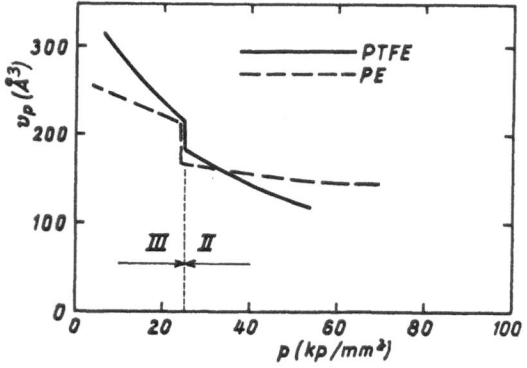

Fig. 9. The pressure dependence of the pressure-activation volume of polyethylene (8, 9) and polytetrafluorethylene (10) at room temperature

(b) v_p is pressure dependent. The strain dependence, registered under (a), is only a small portion of the change due to pressure effects, and is neglected in Fig. 9. At medium pressure, an abrupt transition occurs as a consequence of a change in deformation mechanisms.

The Stress Dependence of Activation Parameters

An essential result is the non-linear stress dependence of the stress-activation volume. As a prerequisite, the force-distance curve must be non-linear, too (this has been assumed already in Fig. 1). The total external work, done by a force $K = \tau A$ during the thermally activated event cannot be expressed by τv_τ alone. This value neglects the portion of the work which was done between the minimum potential energy and the equilibrium position without the help of thermal activation. The correct way to write the external work is as an integral $\int v_\tau(\tau)\, d\tau$ (1—3). The activation enthalpy ΔH is connected with $v_\tau(\tau)$, and also stress dependent. Therefore eq. [7] has to be corrected:

$$[\Delta H(\tau)]_p = \Delta H_0 - \int_0^\tau v_\tau(\tau)\, d\tau \,. \qquad [35]$$

For one more reason it is obvious that the integral has to be used: namely, in order to maintain the validity of the original simple definitions [13] or [16] for the stress-activation volume. With a stress-dependent v_τ, eq. [6] or [7], when differentiated with respect to τ, would yield a more complex relationship between ΔH and v_τ.

It is necessary to rewrite eq. [25] in the same way. Eq. [26] for the total activation enthalpy reads now (9)

$$\Delta H(\tau, p) = \Delta H_0 - \int_0^\tau v_\tau(\tau)\, d\tau + \int_0^p v_p(p)\, dp \,. \qquad [36]$$

The Relation between v_τ and v_p

The interrelation between stress- and pressure-activation volumina is given by eq. [29]. At present, only for PE are data available for v_τ as well as for v_p (8, 9). Let us consider the conditions for yielding at room temperature, i.e. for region III according to the definition for PE (8). τ is about 1 kp/mm², the pressure-hardening coefficient $d\tau/dp$ near atmospheric pressure 0.043, v_p is about 260 Å³ (Fig. 9). According to [29] the stress-activation volume should be $v_\tau \approx 6000$ Å³: a plausible value judging from extrapolating the curve in Fig. 5 to $\tau = 1$ kp/mm².

The Temperature Dependence of τ

By means of eq. [12], which is changed to agree with [35] it should be possible to compute the theoretical temperature dependence of the stress at constant pressure:

$$\int_0^\tau v_\tau \, d\tau = \Delta H_0 + kT \ln (\dot\varepsilon / \dot\varepsilon_0) . \qquad [37]$$

ΔH, $\dot\varepsilon$ and $\dot\varepsilon_0$ have fixed values. The integral on the left side can be obtained from $v_\tau - \tau$-diagrams (Fig. 5), and is of course also related to τ (Fig. 10). Since the portion $0 < \tau < 2 \, \mathrm{kp/mm^2}$ is not available from Fig. 5, the curve can only be calculated between the limits $\int_2^\tau v_\tau \, d\tau$ (the curve for $C = 0$ in Fig. 10). In a previous publication it has been shown that a constant $C = C'$ has to be chosen in order to procure agreement with the experimental τ (7). This adjustment was done only once for a certain $\tau_{T,\varepsilon}$, but in due course the choice of $C = C'$ provided further agreement between theory and experiment for all temperatures and strains (Fig. 11). We would like to emphasise especially the increased temperature dependence of flow stresses at high strains.

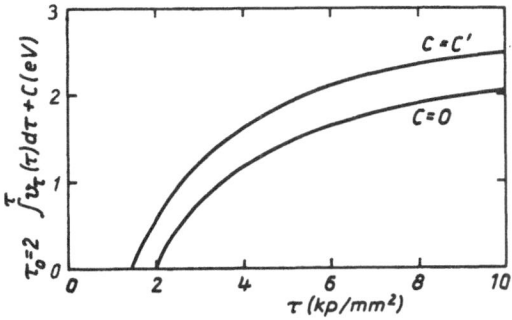

Fig. 10. The stress dependence of external work, as calculated from Fig. 5, of epoxy in deformation range II (7)

A simplified relation between stress and temperature is drawn from the original eq. [7]: τ is inversely proportional to the stress-activation volume. Although this form, too, establishes qualitatively the non-linear relation, the quantitative result is not satisfying (7). This discrepancy is a further indication that eq. [35], and not [7], is the correct form of the activation enthalpy.

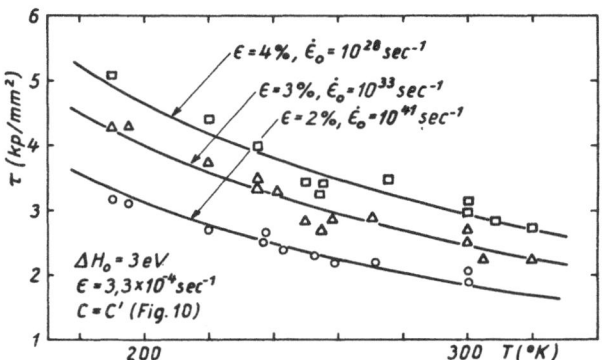

Fig. 11. Theoretical and experimental temperature dependences of flow stresses of epoxy in deformation range II (7)

Possible Mechanisms and Significance of Activation Parameters

The given examples are characteristic of amorphous (EP) and partially crystalline substances (PE). One prerequisite for the validity of the rate theory is the exclusive occurrence of one single mechanism. Conditions in PE may become difficult to survey when the deformation of the crystalline portion contributes also (the deformation of PE-crystals by dislocation movement has been discovered recently (14). No preceding work has attempted to separate these two possible modes; on the contrary, it was claimed that it is only in the amorphous parts that deformation takes place (9, 13).

A further difficulty in discussing EP, PE and PTFE together are their different constitutions. The glass-transition temperature of EP, PC, PVC, PTFE and PMMA is 350 to 400 K. The quoted results for EP and PTFE are thus representing the glassy state. The examples for PE, on the other hand, are for a material with a transition at 175 K, and the activation parameters have been determined in the viscoelastic condition. Astounding is therefore the similarity of a few activation parameters. Especially the agreement of pressure effects appears to be perfect: the pressure-hardening coefficient $d\tau/dp$ for yield points[3]) of PE is 0.038 (9), of PTFE 0.038 (10), (the flow stresses behave similarly); the pressure dependence of the pressure-activation volume of viscous PE and glassy PTFE is almost identical (cf. Fig. 9). The similarities are

---

[3]) The yield point is defined as "upper yield stress" ($d\sigma/d\varepsilon = 0$) for viscous polymers, as proportional limit for glassy polymers.

not as convincing when the temperature dependences of strength and activation characteristics are compared.

Earlier, data of the three deformation ranges have been quoted: especially the II/III transitions of EP and PE are wide apart. Also disagreeing with the results for EP in Fig. 6 is the jump at the II/III transition of 0.5 eV which has been calculated for PE, and the stress-activation volumina of EP which are on an average twice as high as those of PE. At present these similarities or differences between EP and PE cannot be interpreted.

The theoretical part of this work has been aimed at the effects of a shear stress τ. The stress-activation volume v_τ is, according to eq. [6], defined as the product of activation area A and distance d. τ operates in the area A. This model may be relevant, when the deformation is interpreted as a translation of chains against the resistance of side groups, or as stretching of kinked chains through rotation or through the movement of disclinations (16). A description of the pressure-activation volume v_p follows from eq. [25]: it is the volume difference of the deformation element between the non-activated and activated state, caused by external pressure[4]. According to eq. [29] there is a simple relation $v_p \propto v_\tau$. Whether the proportionality factor (the pressure dependence of the shear stress) varies with temperature, is not yet known. Should it turn out to be temperature dependent, v_p as well as v_τ would be a function of stress.

The question how the experimental ΔH is related to ΔG, appears at first of minor interest, since at the present stage it is in many cases sufficient just to compare the enthalpies as determined in different methods (e.g. oscillatory measurements). In the case of PC, both the mechanism which causes the internal friction due to the β-relaxation, and the yield process below 200 K have an activation enthalpy of 0.42 eV (18). For the deformation above 200 K, 3.3 eV has been measured for PC and 3 eV for PVC (19, 20). These values agree with those reported for EP (7). Due to the structural similarities of PC and epoxide resins the comparison seems justified, and the assumption that

the loss peak reflects the mobility of the diether linkage (21) may be valid for both materials.

The surprising agreement of activation parameters for an elastic and macroplastic behaviour should — if the mechanisms were the same — lead to further identities. The temperature dependence, for instance, can in fact be explained in two different ways (7). The activation analysis supplies curves, as they are plotted in Fig. 11. Considering, on the other hand, that *Young*'s modulus of polymers is strongly influenced by temperature, slopes $d\sigma/dT$ can be calculated for stresses at certain temperatures:

$$\left[\frac{d\sigma}{dT} \right]_\varepsilon = \frac{\sigma_\varepsilon}{E} \frac{dE}{dT} . \qquad [38]$$

It has been shown that both methods furnish consistent results as long as the stresses are within the "elastic" range or at its end (determined by a proportional limit). But $[d\sigma/dT]_\varepsilon$ increases when the elastic range is exceeded.

In the calculation of the theoretical curves of Fig. 11 it was found that it is not ΔH_0 which determines the slope $d\sigma/dT$, but only the frequency factor $\dot{\varepsilon}_0$ which has proved to be a function of ε. The difference between a slope for a certain $\sigma_{T,\varepsilon}$ and a slope which stems from the "elastic" behaviour has to be traced back to the frequency factor. In the case of metals the frequency factor is the theoretically attainable maximum strain rate, and related to dislocation density and velocity of sound in the crystal. The high values of EP (10^{25} to 10^{45} sec^{-1}; a similar value was found for PC (20)) cannot be explained on the same basis. According to eq. [11] $\dot{\varepsilon}_0$ consists of two parts: $\dot{\varepsilon}_0'$ may contain the number and density of glide elements, the area which is covered during the thermal event, and a jump frequency; the second part may be determined by the entropy of deformation. Both partials, and especially the second, may be sensitive to the degree of deformation. Every attempt to establish a model for $\dot{\varepsilon}_0$ will have to take into account this extreme variation of $\dot{\varepsilon}_0$ with ε.

Conclusion

The well-known rate theory which relates strain rate, temperature and activation enthalpy through an *Arrhenius*-type equation was used as the basis for a thermo-mechanical activation analysis of the deformation process. It can be

---

[4]) So far it is, however, impossible to relate v_p with macroscopic volume changes which occur during deformation (17).

applied on any kind of mechanical test. The theory has been used successfully to study the deformation of metals. Applied on polymers, it must, however, be modified to accommodate the influence of external hydrostatic pressure on the strength properties. The results differ in an important respect from the related *Eyring* analysis (22): the activation enthalpy, as well as stress- and pressure activation volumes are stress or pressure dependent. A pronounced dependence of the frequency factor on the degree of deformation may be explained by possible changes of the molecular structures, and by entropy variations.

Another result is the existence of temperature regions where the deformation is controlled by different mechanisms. Also of interest is the consistency of experimental findings and their theoretical explanation (e.g. the non-linearity in stress-temperature curves).

Some experimental enthalpies can be compared with those derived with oscillatoric methods. The conclusion may be drawn that the same mechanism is rate controlling during both anelastic and macro-deformation. This is supported by other findings apart from the comparison of enthalpies.

Since the analysis, as one application of phenomenological thermodynamics, is generally valid and not bound to one special mechanism, it may not be correct to judge alone on grounds of equal enthalpies: it is conceivable that different mechanisms with similar enthalpies could exist. Future models will have to predict mechanisms whose theoretical temperature, rate, and pressure dependences conform with the experiments.

Acknowledgement

I would like to thank Professor Dr. W. *Knappe*, Institut für Kunststoffverarbeitung, Montanistische Hochschule Leoben, whose encouragement was valuable in the writing of the paper.

Summary

The model for macro-deformation is based on an *Arrhenius*-type equation. Using a thermodynamic approach, an analysis is described which characterizes the deformation behaviour by activation enthalpies, stress- and pressure-activation volumes and frequency factors, and their relation to stress, pressure and strain. The application of the method in recent experiments on EP, PE, PMMA, etc. is reviewed, and the consistency of theoretical and experimental results is demonstrated.

Zusammenfassung

(Eine Analyse der thermisch aktivierten Makroverformung von Polymeren.) Das Modell für die Makroverformung ist auf einer *Arrhenius*gleichung aufgebaut. Mit Hilfe der Thermodynamik wird eine Analyse abgeleitet, welche das Verformungsverhalten durch Aktivierungsenthalpien, Spannungs- und Druck-Aktivierungsvolumina, Frequenzfaktoren und ihre Abhängigkeiten von Spannung, Druck und dem Verformungsgrad beschreibt. Die Methode ist bisher in Versuchen mit EP, PE, PMMA etc. angewendet worden. Die Übereinstimmung von theoretischen und experimentellen Resultaten wird gezeigt.

References

1) *Conrad, H.*, J. Metals **16**, 582 (1964).
2) *Kronmüller, H.*, Moderne Probleme der Metallphysik, Band I, p. 126, (Berlin, 1965).
3) *Evans, A. G.* and *R. D. Rawlings*, Phys. stat. sol. **34**, 9 (1969).
4) *Ward, I. M.*, Mechanical Properties of Solid Polymers (London, 1971).
5) *Shen, K. H.* and *J. L. Rutherford*, Mater. Sci. Eng. **9**, 323 (1972).
6) *Pink, E.* and *J. D. Campbell*, Report No. 1040/72, Dept. Engineering Science (Oxford, 1972).
7) *Pink, E.* and *J. D. Campbell*, Mater. Sci. Eng. **15**, 187 (1974).
8) *Pampillo, C. A.* and *L. A. Davis*, J. Appl. Phys. **43**, 4277 (1972).
9) *Davis, L. A.* and *C. A. Pampillo*, J. Appl. Phys. **42**, 4659 (1971).
10) *Davis, L. A.* and *C. A. Pampillo*, J. Appl. Phys. **43**, 4285 (1972).
11) *Holt, D. L.*, J. Appl. Polymer Sci. **12**, 1653 (1968).
12) *Duckett, R. A.*, *S. Rabinowitz* and *I. M. Ward*, J. Mater. Sci. **5**, 909 (1970).
13) *Brady, T. E.* and *G. S. Y. Yeh*, J. Appl. Phys. **42**, 4622 (1971).
14) *Gleiter, H.* and *A. S. Argon*, Phil. Mag. **24**, 71 (1971).
15) *Rabinowitz, S.*, *I. M. Ward* and *J. S. C. Parry*, J. Mater. Sci. **5**, 29 (1970).
16) *Argon, A. S.*, Lecture given at Oxford (1972).
17) *Pampillo, C. A.* and *L. A. Davis*, J. Appl. Phys. **42**, 4674 (1971).
18) *Bauwens, J. C.*, J. Mater. Sci. **7**, 577 (1972).
19) *Bauwens, J. C.*, *C. Bauwens-Crowet* and *G. Homès*, J. Polymer Sci. **7-A2**, 1745 (1969).
20) *Bauwens-Crowet, C.*, *J. C. Bauwens* and *G. Homès*, J. Mater. Sci. **7**, 176 (1972).
21) *Cuddihy, E.* and *J. Moacanin*, Epoxy Resins. Amer. Chem. Soc., p. 96 (1970).
22) *Eyring, H.*, J. Chem. Phys. **4**, 283 (1936).
23) *Cottrell, A. H.*, An Introduction to Metallurgy (London, 1971).
24) *Schoeck, G.*, Phys. stat. sol. **8**, 499 (1965).

Author's Address:

Erwin Pink
Erich-Schmid-Institut für Festkörperphysik
der Österreichischen Akademie der Wissenschaften
Leoben, Austria

Progr. Colloid & Polymer Sci. **58**, 90–101 (1975)

*Institut für Makromolekulare Chemie der Technischen Hochschule Darmstadt
und Sonderforschungsbereich 41 Physik und Chemie der Makromoleküle*

Rheologische Messungen an Polyisobutylenlösungen im Weissenberg-Rheogoniometer

S. Krozer und *J. Schurz*

Mit 15 Abbildungen und 6 Tabellen

(Eingegangen am 17. September 1974)

1. Einleitung

In dieser Arbeit soll über unsere Erfahrungen mit dem *Weissenberg*-Rheogoniometer (WRG) berichtet werden. Als Untersuchungsobjekt haben wir Polyisobutylen (PIB), gelöst in Toluol gewählt (unfraktionierte Handelsprodukte), einerseits weil dieses System zugleich in anderen Instrumenten an unserem Institut studiert wurde und andererseits weil gerade über dieses System sehr viele Arbeiten in der Literatur vorliegen. Unserer Ansicht nach ist es beim gegenwärtigen Stand der phänomenologischen Rheologie unbedingt notwendig, Messungen in verschiedenen Geräten an ein und derselben Substanz zu vergleichen. Nur auf diese Weise werden die derzeit noch bestehenden beträchtlichen Diskrepanzen, die man in den Meßgeräten verschiedenen Typs erhält, beseitigt werden können und werden wir dem Ziel näherkommen, objektive Meßergebnisse zu erhalten, die frei von Apparateeinflüssen sind. Überdies stellen solche vergleichende Messungen in verschiedenen Apparaten sehr kritische Tests für die zugrunde liegenden Theorien dar.

Zugleich soll ein Vorschlag für eine einheitliche Darstellung der Meßresultate gegeben werden. Zuletzt endlich sollte dann noch versucht werden, die erhaltenen Meßergebnisse mit in der Literatur vorhandenen Theorien zu vergleichen, wobei wir uns in diesem Fall nur auf die Theorie von *Spriggs* beschränken wollen. In späteren Arbeiten sollen dann unsere Versuche mit anderen Meßgeräten beschrieben und mit den hier mitgeteilten Resultaten verglichen werden.

2. Darstellungsweise

Wir wollen bei der Darstellung der Meßergebnisse von der Erfahrung ausgehen, die man an den Schubviskositäts-Fließkurven im Laufe der letzten Jahrzehnte gewonnen hat. Es hat sich hier als notwendig erwiesen, mit einer scheinbaren Viskosität η' zu arbeiten, die einfach als $\eta' = \tau/D$ definiert ist (τ bzw. D: Schubspannung bzw. Scherrate (1) an einer bestimmten Stelle, meist an der Begrenzung des Fließfeldes). Diese scheinbare Viskosität stellt eigentlich eine Äquivalentviskosität dar, nämlich jene, die eine *Newton*sche Flüssigkeit bei gleichem τ und gleichem D haben würde. Sie ist daher selbstverständlich eine Funktion von D und kann daher nur Punkt für Punkt ausgewertet werden (2). Die Darstellung erfolgt im allgemeinen in Form der doppellogarithmischen Fließkurven, in denen $\log D$ gegen $\log \tau$ aufgetragen ist. Die Steigung dieser Fließkurven ist

$$s = d \log D / d \log \tau \,.$$

In ganz ähnlicher Weise schlagen wir nun vor, die Normalviskositäten darzustellen, indem wir ebenfalls wieder scheinbare Normalviskositäten einführen, die ganz analog ebenfalls Äquivalentwerte darstellen sollen. Nimmt man sie nach der Theorie 2. Ordnung als D^2 proportional an, so kommt man damit auf folgende Beziehungen:

Schubviskosität η':

$$\eta' = \frac{\tau_{12}}{D} = \frac{\tau}{D}$$

(τ ohne Indizes bedeutet immer τ_{12}).

Normalviskositäten:

$$\eta_1' = \frac{\tau_{11} - \tau_{33}}{D^2} = \frac{\sigma_1}{D^2} \qquad \eta_2' = \frac{\tau_{22} - \tau_{33}}{D^2} = \frac{\sigma_2}{D^2}$$

wenn wir, wie üblich, abkürzen:

$$\sigma_1 = \tau_{11} - \tau_{33} \quad \text{und} \quad \sigma_2 = \tau_{22} - \tau_{33}.$$

Für die Normalspannungsdifferenz $\sigma_1 - \sigma_2 = \tau_{11} - \tau_{22}$ wollen wir eine analoge Normalviskosität η_N' einführen:

$$\eta_N' = \frac{\sigma_1 - \sigma_2}{D^2} = \frac{\tau_{11} - \tau_{22}}{D^2} = \eta_1' - \eta_2'.$$

Selbstverständlich sind alle definierten scheinbaren Viskositäten Funktionen von D bzw. D^2.

Führt man weiterhin den Schubmodul G ein:

$$G = \frac{\tau_2}{\tau_{11} - \tau_{22}},$$

so erhält man den interessanten Zusammenhang:

$$G = \frac{\eta'^2}{\eta_N'} \quad \text{bzw.} \quad \eta'^2 = G \cdot \eta_N'.$$

Die so gewonnenen Werte der scheinbaren Normalviskositäten wollen wir in Form von phänomenologischen Fließkurven darstellen, indem wir $\log D$ gegen den Logarithmus der jeweiligen Normalspannungsdifferenz bzw. gegen den Logarithmus von η_N' auftragen. Diese Fließkurven unterscheiden sich von jenen der Schubviskosität insofern, als bei ihnen *Newton*sche Flüssigkeiten infolge der quadratischen Abhängigkeit der Normalspannungen von D sich als Geraden mit der Steigung $1/2$ darstellen werden, während bekanntlich in der Schubviskosität die *Newton*sche Flüssigkeit eine Ge-

rade mit der Steigung Eins repräsentiert. Das entspricht den Formeln

Schubviskosität:

$$\log D = \log (1/\eta') + \log \tau,$$

Normalviskosität:

$$\log D = \tfrac{1}{2} \log (1/\eta_N') + \tfrac{1}{2} \log \tau.$$

Die Steigung der Normalviskositäts-Fließkurven ist definiert als (für η_N'):

$$s_N = d \log D/d \log (\tau_{11} - \tau_{22}).$$

Wir wollen nochmals festhalten, daß wir unter Parametern ohne Indizes stets die Werte für den Schubversuch verstehen wollen, während die Werte für Normalkräfte immer mit Indizes versehen sind, wie sie oben definiert wurden. Insbesondere bedeutet der Index N stets die Differenz der beiden ersten Normalspannungsdifferenzen bzw. der daraus erhaltenen Viskositäten.

Durch diesen Schematismus glauben wir die Darstellung der Meßergebnisse auf eine vernünftige Basis gestellt zu haben, insofern als wir uns sehr eng an die bei der Schubviskosität verwendeten Formalismen gehalten haben. Es wäre zu wünschen, daß alle Autoren sich diesen Vorschlägen anschließen; man könnte dadurch publizierte Ergebnisse viel besser miteinander vergleichen, um so mehr als die Anschaulichkeit der Fließkurven im genannten Sinn ja seit Jahrzehnten erprobt ist.

3. Experimentelles

Die Messungen wurden mit dem *Weissenberg*-Rheogoniometer durchgeführt. Es wurden gemessen: Schubspannungen und Normalspannungen als Funktion der

Tabelle 1. Meßbedingungen

Temperatur °C	Geometrie Durchmesser $2R$ (mm)	Winkel Kegel-Platte	Spalt Platte-Platte (mm)	Torsions-federkonst. k_T (dyn/u)	Normal-federkonst. k_N (dyn/u)	Eigenschwingungsfrequenz f [s^{-1}] oberer Teil Torsion	unterer Teil Normalkraft
25	50	2°	1	980	86,5	48	1,66

Tabelle 2. Meßgrenzen

Geschwindigkeits-gefälle-Bereich (s^{-1})	Frequenz-Bereich (s^{-1})	Spannungsbereiche Scher (dyn/cm²)	Normal (dyn/cm²)
$0,29 - 1,9 \times 10^3$	$10^{-2} - 10^2$	$10^1 - 4 \times 10^4$	$10 - 2,5 \times 10^4$

Scherrate mit Kegel-Platte (KP) und Platte-Platte (PP) Geometrie. Ferner wurde der komplexe Modul bei Drehoszillationen gemessen.

Die Konstruktion unseres Gerätes (Modell R 18, Fa. Sangamo, England) ist in der Literatur beschrieben (3).

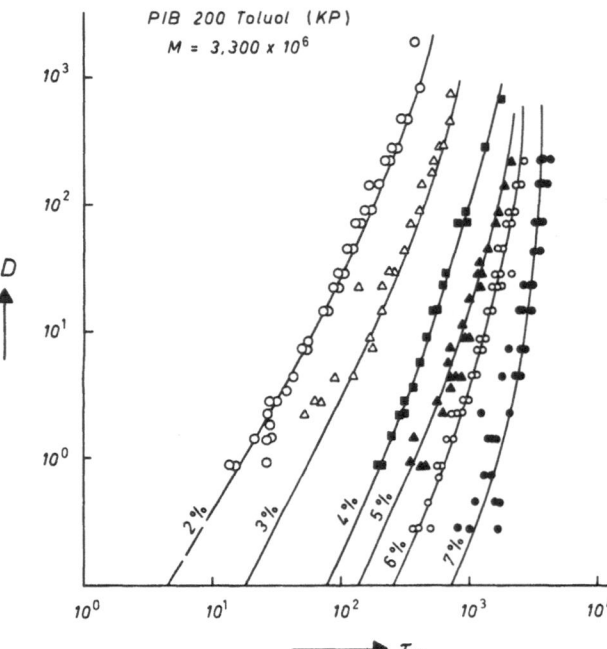

Abb. 1. Fließkurve von PIB 200 in Toluol (Schubviskosität)

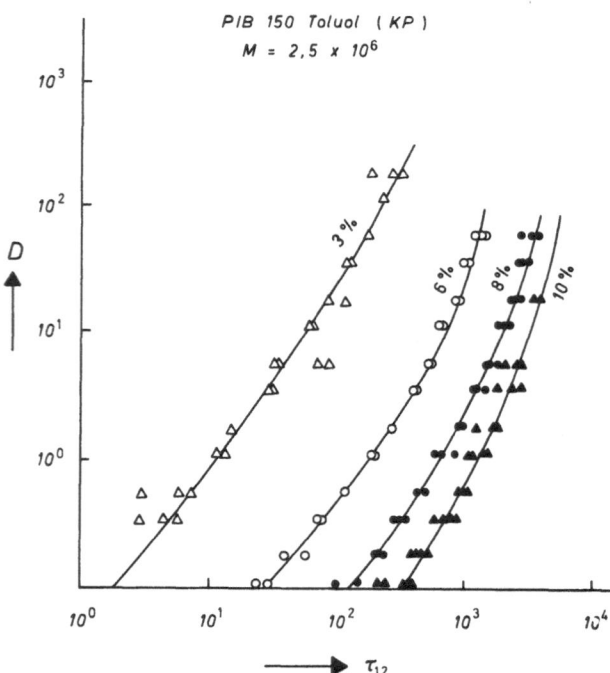

Abb. 2. Fließkurve von PIB 150 in Toluol (Schubviskosität)

Die Meßgrenzen und Meßbedingungen sind in den Tabellen 1 und 2 zusammengestellt. Die Meßmöglichkeiten bei Schwingungen sind beschränkt durch die Eigenfrequenzen der Meßsysteme (Torsion 48 s^{-1}, Normalkraft 1,7 s^{-1}), da die verwendete Frequenz viel kleiner sein muß als die Eigenfrequenz. Das ist nur bei Torsionsmessungen realisierbar. Für die Normalkraftmessungen (4) muß die Frequenz sehr niedrig sein, wodurch die Normalkraft sehr klein wird.

Um die Temperatur konstant zu halten, befinden sich die Meßsysteme mit der Probe in einem Mantel, in welchen Röhrchen mit durchfließendem Wasser eingebaut sind. Die Eichuntersuchungen haben gezeigt, daß die Temperaturschwankungen bei einer Umgebungstemperatur von ca. 20 °C zu 25 ± 0,5 °C angenommen werden können. Eine wesentliche Fehlerquelle stellt die „Austrocknung" der Probenschichten dar, die sich an der Phasengrenze Lösung/Luft befinden. Infolge der Verdunstung des Lösungsmittels bildet sich am Spaltrand eine relativ harte Schicht, welche die beiden Platten gewissermaßen verklebt und damit zur Verfälschung des Meßergebnisses führen kann. Um die Verdunstung geringer zu machen, haben wir in der Nähe des Spaltes rund um die Platten Gefäße mit Lösungsmittel angebracht, wodurch die Resultate verbessert wurden. Es wurde auch Lösungsmittel auf den Spaltenrand aufgespritzt. Diese Maßnahmen haben zur Verbesserung der Konstanz der Nullposition der Normalkraftanzeige geführt.

Die Messungen wurden mit Lösungen kommerzieller Polyisobutylen(PIB)-Proben (Oppanol, Fa. *BASF*) durchgeführt. Als Lösungsmittel wurde Toluol verwendet. Meßtemperatur war 25°C. Die vermessenen Proben und der Meßbereich sind in Tabelle 3 charakterisiert.

Tabelle 3. Die Charakteristik der angewandten Proben

PIB (Oppanol)	M_w*)	$[\eta]$*) Toluol 25 °C	c g/100 ml
B 50	328 000	1,05	8, 10, 12, 14
B 100	1 250 000	2,30	3, 6, 8, 10, 12
B 150	2 500 000	3,85	3, 6, 8, 10
B 200	3 550 000	5,75	2, 3, 4, 5, 6, 7

*) Meßergebnisse aus (8). $[\eta]$ in 100 ml/g.

Die Lösungen wurden wie üblich hergestellt, indem zerkleinerte PIB-Stückchen nach ein paar Tagen Quellung in Toluol am Wälzrad belassen wurden, bis vollständige Lösung eintrat (3—4 Wochen). Alle Messungen wurden 2—4mal wiederholt.

4. Resultate

Die erhaltenen Resultate sind in der im Abschnitt 2 beschriebenen Weise dargestellt. Zunächst zeigen die Abb. 1 bis 4 die Fließkurven für die Schubviskosität, indem log D gegen log τ aufgetragen ist. Man sieht an der graphischen Darstellung, daß die Streuung der Punkte relativ groß ist, jedenfalls wesentlich höher, als man sie beim Arbeiten mit sonstigen Rotationsviskosimetern oder gar mit dem Kapillarhochdruckviskosimeter gewohnt ist. Die Steigungen erscheinen sehr hoch, insbesondere bei den Proben mit den höheren Molekulargewichten. Bei den Proben mit niedrigeren Molekulargewichten (Abb. 3 und Abb. 4) wird schon das Gebiet der *Newton*schen Viskosität mit den 45°-Geraden angenähert. Alle in den Abb. 1 bis 4 gezeigten Fließkurven wurden mit Kegelplatte-Geometrie erhalten, eine Korrektur für nicht *Newton*sche Flüssigkeiten (5) wurde nicht angebracht. In den Abb. 5 bis 8 sind die analogen Normalviskositätsfließkurven dargestellt, hier ist log D gegen log ($\tau_{11} - \tau_{22}$) aufgetragen. Man sieht hier zweierlei. Wiederum fällt auf, daß die Streuung der Meßpunkte hoch ist, in diesem Fall noch bedeutend höher als bei der Schubviskosität. Wir haben uns zu helfen gesucht, indem wir möglichst viele Meßpunkte aufnahmen (jede Fließkurve ist aus 3 bis 4 verschiedenen Meßserien zusammengesetzt) und daß wir dann versuchten, durch die erhaltenen streuenden Meßpunkte die beste Fließkurve durchzulegen. Weiters fällt auf, daß auch die Normalviskositätsfließkurven durch keine einfache analytische Formel beschrieben werden können, so daß auch hier nichts anderes übrig bleibt, als mit der scheinbaren Normalviskosität $\eta'_N (D)$ zu operieren und auch die Normalviskositäts-Fließkurven Punkt für Punkt zu vermessen. Im großen und ganzen finden wir also auch hier ein ähnliches Bild wie bei den Schubviskositäts-Fließkurven, und wir glauben, daß man bei dieser Darstellungsweise sofort ein übersichtliches und anschauliches Bild des Normalviskositäts-Fließverhaltens für eine Serie von Lösungen verschiedener Konzentration erhält. Eine genauere Betrachtung zeigt, daß bei den großen Molekulargewichten (Abb. 5 und 6) wiederum sehr hohe Steigungen erhalten werden. Der *Newton*sche Bereich, in dem die Normalviskosität dem Quadrat der Scherrate proportional ist, wird nur bei den niedermolekularen Proben (Abb. 7 und Abb. 8) erreicht. Eine geringere Steigung als 1/2 wird, in Übereinstimmung mit der theoretischen Erwartung, nirgends gefunden.

Im großen und ganzen glauben wir damit gezeigt zu haben, daß die von uns vorgeschlagene Darstellung mit Hilfe einer „scheinbaren Normal-

viskosität" nützlich ist und praktische Vorteile bringt. Unbedingt notwendig ist dagegen eine Erhöhung der Präzision der Normalkraftmessungen.

Bemerkenswert ist übrigens auch, daß die größte Streuung der Meßpunkte bei den höch-

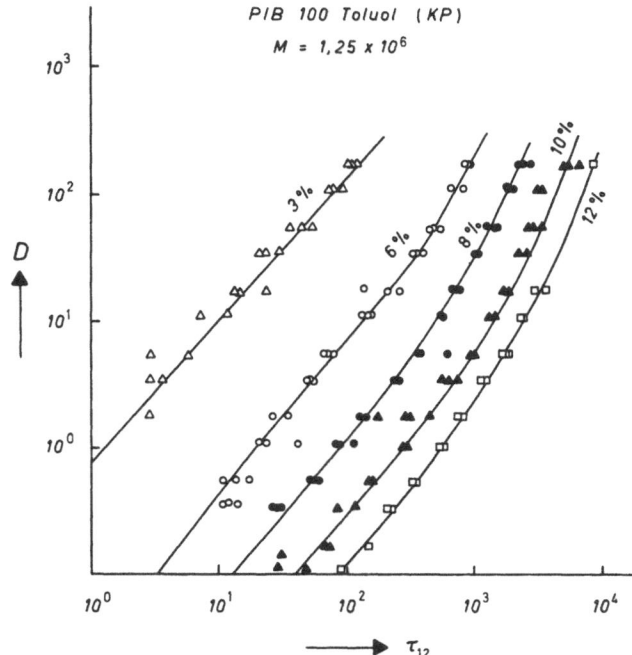

Abb. 3. Fließkurve von PIB 100 in Toluol (Schubviskosität)

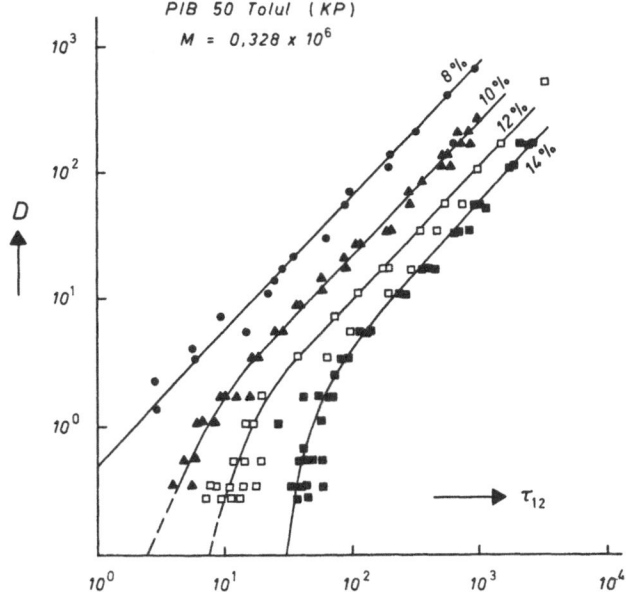

Abb. 4. Fließkurve von PIB 50 in Toluol (Schubviskosität)

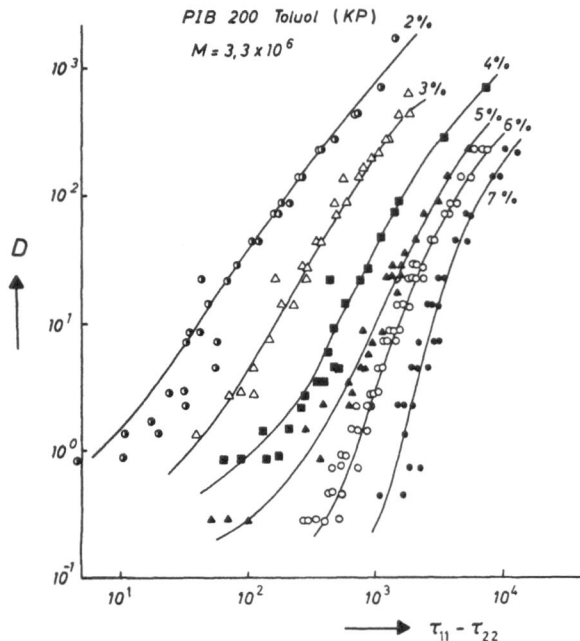

Abb. 5. Fließkurve von PIB 200 in Toluol (Normal-viskosität, KP-Geometrie)

sten Konzentrationen beobachtet wird, was wahrscheinlich mit der Austrocknung der Proben verbunden ist.

Bei der Kegel-Platte-Geometrie erhält man aus den Normalkraftmessungen nur die Normalspannungsdifferenz als Funktion der Scherrate.

Es gilt (6):

$$\sigma_1 - \sigma_2 = \tau_{11} - \tau_{22} = \frac{2 N_{KP}}{\pi \cdot R^2}$$

N_{KP}: Normalkraft für KP,
R: Plattenradius.

Um die Funktionen $\sigma_1(D)$ und $\sigma_2(D)$ getrennt zu erhalten, muß man als Ergänzung Messungen mit einer anderen Geometrie ausführen. In dieser Arbeit wurden dazu Messungen der Normalkraft mit Platte-Platte-Geometrie (PP) herzugezogen. In diesem Fall gilt die Formel (7):

$$\sigma_1 - 2\sigma_2 = \frac{N_{PP}}{\pi \cdot R^2}\left(2 + \frac{d \ln N_{PP}}{d \ln D}\right)$$

N_{PP}: Normalkraft für PP,
R Plattenradius.

Der Radius der angewandten Platten war 2,5 cm, Spaltenbreite 1 mm. Die Genauigkeit der parallelen Einstellung betrug ca. 2 μm.

Die Meßergebnisse für das PIB B 200 sind auf der Abb. 9 wiedergegeben, auf der $\log D$ gegen $\log(\sigma_1 - 2\sigma_2)$ aufgetragen ist. Wir finden wiederum normale Fließkurven, deren Steigung allerdings beträchtlich über jener ist, die für das *Newton*sche Verhalten zu erwarten wäre.

Zuletzt wurden auch komplexe Viskositäten bei Oszillationsbeanspruchung gemessen. Die Meßergebnisse für PIB B 200 sind in Abb. 10 dargestellt. Es sind dort $\eta'(\omega)$ sowie $\eta''(\omega)$ als

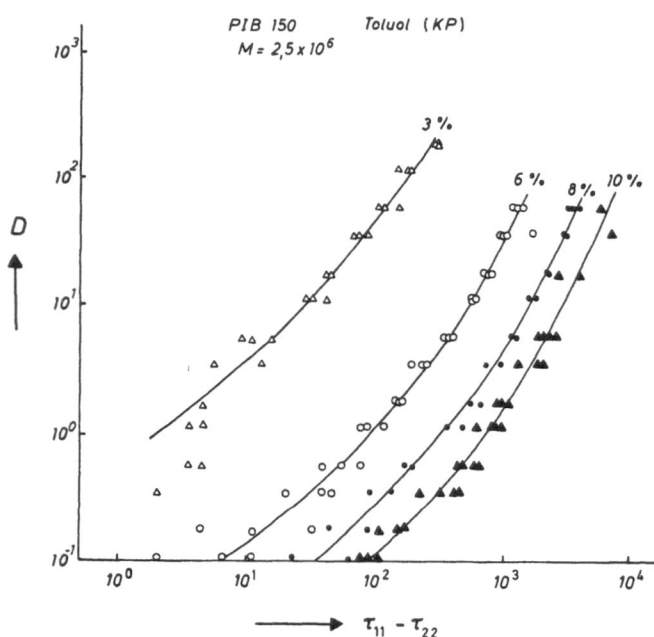

Abb. 6. Fließkurve von PIB 150 in Toluol (Normalviskosität, KP-Geometrie)

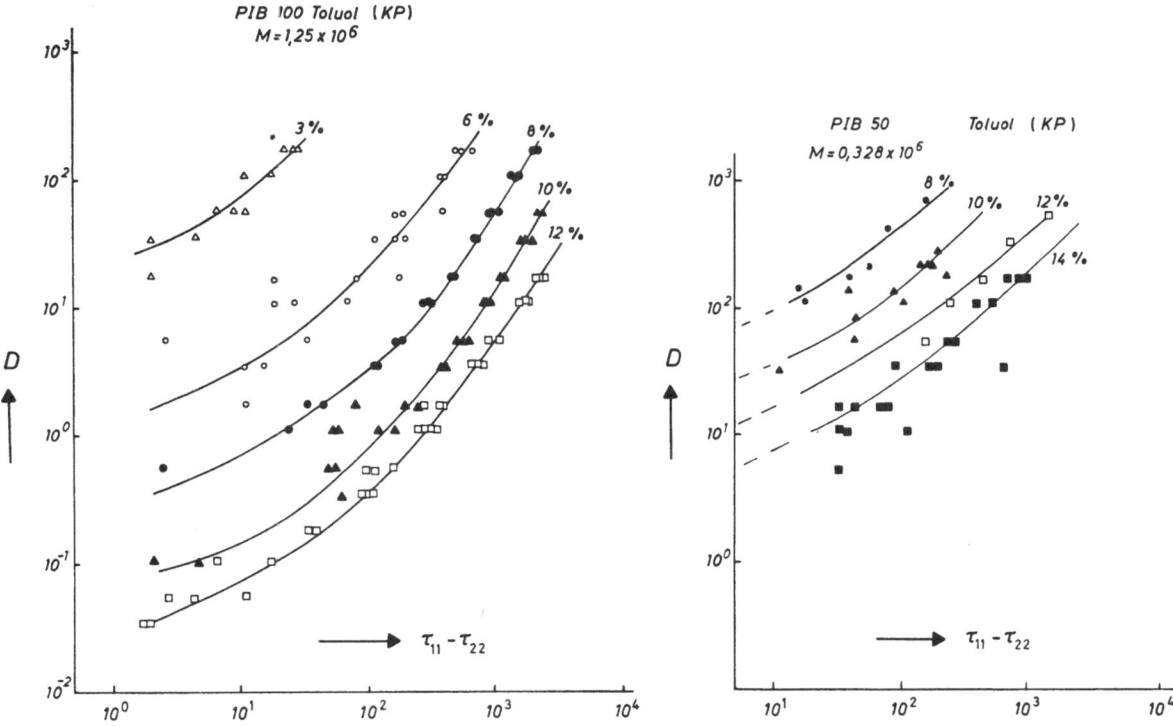

Abb. 7. Fließkurve von PIB 100 in Toluol (Normalviskosität, KP-Geometrie)

Abb. 8. Fließkurve von PIB 50 in Toluol (Normalviskosität, KP-Geometrie)

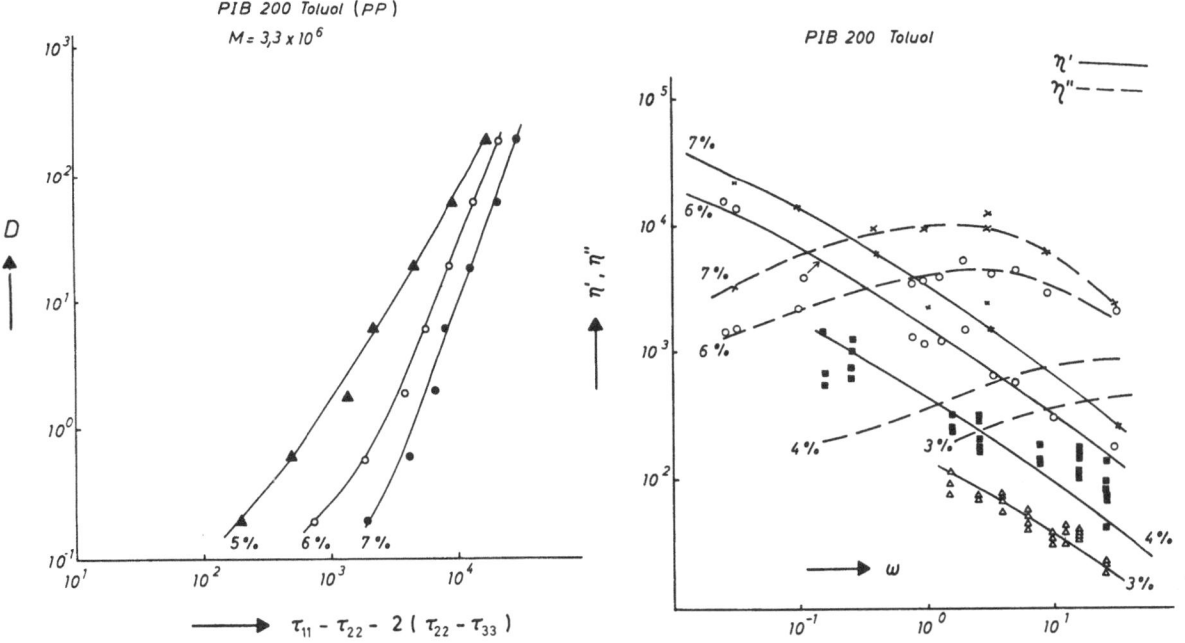

Abb. 9. Fließkurve von PIB 200 in Toluol (Normalviskosität, PP-Geometrie)

Abb. 10. Dynamische Fließkurven von PIB 200 in Toluol

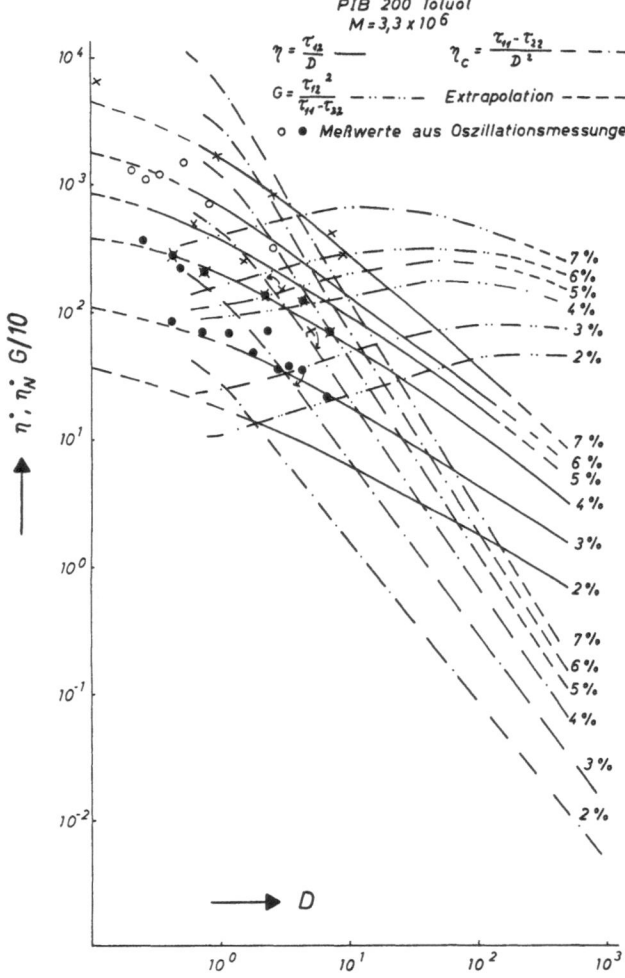

PIB 200 Toluol
M = 3,3 x 10⁶

$\eta = \dfrac{\tau_{12}}{D}$ ———　　$\eta_c = \dfrac{\tau_{11} - \tau_{22}}{D^2}$ — — —

$G = \dfrac{\tau_{12}^2}{\tau_{11} - \tau_{22}}$ —··—··—　　Extrapolation — — — —

o • Meßwerte aus Oszillationsmessungen

Abb. 11. Vergleich von dynamischen und stationären Meßwerten für PIB 200 in Toluol (Linien: stationäre Messungen; Punkte: Oszillationsmessungen)

Über die Bedeutung dieses Maximums wird noch zu sprechen sein (aus den dynamischen Viskositäten erhält man die dynamischen Moduln nach $G' = \omega \cdot \eta''\,(\omega)$ und $G'' = \omega \cdot \eta'(\omega)$, wobei G' der Speichermodul und G'' der Verlustmodul ist).

Aufgrund der durchgeführten Messungen war es nun möglich, die Schubviskosität η', die Normalviskosität η'_N und auch den Schermodul G zu bestimmen, wobei die genannten Größen wie folgt definiert sind:

$$\eta' = \frac{\tau}{D}\;;\quad \eta'_N = \frac{\tau_{11} - \tau_{22}}{D^2}\;;\quad G = \frac{\tau^2}{\tau_{11} - \tau_{22}}\,.$$

Die erhaltenen Resultate sind in den Abb. 11, 12, 13 und 14 jeweils für die Proben PIB B 200, B 150, B 100 und B 50 dargestellt. Der besseren Übersicht halber wurde in diesen Abbildungen auf die Angabe von Meßpunkten verzichtet. Schließlich sind zuletzt noch die getrennten Werte für σ_1, σ_2 sowie $\eta'_1 = \sigma_1/D^2$ und $\eta'_2 = \sigma_2/D^2$ in der Abb. 15 für PIB B 200 aufgetragen worden. Man sieht, daß σ_1 ansteigende Kurven zeigt, während σ_2 ein Plateau ausbildet, um dann bei einem D-Wert von etwa 10^2 plötzlich anzusteigen. Die Normalspannungsviskositäten η'_1 und η'_2 stellen mit der Scherrate monoton fallende Funktionen dar, die in der doppellogarithmischen Auftragung als lineare Geraden anfallen, so daß also hier die Normalviskositäten durch ein Potenzgesetz approximiert werden können.

5. Diskussion der Resultate

5.1. Die Theorie von *Spriggs*

Um das rheologische Verhalten von Stoffen zu beschreiben, wurden verschiedene konstitutive Gleichungen differentieller (8) und integraler (9) Art vorgeschlagen. *Spriggs* hat eine differentielle Theorie veröffentlicht (10), in welcher der konstitutiven Gleichung ein Vier-Parameter-Modell zu Grunde gelegt ist. Die gleichen Resultate bringt auch die integrale Theorie von *Spriggs* et al. (11).

Spriggs' Theorie geht von *Oldroyds* Zustandsgleichung (12) aus. Um das experimentell gefundene Verhalten von Viskosität und Normalspannungen zu beschreiben, hat *Oldroyd* in die Gleichung für die *Maxwell*sche Flüssigkeit (in mitgeführten Koordinaten) zusätzliche Glieder eingeführt. Er erhält:

$$\boldsymbol{\tau} + \lambda\,\mathscr{F}\boldsymbol{\tau} = \eta\,\boldsymbol{D}\,. \tag{1}$$

Funktion der Kreisfrequenz ω aufgetragen[1]. Man sieht, daß die Verlustviskosität $\eta'(\omega)$ in dieser Darstellung linear abfallende Kurven gibt, während die Speicherviskosität $\eta''(\omega)$ ansteigende Kurven ergibt, die bei den höheren Konzentrationen sogar ein Maximum ausbilden.

---

[1] Anmerkung: Wir sind uns des Umstandes bewußt, daß hier die übliche Bezeichnungsweise unklar und doppelsinnig ist, da η' entweder die scheinbare Schubviskosität oder die dynamische Verlustviskosität (Imaginärteil) im Schwingungsversuch sein kann. Wir schlagen daher vor, in allen Fällen, in denen eine Verwechslung möglich ist, die dynamischen Größen als Funktion von ω anzuschreiben, also:

$$\eta^*(\omega),\quad \eta'(\omega),\quad \eta''(\omega).$$

In der Literatur wird das ω vielfach weggelassen. Dies erscheint uns jedoch nur dann erlaubt, wenn eine Verwechslung ausgeschlossen ist.

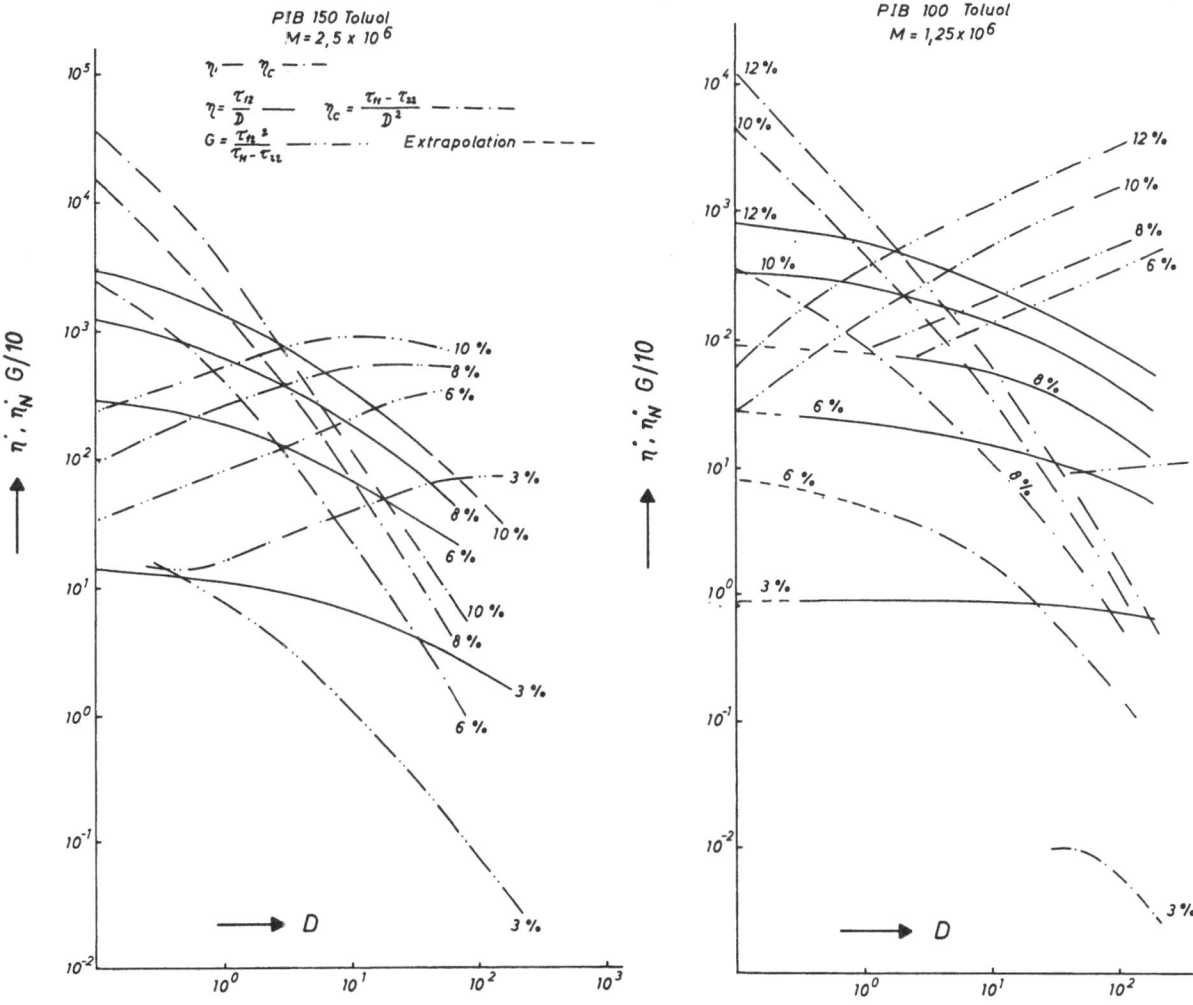

Abb. 12. Schub- und Normalviskosität sowie Schub-modul als Funktion von D für PIB 150 in Toluol

Abb. 13. Schub- und Normalviskosität sowie Schub-modul als Funktion von D für PIB 100 in Toluol

Hier sind:

$\boldsymbol{\tau} = \tau_{ik}$ der Spannungstensor,

$\boldsymbol{D} = D_{ik}$ der Scherratetensor,

λ, η sind Materialkonstanten (Relaxationszeit und Viskosität).

$\mathscr{F}$ bezeichnet den Operator:

$$\mathscr{F}\tau_{ik} = \frac{\mathscr{D}\tau_{ik}}{\mathscr{D}t} + \frac{1+\varepsilon}{2}$$
$$\times (\tau_{ij}D_{jk} + \tau_{jk}D_{ij} + \tfrac{1}{3}\tau_{jm}D_{jm}\delta_{ik}), \qquad [2]$$

$\dfrac{\mathscr{D}}{\mathscr{D}t}$: die Ableitung nach *Jaumann*,

ε: eine Materialkonstante,

δ_{ik}: *Kronekers* Symbol.

Spriggs beschreibt das rheologische Stoffver-halten durch ein System von Gleichungen (11):

$$\boldsymbol{\tau}_p + \lambda_p \mathscr{F}\boldsymbol{\tau}_p = \eta_p \boldsymbol{D}, \qquad [3]$$

wobei:

$$\boldsymbol{\tau} = \sum_p \boldsymbol{\tau}_p$$

mit $p = 1, 2, 3, \ldots$

Für Relaxationszeiten λ_p und die Viskosi-täten η_p werden folgende empirische Gleichun-gen verwendet:

$$\lambda_p = \lambda \cdot p^{-\alpha}$$

$$\eta_p = \eta_0 \Big/ \Big(p^\alpha \sum_{p=1}^\infty p^{-\alpha}\Big) = \eta_0/[p^\alpha \cdot Z(\alpha)],$$

wobei:

λ: eine charakteristische Relaxationszeit,

η_0:   die Viskosität bei der Scherrate $D = 0$ und

$Z(\alpha)$: die *Riemann*sche Zetafunktion.

Aufgrund dieser Voraussetzungen ist es möglich, ein Vier-Parameter-Modell zu betrachten. Die

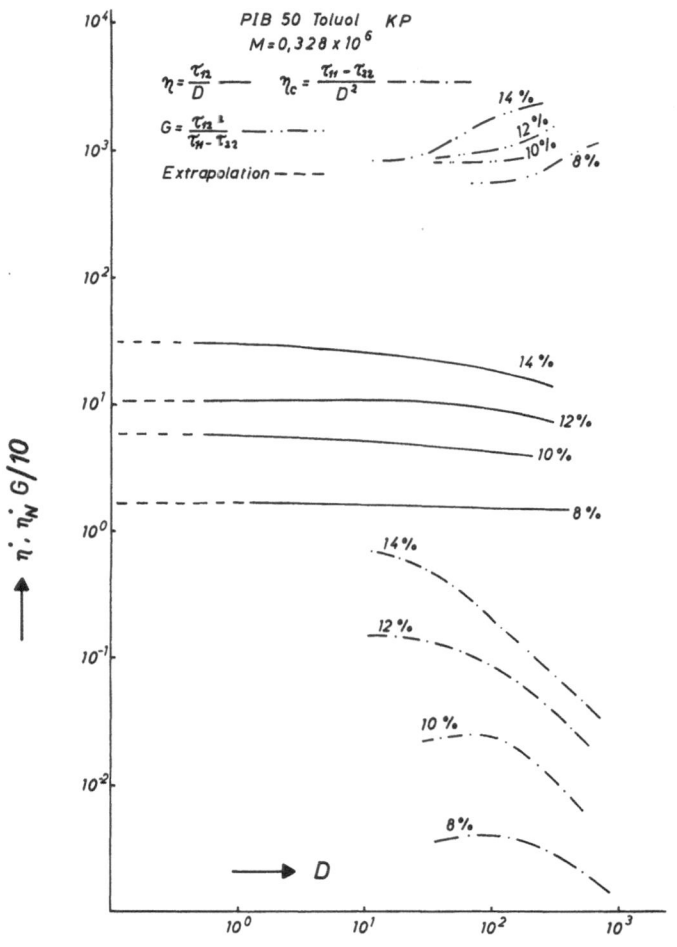

Abb. 14. Schub- und Normalviskosität sowie Schubmodul als Funktion von D für PIB 50 in Toluol

Parameter sind:

η_0, α, λ, ε.

Aus der Gleichung [1] bekommt man dann:
Für einfaches Scherfließen:

a) $\quad \eta(D) = \dfrac{\eta_0}{Z(\alpha)} \displaystyle\sum_{p=1}^{\infty} \dfrac{p^\alpha}{p^{2\alpha} + (\lambda CD)^2}$

b) $\quad \eta_N(D) = \dfrac{\sigma_1 - \sigma_2}{D^2}$

$\qquad\qquad = -\dfrac{2\lambda\eta_0}{Z(\alpha)} \displaystyle\sum_{p=1}^{\infty} \dfrac{1}{p^{2\alpha} + (\lambda CD)^2}$ [4]

c) $\quad \dfrac{\sigma^2}{D^2} = -\dfrac{\varepsilon}{2} \cdot \eta_N$

mit $\sigma_1 = \tau_{11} - \tau_{33}$ und $\sigma_2 = \tau_{22} - \tau_{33}$.

Für die dynamische Viskosität bei der Winkel-geschwindigkeit ω erhält man:

$\left. \begin{array}{l} \eta'(\omega) = \dfrac{\eta_0}{Z(\alpha)} \cdot \displaystyle\sum_{p=1}^{\infty} \dfrac{p^\alpha}{p^{2\alpha} + (\lambda\omega)^2} \\[3ex] \eta''(\omega) = \dfrac{\eta_0}{Z(\alpha)} \cdot \displaystyle\sum_{p=1}^{\infty} \dfrac{1}{p^{2\alpha} + (\lambda\omega)^2} \end{array} \right\}$ [5]

Der Parameter C ist mit ε durch die Gl. [6] ver-knüpft:

$$C = \frac{2 - 2\varepsilon - \varepsilon^2}{\varepsilon}, \qquad [6]$$

wobei ε gegeben ist als:

$$\varepsilon = -\frac{2(\tau_{22} - \tau_{33})}{\tau_{11} - \tau_{22}} = -\frac{2\sigma_2}{\sigma_1 - \sigma_2}.$$

Bekanntlich ist nach der *Weissenberg*-Hypothese $\tau_{33} = \tau_{22}$, daher $\sigma_2 = \tau_{22} - \tau_{33} = 0$ und deshalb auch $\varepsilon = 0$.

5.2. Vergleich mit den Meßwerten

Wir wollen nun die von uns erhaltenen Meß-werte mit der *Spriggs*schen Theorie vergleichen. Dazu haben wir zunächst die Grenzwerte für die Viskosität bei verschwindender Scherrate, η_0, in der Tabelle 4 zusammengestellt. Sie wurde aus unseren Meßwerten durch Extrapolation er-halten, was natürlich wiederum eine gewisse Unsicherheit mit mit sich bringt. Die Größe η_0

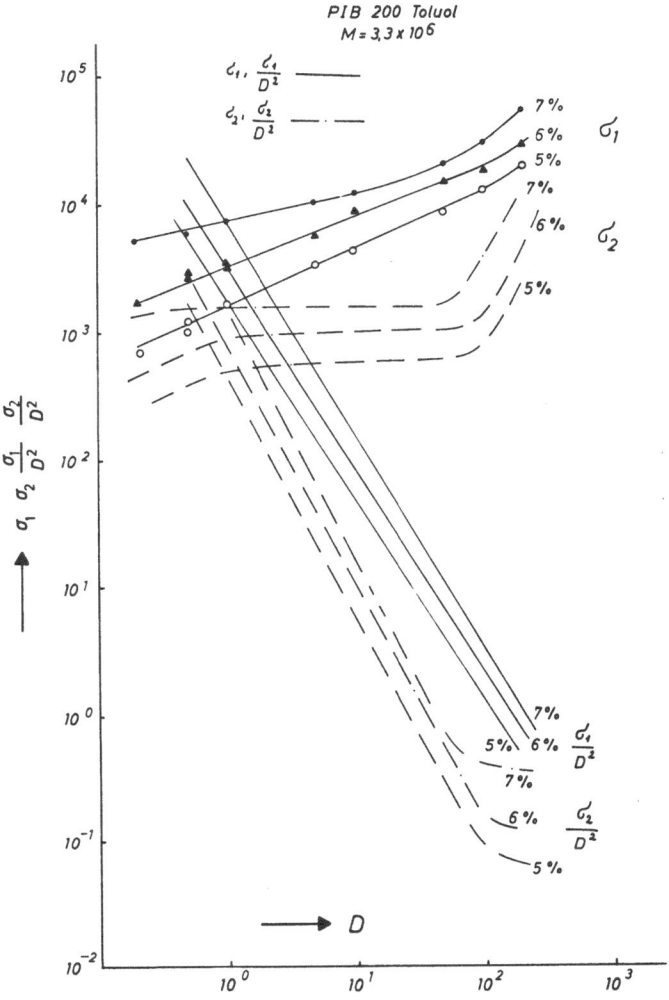

Abb. 15. Normalspannungen und Normalviskositäten als Funktion von D für PIB 200 in Toluol

ist eine der vier Materialkonstanten, die in der *Spriggs*-Theorie auftreten. Weiters ergibt die *Spriggs*-Theorie, daß bei sehr hoher Scherrate η' und η'_N der Scherrate D mit den Potenzen n und n' proportional sein sollen, wobei $n + n' = 2$.

Wie aus unseren Meßergebnissen, wie sie in Tabelle 5 dargestellt sind, zu ersehen ist, sind die von uns gefundenen Zahlen tatsächlich dem Wert 2 sehr nahe. Man beobachtet jedoch, daß die Summe mit wachsender Konzentration ansteigt. Es gibt eine Konzentration c^*, bei der $n + n'$ genau 2 ist. Diese Konzentration c^* sinkt mit wachsendem Molekulargewicht der Probe. Die zweite Materialkonstante der *Spriggs*-Theorie ist $\alpha = 1/n$. α^* soll der Wert von α sein, der der Konzentration c^* entspricht. Nach unseren Messungen ist α^* unabhängig vom Molekulargewicht und gleich 1,56 mit einer Genauigkeit von 3%. Weiterhin soll ent-

sprechend der *Spriggs*schen Theorie der Ausdruck σ_2/D^2 dem Wert von η'_N proportional sein nach:

$$\frac{\sigma_2}{D^2} = -\frac{\varepsilon}{2} \cdot \eta'_N \, .$$

Wie man aus der Abb. 15 sieht, sind die Funktionen $\eta'_N(D)$ vs. D und $\sigma_2(D)/D^2$ vs. D im doppelt-logarithmischen Maßstab nahezu parallele Geraden. Die Steigungen dieser Geraden haben sehr ähnliche Werte, wie man in Tabelle 6 sieht. Jedoch stehen die gefundenen ε-Werte im Widerspruch mit der theoretischen Bedingung, daß $-\varepsilon$ kleiner als 1,73 sein soll.

Eine weitere wichtige Voraussagung der *Spriggs*schen Theorie ist, daß die Schubviskosität $\eta'(D)$ als Funktion von D bei stationären Messungen ähnlich verlaufen sollte wie die dynamische Viskosität $\eta'(\omega)$ als Funktion von ω

Tabelle 4. η_0 und $\eta_{N,0}$ bei niedrigem Geschwindigkeitsgefälle

PIB 50			PIB 100			PIB 150			PIB 200		
c %	η_0	$\eta'_{N,0}$	c %	η_0	$\eta'_{N,0}$	c %	η_0	$\eta'_{N,0}$	c %	η_0	$\eta'_{N,0}$
8	1,7	0,0042	3	0,89	0,001	3	14	38,0	2	17	32
10	5,8	0,025	6	28	8,0	6	280	3800 *)	3	128	91
12	10,5	0,15	8	89	80 *)	8	1400	50 000 *)	4	520	1 350
14	31	0,75	10	345	180 *)	10	3800	75 000 *)	5	1000	4 000
			12	860	900 *)				6	2500	6 300
									7	6500	30 000

*) Der Bereich der Proportionalität $\tau_{11} - \tau_{22} \sim D^2$ war nicht erreicht. Die Werte sind aus Extrapolation gefunden.

Tabelle 5. Exponent in den Gleichungen $\eta = AD^n$ und $\eta_N = BD^{n1}$ bei hohem Schergefälle ($> 200 \text{ s}^{-1}$) Lösungen in Toluol

PIB 50	$-n$	$-n'$	$-(n + n^1)$	c^*	$-\alpha^*$
8	—	0,86			
10	—	1,24			
12	—	1,22			
14	—	1,01			
PIB 100					
3	—	—			
6	0,570	1,262	1,832		
8	0,683	1,559	2,242	6,8	1,563
10	0,701	1,582	2,283		
12	0,675	1,569	2,244		
PIB 150					
3	0,555	1,325	1,880		
6	0,623	1,451	2,074	5	1,613
8	0,822	1,458	2,280		
10	0,932	1,640	2,572		
PIB 200					
2	0,604	1,201	1,805		
3	0,614	1,369	1,983		
4	0,750	1,421	2,171	3,2	1,515
5	0,789	1,539	2,328		
6	0,791	1,530	2,321		
7	0,887	1,536	2,473		

bei Oszillationen. Trägt man sie daher doppellogarithmisch auf, so sollte man für beide Funktionen übereinstimmende Kurven erhalten, wenn man die D-Achse um einen konstanten Wert C verschiebt. Aus einem Vergleich der $\eta'(\omega)$ vs. ω-Kurven (Abb. 10) mit den $\eta'(D)$ vs. D-Kurven (Abb. 5) bekommt man als Verschiebungsfaktor für unsere Messungen den Wert $C = 0,28$. Die in der Abb. 11 eingezeichneten offenen Kreise sind $\eta'(\omega)$-Werte, die mit diesem Verschiebungsfaktor errechnet wurden. Man sieht, daß die Übereinstimmung sehr gut ist, womit man feststellen kann, daß die *Spriggs*sche Theorie den Übergang von der stationären zur dynamischen Viskosität gut beschreibt. Dagegen scheint die theoretische Voraussage:

$$\frac{\sigma_2(D)}{D^2} \quad \text{prop.} \quad \frac{\eta'_N(D)}{D^2}$$

für unsere Messungen nicht zu gelten, auch die bei hoher Scherrate gefundenen Potenzen n und n' für $\eta'(D)$ und $\eta'_N(D)$ folgen nicht den theoretischen Voraussagen.

Der letzte Parameter in der *Spriggs*schen Theorie ist die Größe λ, die die Rolle einer Relaxationszeit spielt. Sie ist aus der Formel

$$\lambda = \frac{\eta'_{N,0}}{2 f(\alpha) \cdot \eta_0} \quad \text{mit} \quad f(\alpha) = \sum_{p=1}^{\infty} p^{-2\alpha} \Big/ \sum_{p=1}^{\infty} p^{-\alpha}$$

zu berechnen. Wir haben unsere Messungen in dieser Beziehung ausgewertet und die λ-Werte errechnet. Da wir zur Errechnung der λ-Werte

Tabelle 6. Vergleich von Richtungskoeffizienten K und ε-Werten aus der Formel $(\sigma_2/D^2) = - (\varepsilon/2) \eta_N$ für PIB 200 in Toluol (Abb. 11 und 15). Zum Vergleich ist ε aus dem Verschiebungskoeffizient berechnet (letzte Kolonne)

c %	5 σ_2/D^2	η'_N	6 σ_2/D^2	η'_N	7 σ_2/D^2	η'_N	Verschiebungs-koeffizient (shift factor)
K	1,9	1,5	1,8	1,6	1,9	1,6	0,28
$-\varepsilon$		4,4		3,6		4,0	2,6

zum Teil auf extrapolierte Kurven zurückgreifen mußten, kommt ihnen keine sehr große Aussagekraft zu. Man kann daher nur schließen, daß sich die Relaxationszeiten bei hohen Molekulargewichten mit zunehmender Konzentration nicht so stark verändern wie bei niedrigen Molekulargewichten. Generell nimmt λ im vermessenen Bereich mit steigendem Molekulargewicht zu.

Abschließend kann somit gesagt werden, daß die Meßgenauigkeit noch nicht befriedigend ist; eine Reihe von Fehlerquellen wurde erkannt und sollte bald beseitigt werden. Die experimentellen Verbesserungen, mit denen es z. B. *Meißner* (13) gelang, das *Weissenberg*-Rheogoniometer durch nicht einfache Umbauten für Messungen an Schmelzen zu adaptieren, sind uns bekannt. Wir glauben jedoch, daß die im Prinzip einfacheren Messungen an Lösungen auch mit dem *Weissenberg*-Rheogoniometer in der vorliegenden Form möglich sein würden. Aufgrund unserer Messungen können wir feststellen, daß die *Spriggs*sche Theorie, die wir als Beispiel untersuchten, manche unserer Meßresultate richtig wiedergibt, während andere theoretische Voraussagungen durch unsere Messungen nicht bestätigt werden. Es ist dringend von Nöten, daß die Präzision der Normalspannungsmessungen bedeutend erhöht wird. Nur dann wird es uns gelingen, fundierte Aussagen und experimentelle Meßwerte zu liefern, mit denen die vorgeschlagenen Theorien wirklich kritisch überprüft werden können, so daß eine verläßliche Brücke zwischen phänomenologischer Rheologie und Strukturrheologie (14) gespannt werden kann.

Zusammenfassung

Es werden Erfahrungen bei der Vermessung von Polyisobutylen-Lösungen in Toluol im *Weissenberg*-Rheogoniometer beschrieben. Die Resultate zeigen eine relativ große Streuung, deren mögliche Ursache diskutiert wird. Auf einige Fehlerquellen wird hingewiesen.

Nach Einführung von „scheinbaren" Normalviskositäten werden die Ergebnisse nach der Theorie von *Spriggs* interpretiert, was zu teilweiser Übereinstimmung führt.

Summary

Polyisobutylene-solutions in toluene are measured in a *Weissenberg*-rheogoniometer. The performance of the instrument is described. The results show a relatively large scattering, whose origin is discussed. Several sources of error are pointed out. After introduction of "apparent" normal viscosities, the results are interpreted in terms of *Spriggs*' theory, whereby partial agreement is found.

Literatur

1) *Mavrommatakos, A.* und *J. Schurz,* Allg. u. Prakt. Chem. **24,** 174 (1973).
2) *Schurz, J.,* Rheol. Acta **4,** 107 (1965).
3) *Wazer, J. R., J. W. Lyons, K. Y. Kim* und *R. E. Colwell,* Viscosity and Flow Measurements (1966).
4) *Sakai, M., H. Fukaya* und *M. Nagasawa,* Trans. Soc. Rheol. **16** No. **4,** 635 (1972).
5) *Schurz, J.,* Viskositätsmessungen an Hochpolymeren (Stuttgart, 1972).
6) *Middleman, S.,* The Flow of High Polymers (London-Sydney-Toronto, 1968).
7) *Kotaka, T., M. Kurata* und *M. Tamura,* J. Appl. Phys. **30,** 1705 (1959).
8) *Oldroyd, J. G.,* Proc. Roy. Soc. A **245,** 278 (London, 1958).
Spriggs, T. W., Chem. Eng. Sci. **20,** 931 (1965).
9) *Coleman, B. D.* und *W. Noll,* Rev. Mod. Phys. **23,** 239 (1961).
Spriggs, T. W., J. D. Huppler und *R. B. Bird,* Trans. Soc. Rheol. **10** No. **1,** 191 (1966).
Bernstein, B., E. H. Kearsley und *L. Zapas,* J. Res. Natl. Bur. Std. **68 B,** 103 (1964).
10) *Spriggs, T. W.,* Chem. Engng. Sci. **20,** 931 (1965).
11) *Spriggs, T. W., J. D. Huppler* und *R. B. Bird,* Trans. Soc. Rheol. **10,** 191 (1966).
12) *Oldroyd, J. G.,* Proc. Roy. Soc. A **245,** 278 (London, 1958).
13) *Meissner, J.,* J. Appl. Polymer Sci. **16,** 2877 (1972).
14) *Schurz, J.,* Strukturrheologie (Stuttgart, 1974).

Adresse des Autors:

Prof. Dr. *S. Krozer*
c/o Institut für Makromolekulare Chemie
der TH Darmstadt
6100 Darmstadt

Progr. Colloid & Polymer Sci. **58**, 102–107 (1975)

Department of Textile Technology, Indian Institute of Technology, (India)

Deformation mechanisms in anisotropic polymers at small strains

V. B. Gupta

With 3 figures and 1 table

(Received February 12, 1974)

The mechanical properties of semi-crystalline polymeric fibres and films in the small strain region are technically very important and it is of interest to find out the deformation mechanisms that become operative at these small strains. Two deformation mechanisms involving the fibrils and the lamellae are discussed and their effect on the compliance of anisotropic polymers is examined.

Introduction

The *Young*'s modulus of a highly oriented fibre along the fibre axis is at least ten times less than the corresponding theoretical modulus of a fully aligned long chain polymer. It is now accepted that this is due to the folded nature of the long chain molecules because the back-folding allows there to be a high degree of chain alignment without extensive chain continuity along the filament axis (1). It would thus be expected that the stretching of covalent bonds would not be the predominant mechanism of deformation, particularly at room temperature. The straightening of molecular chains, which is another deformation mechanism, will also be relatively unimportant in the case of semi-crystalline polymers because this deformation mechanism results in a modulus which is a few orders of magnitude less than the modulus of anisotropic crystalline polymers.

The question which arises is "What are the actual mechanisms of compliance in polymers which are materials of relatively low modulus?" There can be two broad approaches to this problem. The first approach involves finding out the molecular mechanisms which result in the moduli that are practically observed. As already stated, bond stretching and chain straightening are two molecular mechanisms that have generally been believed to be the predominant mechanisms of deformation in a polymer. The second approach involves finding out the mechanical deformation modes that become operative when a force is applied and to establish how they can be quantitatively related to the moduli. Two such modes of deformation, which were suggested from extensive work on polyethylene, are C-axis shear and interlamellar shear (2, 3). The C-axis shear is a crystalline elastic shear mode with displacement along the C-axis. Interlamellar shear involves slip between the crystal lamellae, which is oblique to the axis of chain alignment. Both these deformation modes have been shown to make important contribution to the total elastic compliance of polymers. In this paper, these two deformation modes will be examined in some detail and some expressions derived to describe their effect on the elastic compliance.

Before discussing the mechanics of the problem, it might be worthwhile to consider how these deformation mechanisms are related to polymer structure.

Previous work has shown that while C-axis shear is important in cold-drawn polymers, interlamellar shear plays an important role in polymers that have been heat-set subsequent to being cold-drawn (2, 3). The structure of cold-drawn polymer may be visualised in terms of the model of *Hosemann* (4), viz. a highly oriented structure in which there are no sharp boundaries between the crystalline and non-crystalline phases. The polymer can therefore be approximated to a single phase. In view of this and the fibrillar nature of cold-drawn polymers the dominant deformation mechanism on the application of a tensile force will be the deformation of the fibrils under combined tension and shear. When a cold-drawn sheet or filament is heat-set at an appropriate temperature, a different structure is obtained and this may be represented by a model evolved by *Fischer et al.* (5). The polymer now has a clear-cut two-phase structure; the oriented crystallites are in form of lamellae with the less oriented, non-crystalline phase sandwiched in the interlamellar regions. The lamellar planes may be perpendicular to the fibre axis, in which case ideally a two-point low-angle X-ray diffraction pattern is obtained or it may be inclined to the fibre axis, by say 45°, in which case ideally a four-point pattern will be obtained (6, 7). In such samples, the deformation may be analysed in terms of the deformation of the interlamellar regions under combined tension and shear (8).

These deformation mechanisms will now be discussed.

Deformation of Fibrils

1. One phase model. The fibre is assumed to be composed of fibrils (Fig. 1) which, in turn, contain the molecular chains either in the folded or extended configuration.

A simplifying assumption may be made at this stage, viz. that the fibril has a much higher *Young*'s modulus along its axis of symmetry than the transverse modulus and the lateral contraction of the fibril, when a stress is applied

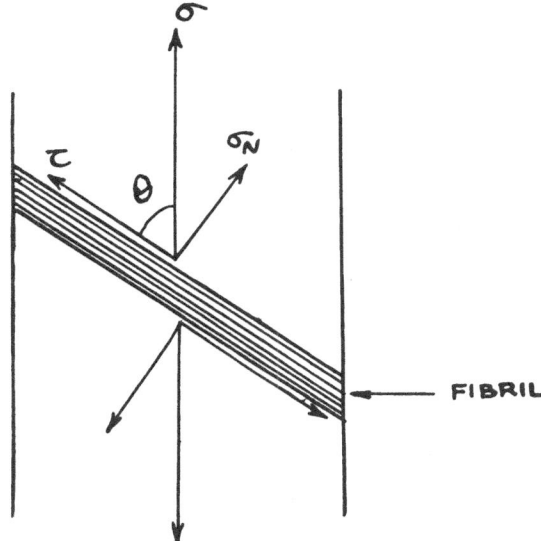

Fig. 1. Resolution of stresses on a fibril in a single-phase model

along the axis of the fibril, is negligible. With this assumption, the stress σ applied to the fibre along the fibre axis may be resolved into the following two components (Fig. 1):

$$\sigma_N = \sigma \sin^2 \theta, \quad \text{and} \quad \tau = \sigma \sin \theta \cos \theta, \qquad [1]$$

where σ_N is the stress normal to the axis of the fibril, θ the angle which the axis of the fibril makes with the fibre axis and τ is the shear stress.

The strain along the normal to the fibrillar axis is $\varepsilon_N = S_{11} \sigma_N$, where S_{11} is the transverse compliance of the fibril. The shear strain, y, is given by $y = S_{44} \tau$, where S_{44} is the shear compliance of the fibril.

The tensile strain, ε, along the fibre axis is therefore

$$\varepsilon = S_{11} \sigma_N \sin^2 \theta + S_{44} \tau \sin \theta \cos \theta$$
$$= S_{11} \sigma \sin^4 \theta + S_{44} \sigma \sin^2 \theta \cos^2 \theta. \qquad [2]$$

The contribution of a single fibril to the compliance along the fibre axis is

$$S'_{33} = \varepsilon / \sigma = S_{11} \sin^4 \theta + S_{44} \sin^2 \theta \cos^2 \theta. \qquad [3]$$

For a number of synthetic polymers like polyethylene terephthalate, nylon, polypropylene and high-density polyethylene, it is found (9) that $S_{11} \approx S_{44}$. With this approximation, eq. [3] may be written as

$$S'_{33} = S_{11} \sin^2 \theta. \qquad [4]$$

Averaging the contribution of all the fibrils in the drawn polymer, we obtain

$$\overline{S'_{33}} = S_{11} \overline{\sin^2 \theta}, \qquad [5]$$

where $\overline{S'_{33}}$ is the tensile compliance of the fibre along the fibre axis. $\overline{\sin^2 \theta}$ is the average value of $\sin^2 \theta$ and is a measure of the distribution of fibrils.

This is an interesting result. It was first derived for sonic compliance of fibres whose axial modulus is much higher than their transverse modulus (10). The present treatment gives a physical basis to this relationship. It shows that for any fibre which has a fibrillar structure and there is no discrete substructure within the fibril, this expression, i.e. eq. [5], will approximately hold. It has been shown that it holds good for some cold-drawn fibres like polyethylene terephthalate (11), polypropylene (11) and high-density polyethylene (12). The interesting conclusion may be drawn that in cold-drawn fibres, this crystallite shear mode plays an important part in determining their elastic compliance.

2. Two-phase model. A two-phase model may be considered to represent a composite in which one phase is comparatively more compliant than the other and that this softer phase is sandwiched between rod- or strip-like units of the stiffer phase. Such structures may be obtained in chemically homogeneous polymers where the amorphous phase may be sandwiched between rod-like crystalline units or in chemically heterogeneous materials like copolymers of butadiene and styrene where the styrene is present in the form of rods or strips.

The simplest approach to this problem is to assume that it is only the soft phase which deforms. The treatment of this problem is indicated in Fig. 2, and is similar to the one described in the case of one phase model and leads to the following expression for the contribution of a single unit to the tensile compliance,

$$S'_{33} = S^a_{11} \sin^4 \theta + S^a_{44} \sin^2 \theta \cos^2 \theta , \qquad [6]$$

where S^a_{11} and S^a_{44} are now the tensile and shear compliance of the amorphous or soft phase. The tensile compliance of the composite is obtained by averaging for all the units and this gives

$$\overline{S'_{33}} = S^a_{11} \overline{\sin^4 \theta} + S^a_{44} \overline{\sin^2 \theta \cos^2 \theta} , \qquad [7]$$

where $\overline{\sin^4 \theta}$ and $\overline{\sin^2 \theta \cos^2 \theta}$ represent the average distribution of the deforming units. For the amorphous phase, S^a_{44} is no longer equal to S^a_{11} but is equal to $3 S^a_{11}$ in case of simple shear

and $4 S^a_{11}$ in case of pure shear. This approach has been employed by *Folkes* and *Keller* (13) to predict the mechanical anisotropy of a three-block copolymer, viz. styrene-butadiene-styrene.

Deformation of Interlamellar Regions

This deformation mechanism involves the deformation of interlamellar material which is sandwiched between the crystalline lamellae.

Polymers, which have been heat-set subsequent to being cold-drawn, show a clear-cut lamellar morphology. A particularly simple case is illustrated in Fig. 3 and the treatment in this case is similar to that for the two phase model considered above. If the normal to the lamellar plane makes an angle of α with the fibre axis along which the tensile stress σ is applied, the expression for tensile compliance, $\overline{S'_{33}}$, comes out to be

$$\overline{S'_{33}} = S^a_{33} \overline{\cos^4 \alpha} + S^a_{44} \overline{\cos^2 \alpha \sin^2 \alpha} . \qquad [8]$$

If the angle α has a fixed value, then this gives

$$\overline{S'_{33}} = S^a_{33} \cos^4 \alpha + S^a_{44} \cos^2 \alpha \sin^2 \alpha . \qquad [9]$$

In a previous paper (3), the mechanical anisotropy of low-density polyethylene sheets, which had been prepared by cold-drawing, rolling and heat-setting, was reported. The X-ray diffraction patterns of these samples show orthorhombic symmetry and their mechanical anisotropy was explained quantitatively on the basis of a simple model (8) in which the crys-

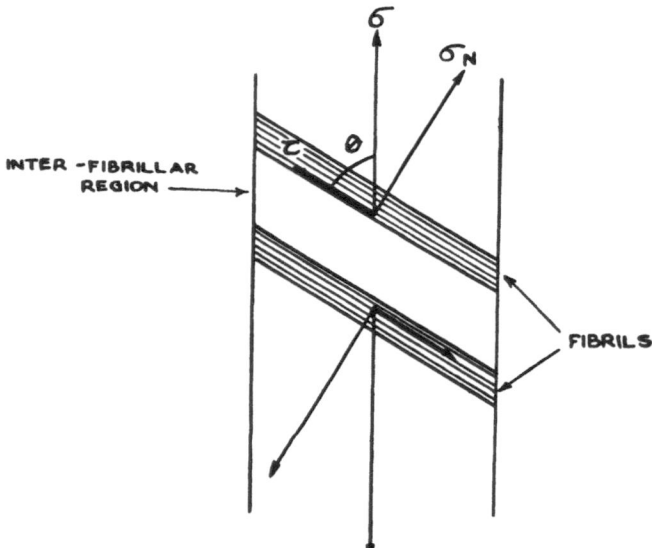

Fig. 2. Resolution of stresses in the interfibrillar region in a two-phase model

Fig. 3. Resolution of stresses in the inter-lamellar region

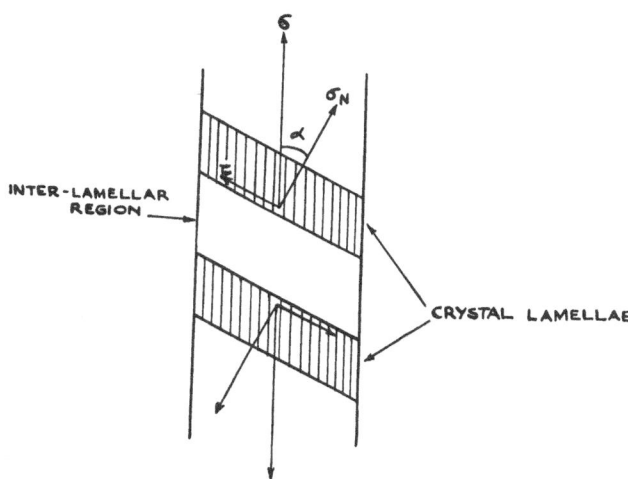

INTER-LAMELLAR REGION

CRYSTAL LAMELLAE

tallites are in the form of lamellae stacked one over the other forming fibrils. The amorphous phase is present either as interlamellar material or is embedded between the fibrils as interfibrillar material. When a tensile force is applied along any of the three crystallographic axes, the interlamellar and/or interfibrillar material will deform. The previous analysis, which was based on geometric analysis of deformation, showed that the model predicted that the modulus along the crystallographic "b" axis is much higher than the moduli along the "a" and "c" axes, the latter two being nearly equal. This was the observed experimental anisotropy in such samples (3), thus showing that the model had some merit.

A more rigorous analysis of the problem can be made using a more general treatment (14) which considers the distribution of lamellar plane normals in a plane normal to the sheet containing the initial draw direction (IDD). Let

β = angle between the lamellar plane normal and the IDD,

φ = angle in the plane of the sheet between the applied stress and the IDD, and

δ = angle between the applied stress and the lamellar plane normal.

The general result for the compliance at an angle φ to the IDD in the plane of the sheet can be written as

$$\overline{S'_\varphi} = S^a_{33} \overline{\cos^4 \delta} + S^a_{44} \overline{\sin^2 \delta \cos^2 \delta} , \qquad [10]$$

where

$$\cos \delta = \cos \varphi \cos \beta . \qquad [11]$$

For a fixed value of δ, eq. [10] may be written as

$$\overline{S'_\varphi} = S^a_{33} \cos^4 \delta + S^a_{44} \sin^2 \delta \cos^2 \delta . \qquad [12]$$

It will be assumed that the model for the polymer is the one used in the earlier paper (8) and already briefly described above. The compliances are then calculated keeping in view the following points:

(i) only that amorphous material which is in series with the crystalline material can deform.

(ii) The individual contributions of the interlamellar or interfibrillar amorphous material to the total compliance are in proportion to the volume fraction of the material which deforms.

(iii) In this model, under the conditions being considered, the interfibrillar material deforms in tension and its tensile compliance is denoted by S^a_{33}.

(iv) The contribution of interlamellar material is described by eqs. [10] and it is assumed that the deformation involves pure shear, the shear compliance being denoted by S^a_{44}, and

(v) The total compliance is obtained by simple summation of the compliances of the deforming regions.

The compliances of the low-density polyethylene sheets along the crystallographic "c", "b" and "a" axes can now be calculated in terms of the volume fractions of the interlamellar and interfibrillar regions and their respective contributions to the compliance. The moduli along the a-, b- and c-axis directions had been experimentally determined (3) by examining two sheets: The first had the "a" axis along the IDD and the "b" axis perpendicular to it but in the plane of the sheet. For this sheet, therefore, $\varphi = 0°$ for the a-axis direction and $\varphi = 90°$ for the "b" axis direction. The second sheet had the "c" axis along the IDD and the "b" axis in the plane of the sheet but perpendicular to the "c" axis. For this sheet, therefore, $\varphi = 0°$ for c-axis direction and $\varphi = 90°$ for the b-axis direction. Thus, for these two sheets, from eq. [11], $\cos \delta = \cos \beta$ for the c- and a-axis directions, while $\cos \delta = 0$ for the b-axis direction. This shows that the deformation of the interlamellar material contributes only to the compliances in the c- and a-axis directions.

The three compliances are thus given by:

$S'_c =$ Fraction of interlamellar material deforming $\times [S^a_{33} \cos^4 \beta + S^a_{44} \sin^2 \beta \cos^2 \beta]$,

$S'_b =$ Fraction of interfibrillar material deforming $\times [S^a_{33}]$,

$S'_a =$ Fraction of interlamellar material deforming $\times [S^a_{33} \cos^4 \beta + S^a_{44} \sin^2 \beta \cos^2 \beta]$
+ fraction of interfibrillar material deforming $\times [S^a_{33}]$.

For the perpendicular lamellar plane model $\beta = 90°$ and for the inclined lamellar plane model $\beta = 45°$. The volume fraction of material deforming in different regions is taken to be the same as in the earlier paper (8). The results so obtained are shown in Table 1, in which compliances have been converted to moduli.

It is seen that the perpendicular lamellar plane model predicts $E_b = E_a$, which is not what is experimentally obtained (3). The inclined lamellar plane model predicts the correct pattern of anisotropy as long as the volume fraction of the interlamellar material is

more than or equal to the volume fraction of interfibrillar material. The current models for polymer morphology support this observation because folds, chain ends, etc. are mostly in the interlamellar region and thus it is very unlikely that the volume fraction of the amorphous material in the interfibrillar region will exceed that in the interlamellar region.

Thus the more rigorous approach presented in this paper supports the earlier conclusion (8) that an inclined lamellar plane model is able to explain the mechanical anisotropy of these samples whose wide angle X-ray diffraction patterns show orthorhombic symmetry and the low-angle patterns show a clear-cut lamellar morphology with the lamellar plane normal at nearly 45° to the fibre axis. This analysis also throws some light on the likely distribution of the amorphous phase between the interlamellar and interfibrillar regions. It also corrects an error in the earlier calculations of the moduli of the inclined lamellar plane model.

Conclusions

It has been shown that if cold-drawn polymers are assumed to be one-phase structures, intra-fibrillar shear is the predominant mechanism of deformation. In heat-set fibres which have a lamellar morphology, interlamellar shear is the predominant mechanism of deformation. Quantitative correlation between the compliance and structure is established for both models of deformation.

Summary

The compliances of an anisotropic polymer are calculated in terms of two mechanical modes of deformation. The first is a crystallite shear mode with displacement along the c-axis and is important in cold-drawn

Table 1. The tensile moduli predicted on the basis of the two models

Distribution of amorphous material			Moduli predicted for perpendicular lamellar plane model (X E_{amor})			Moduli predicted for inclined lamellar plane model (X E_{amor})		
Total	Volume Fraction Inter-fibrillar	Inter-lamellar	E_c	E_b	E_a	E_c	E_b	E_a
0.4	0.1	0.3	3.33	20	20	2.67	20	3
	0.2	0.2	5	10	10	4	10	3.5
	0.3	0.1	10	6.7	6.7	8	6.7	4.1

polymers which have a fibrillar structure with no discrete sub-structure within the fibrils. The second is interlamellar shear which involves slip between the crystal lamellae, which are oblique to the axis of chain alignment. It is the predominant mode of deformation in anisotropic polymers which have been heat-set subsequent to being cold-drawn.

Zusammenfassung

Die Kompliancen eines anisotropen Polymeren werden mit Hilfe von zwei mechanischen Deformationsarten ausgedrückt. Der erste Mode ist eine Kristallitscherung mit Verschiebung entlang der *C*-Achse und ist wichtig in kaltverstreckten Polymeren mit fibrillärer Struktur ohne diskrete Substrukturen innerhalb der Fibrillen. Der zweite Mode ist zwischen lamellarer Scherung, die ein Gleiten zwischen Kristall-Lamellen einschließt, wobei letztere schräg zu den Längsachsen der Kette liegen. Dieser Mode ist vorherrschend für eine Deformation anisotroper Polymerer nach einer Hitzebehandlung auf Kaltverstreckung.

References

1) *Frank, F. C.*, Proc. Roy. Soc. London A **319**, 127 (1970).
2) *Gupta, V. B.* and *I. M. Ward*, J. Macromol. Sci. (Phys.), B(I) **2**, 373 (1967).
3) *Gupta, V. B.* and *I. M. Ward*, J. Macromol. Sci. (Phys.), B. **2**, 89 (1968).
4) *Hosemann, R.*, J. Appl. Phys. **34**, 25 (1963).
5) *Fischer, E. W., H. Goddar* and *G. F. Schmidt*, Proc. of Conference on Polymer Structure and Mechanical Properties, April 19—21 (Natick, Mass, U.S.A., 1967).
6) *Seto, T.* and *T. Hara*, Rept. Prog. Polymer Physics (Japan) **7**, 63 (1963).
7) *Hay, I. L.* and *A. Keller*, J. Materials Sci. **1**, 41 (1966).
8) *Gupta, V. B.*, Kolloid-Z. u. Z. Polymere **251**, 117 (1973).
9) *Hadley, D. W., P. R. Pinnock* and *I. M. Ward*, J. Materials Sci. **4**, 152 (1969).
10) *Moseley, W. W.*, J. Appl. Pol. Sci. **3**, 266 (1960).
11) *Ward, I. M.*, Mechanical Properties of Solid Polymers (London, 1971).
12) *Gupta, V. B., P. N. Khanna* and *T. H. Somashekar*, Colloid & Polymer Sci. (in press).
13) *Folkes, M. J.* and *A. Keller*, Polymer, Vol. **12**, 222 (1971).
14) *Owen, A. J.* and *I. M. Ward*, J. Materials Sci. **6**, 485 (1971).

Author's address:

Dr. *V. B. Gupta*
Textile Technology Department
Indian Institute of Technology
Hauz Khas, New Delhi-110 029 (India)

Progr. Colloid & Polymer Sci. **58**, 108—113 (1975)

Textile Research Laboratory, Asahi Chemical Industry Co., Ltd., Takatsuki (Japan)
and Institute for Chemical Research, Kyoto University, Uji (Japan)

Comments on the morphology of polymers crystallized from oriented melts

T. Amano, S. Kajita and *K. Katayama*

With 4 figures

(Received May 29, 1974)

Introduction

The character of polymer crystallization is highly dependent on the amount of molecular orientation which exists in the amorphous state. The morphology of polymers crystallized from oriented melts is distinctly different from classical solution-grown single crystals or from the spherulitic textures growing out of quiescent melts. Also a striking increase in the rate of crystallization is observed with increasing orientation. In practice, most polymer processing operations involve crystallization from an oriented state. Melt spinning, blowing, and injection molding of crystalline polymers are good examples of such processes. From a technological point of view, therefore, these phenomena are important ones for study because they are closely related to the final properties of the products.

Pennings and *Kiel* (1) first reported the so-called shish-kebab structure which is formed in a sheared dilute solution. Their observations have stimulated a number of investigations on this kind of crystallization. Theoretical studies (2, 3) on how molecular orientation influences the crystallization have also been made. A number of papers (4) have been published on this topic to date. The interest has centered on melt- or solution-grown crystalline structure, although crystalline structure growing out of an oriented glassy state has recently been investigated.

The present paper deals with the development of morphology in the crystalline structure associated with flow-induced molecular orientation. A fundamental picture for describing such crystalline morphology will be proposed.

Models of the lamellar growth mechanism

Studies using electron microscopy and small angle x-ray techniques have shown that a polymer crystallized under molecular orientation consists of distinct lamellae arranged normal to the flow direction. Molecular chains in the lamellae are folded in wafer-shaped structures about $100-200$ Å thick. Fig. 1 shows a typical surface structure obtained under conditions of applied stress. This morphology is quite common in polymers crystallized from an oriented amorphous state. In the present paper such morphology will be designated "stacked lamellar" structure. Also it is known that certain other textures, which are peculiar to particular polymers, are combined with the common lamellar morphology (5). *Keller and Machin (Hill)* (6, 7) have published detailed reports on certain aspects of oriented textures in polyethylene and on the more general problem of crystallization under stress. The important concept of row structure was introduced to describe the morphology of bulk polymers crystallized under stress. (The characteristics of row structure are fibrillar nucleation with extended chains parallel to the flow direction, and lamellae with folded chains overgrowing the fibrillar nuclei epitaxially.) This structural model has played an important role in understanding oriented crystallization and a large number of observations have been explained in terms of it.

However, direct evidence for fibrillar nuclei with extended chains is not yet available. Certain observations can hardly be explained with such a picture. *Yeh* (8) raised a basic

Fig. 1. Typical "stacked lamellar" structure at the surface of as-spun polyethylene filament

question in his studies on polymers crystallized from an oriented glassy state, using electron microscopy and diffraction techniques. The "stacked lamellar" structure shown in Fig. 1 is easily observed even in polymers crystallized under conditions of extremely low molecular orientation. Under such conditions, the molecular chains cannot be imagined to extend to form fibrillar nuclei, unless an extraordinary concentration of stress occurs. Similar morphology is also formed in as-spun fibers (5). In melt spinning the stress applied over the cross section before solidification is, in general, not so high — of the order of 10^6 dyne/cm^2. The average birefringence is of one or two magnitudes less than that of drawn fibers. On the other hand, in order to extend molecular chains in melts, the stress must be concentrated to a value $10^2—10^3$ times larger than that under ordinary conditions (9).

There are some other arguments against the picture of row nucleation with extended chains. A spiral growth mechanism involving folded

lamellae to form a "stacked lamellar" structure was proposed by *Kobayashi* (10). The screw dislocations are parallel to the stress direction and play an important part in lamellar growth. The mechanism is quite analogous to that in the crystal growth of low molecular weight materials. *Yeh* (8) suggested that "stacked lamellar" structure is due to the alignment and orientation of nodular grains which are present originally in the melts or glassy state. Another idea was advanced by *Baranov et al.* (11). They showed that the usual radial growth of lamellae, as seen in spherulitic crystallization, suffers interference in a stressed field. The anisotropy in the rate of lamellar growth causes the spherulites to flatten along the flow direction. The "stacked lamellar" structure of Fig. 1 results from aggregates of extremely flattened spherulites. All descriptions are based on the picture of a folded chain structure to explain the growth mechanism.

The two extreme cases, that is, crystallization under quiescent conditions and from a highly oriented state, have been fairly well-studied. The intermediate structure, however, is less well-understood and a generalized concept for connecting the extreme cases is lacking. The detailed relationship between morphology and molecular orientation has yet to be studied. Specifically one needs to understand how the change from spherulitic structure to "stacked lamellar" structure is related to changes in molecular orientation before crystallization begins.

The following observations serve, then, to provide some understanding of these intermediate structures.

Experimental

Polyethylene oxide (viscosity average molecular weight, $M\eta = 450,000$) was used as a sample for the experiments. This material is a very suitable one because it easily produces very large spherulites. A 3% chloroform solution of polyethylene oxide was placed on a thin silicon rubber strip, 0.3 mm in thickness, and the solvent was evaporated at room temperature. The polymer film obtained on the strip was melted in an oil bath and then quickly immersed into a crystallization bath, whose temperature was previously controlled. The rubber strip is uniaxially stretched at a certain ratio when its temperature becomes equal to the bath temperature, but crystallization of the sample does not take place at that time. The crystallization temperature must be carefully selected so as to satisfy the above condition. Crystallization proceeds, however, during the course of stretching. Polymer chains are oriented during the deformation, but relax partially. At most, the melt was strained to the same extent as the strip. After stretching, the sample was quenched in a cold bath. The deformation ratio of the rubber strip was in the range between 1 and 3.5. Most of the observations on the specimens were made with a polarizing microscope.

Results and Discussion

Fig. 2 shows polarized light micrographs of samples crystallized by the above method. The light scattering pattern of the samples is quite similar to that obtained by *Baranov et al.* It is therefore presumed that their samples had the same textures as ours. From the results of the light scattering experiments, they deduced that the spherulites flatten during crystallization under molecular orientation, since the rate of lamellar growth is dependent on the direction (11). However, the results of the present study suggest a somewhat different texture from the flattened spherulites they proposed. In a flattened

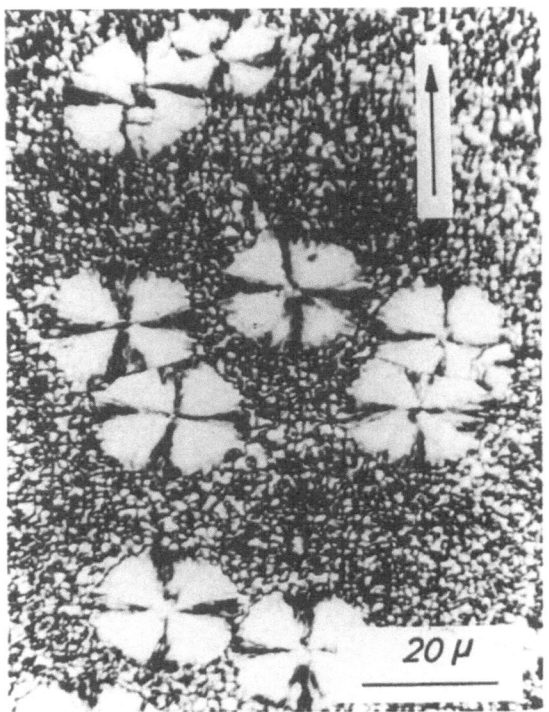

b) stretch ratio 2.3,

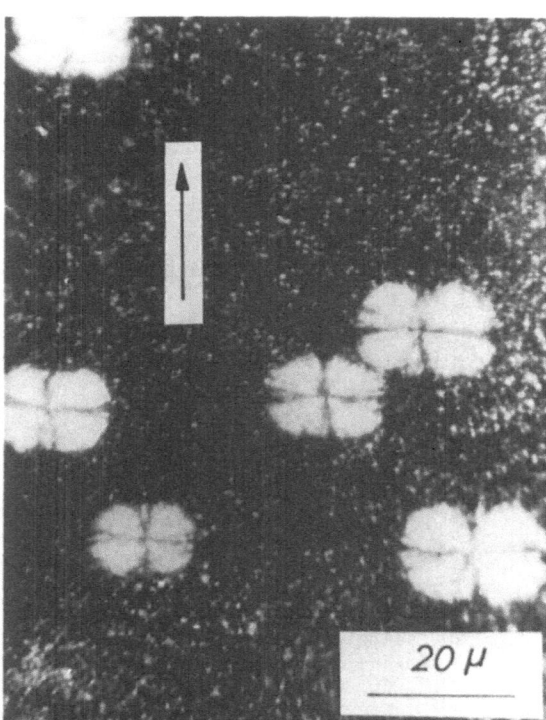

a) stretch ratio 1.8,

Fig. 2. Polarized light micrographs (crossed Nicols) of polyethylene oxide, in which the sample was crystallized from oriented melts on a rubber strip and then stretched at various ratios. The stretching axis coincides with the vibration direction of the polarizer.

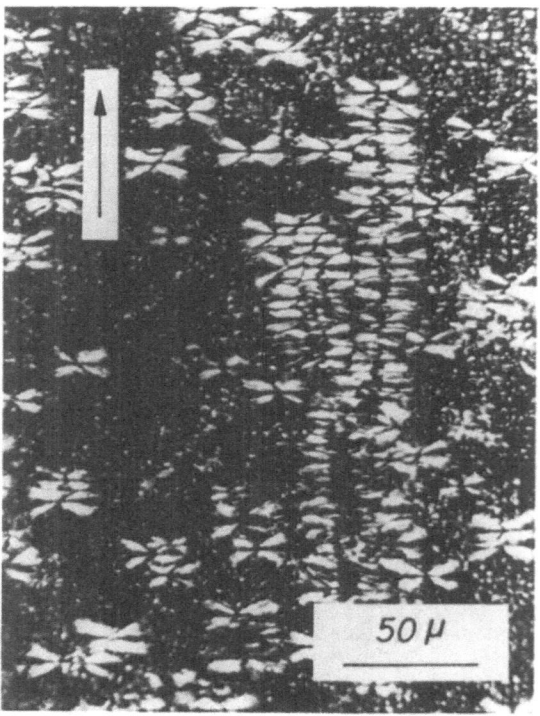

c) stretch ratio 3.0

spherulite aligned along a flow line, one expects a radial texture to remain. Therefore, the illuminated parts of the *Maltese* cross would intersect at a right angle to each other. The observed angle, however, is considerably different from a right angle, as shown in Fig. 2. The angle of the cross is closely dependent on the molecular orientation, but the relationship between these factors is merely qualitative and cannot be stated precisely in a quantitative way. *Keller* pointed out (12) that nucleation tends to occur along cracks, edges, and flow lines, and that the rate of nucleation can be so high as to cause steric interference between the resulting degenerate spherulites. It can be seen in Fig. 2 that the difference in the angle of the cross is not due to steric interference; the same feature is observed even in an isolated spherulite. If one rotates a specimen on the stage of a polarizing microscope, the *Maltese* cross due to the spherulites is observed at all intermediate positions throughout the rotation. However when the stretching axis of the sample is located 45° from the vibration direction of the polarizer, i.e. also 45° from the analyzer, extinction occurs

in the quadrant involving the stretching axis. This is shown in Fig. 3. This evidence indicates that the characteristic radial structure of the spherulites does not appear along the flow direction. That is to say, the structure along the flow direction is much more disordered compared to that in the region normal to the flow direction. This fact is not explained by a simple flattening model for the spherulites and is contradictory to the prediction of fibrillar nucleation theory. The structure along the stretching direction is so irregular that it is improbable that couplings between nuclei exist along that direction, such as extended chain crystals or screw dislocations. *Yeh* and *Lambert* (12) also suggested, from their dark field studies, that extended chain crystals are absent, at least there are no such crystals with diameters greater than about 20 Å nor lengths greater than about 75—150 Å.

Fig. 4 schematically shows the two modes of lamellar growth: a) a mode due to anisotropy of the lamellar growth rate, and b) a mode with restricted lamellar growth directions. The experimental results lead to a conclusion that the lamellar growth must be restricted to certain

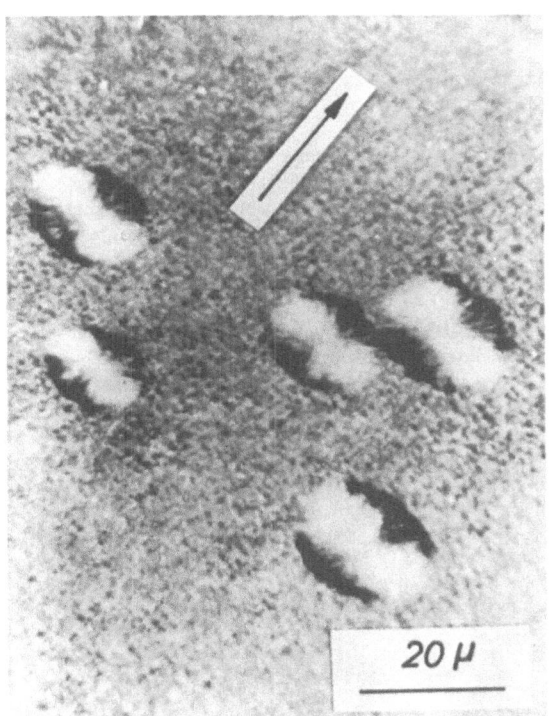

a) the same sample as in Fig. 2 a),

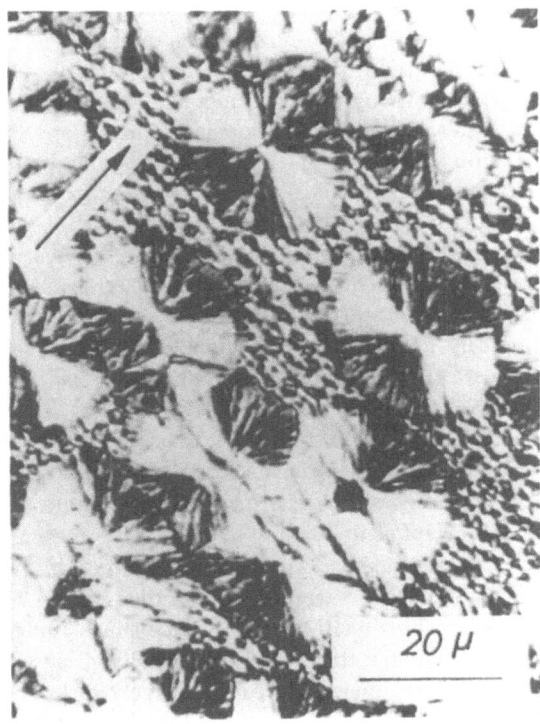

b) the same sample as in Fig. 2 b)

Fig. 3. Polarized light micrographs (crossed Nicols) of the same samples as in Fig. 2 a) and b). The stretching axis makes an angle of 45° with the vibration direction of the polarizer.

 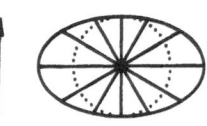 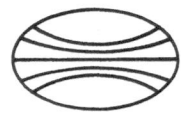

Fig. 4. Schematic illustration of lamellar growth.
a) anisotropy of lamellar growth rate,
b) restriction of lamellar growth to a preferred direction

directions in stressed systems. The intermediate texture between "stacked lamellar" structure and spherulitic structure, as shown in the figures, has been seen frequently in photographs already published (13, 14). This texture is quite similar to the central portion of a spherulite, as observed with an electron microscope. In the proposed concept, the restriction of the lamellar growth to preferred directions is emphasized as opposed to a directional dependence of the growth rate. In addition, the orientation of the primary nuclei is likely to play an important role in the development of the whole texture. With this point of view, the continuous process from spherulites to "stacked lamellar" structure can be well accounted for without considering any particular structure for the nuclei.

As pointed out by *Keller* primary nuclei tend to line up along the flow line. The appearance of a nucleus probably increases the chance of forming another nucleus in the same neighbourhood. Under a high molecular orientation, the anisotropy of the lamellar growth is much increased because of the steric interference due to the alignment of nuclei. It is likely that some structural interrelations exist among these densely lined-up nuclei. Their structure in such cases would be much more complicated than that of an isolated spherulite. While these effects may be present, they are not felt to be essential to the basic mechanism of growth.

Even in the case of growth from the quiescent state, a spherulite lacks spherical symmetry in the neighbourhood of its nucleus, and in fact many lamellae seem to be bundled up in the nucleus. This fact suggests that the structure of the primary nucleus is different from that of a secondary nucleus. Strong interrelations, such as screw dislocations, can probably be expected among the lamellae in a nucleus. The large increase in crystallization rate under orientation is thus mainly attributed to an increasing number

of primary nuclei. And other properties may qualitatively be described by considering that an increasing number of primary nuclei are involved.

Acknowledgements

The authors wish to express their sincere thanks to Professor *D. C. Bogue* of the University *of Tennessee* who critically read the manuscript and made a number of valuable suggestions. Their thanks are also due to Mr. *K. Kagawa* for his kind assistance during the experiment.

Summary

Molecular orientation due to melt flow has a profound effect on the nature of subsequent crystallization from the melt. This resulting crystalline morphology was examined with a polarizing microscope to understand the cases intermediate between quiescent melts on the one hand and highly oriented melts on the other. It seems unlikely that fibrillar nucleation along extended molecules occurs in general. Rather a more satisfactory picture is one in which the dominant orientation-dependent factor is a restriction of the lamellar growth to certain directions. This mechanism is emphasized as opposed to the concept of a direction-dependent growth rate or the earlier concept of fibrillar nucleation. A picture developed along these lines allows one to explain the entire spectrum of structure, ranging from spherulites on the one hand to "stacked lamellar" structure on the other.

Zusammenfassung

Molekularorientierung aus dem Fließen einer Schmelze hat einen nachhaltigen Einfluß auf die Natur der nachfolgenden Kristallisation aus der Schmelze. Diese resultierende kristalline Morphologie wurde mit einem Polarisationsmikroskop geprüft, um die Zwischenzustände zwischen ruhenden Schmelzen auf der einen Seite und hochorientierten Schmelzen auf der anderen zu verstehen. Es erscheint unwahrscheinlich, daß im allgemeinen eine fibrillare Keimbildung an gestreckten Molekülen eintritt. Ein viel befriedigenderes Bild ergibt es, daß der vorherrschende orientierungsabhängige Faktor in einer Hemmung des Lamellenwachstums in bestimmten Richtungen besteht. Dieser Mechanismus ist betont entgegengesetzt dem Konzept der richtungsabhängigen Wachstumsgeschwindigkeit oder dem früheren Konzept der fibrillaren Keimbildung. Ein gemäß diesen Linien entwickeltes Bild erlaubt eine Erklärung des gesamten Spektrums von Strukturen, von Sphärolithen einerseits zu der gestapelten lamellaren Struktur andererseits.

References

1) *Pennings, A.* and *A. M. Kiel*, Kolloid-Z. u. Z. Polymere **205**, 160 (1965).
2) *Kobayashi, K.* and *T. Nagasawa*, J. Macromol. Sci.-Phys. **B 4**, 331 (1970).

3) *Ziabicki, A.*, Colloid & Polymer Sci. **252**, 207, 433 (1974).

4) *Tucker, P.* and *W. George*, Polym. Eng. Sci. **12**, 364 (1972).

5) *Katayama, K., T. Amano* and *K. Nakamura*, Kolloid-Z. u. Z. Polymere **226**, 125 (1968).

6) *Keller, A.* and *M. J. Machin*, J. Macromol. Sci.-Phys. **B 1**, 41 (1967).

7) *Hill, M. J.* and *A. Keller*, J. Macromol. Sci.-Phys. **B 5**, 591 (1971).

8) *Yeh, G. S. Y.*, J. Macromol. Sci.-Phys. **B 6**, 465 (1972).

9) *Katayama, K., T. Amano* and *T. Nakamura*, Appl. Polym. Symp., No. **20**, 237 (1973).

10) *Kobayashi, K.* in *P. H. Geil*, Polymer Single Crystals (New York, 1963).

11) *Baranov, V. G., T. I. Volkov, G. S. Farshyan* and *S. Ya. Frenkel*, J. Polym. Sci., C No. **30**, 305 (1970).

12) *Yeh, G. S. Y.* and *L. Lambert*, J. Appl. Phys. **42**, 4614 (1971).

13) *Keller, A.*, J. Polym. Sci. **15**, 31 (1955).

14) for example, *Andrews, E. H.*, Proc. Roy. Soc. **A 277**, 562 (1964).
Haas, T. W. and *B. Maxwell*, Polym. Eng. Sci. **9**, 225 (1965).

Authors' address:

Dr. *T. Amano*
Technical Research Laboratory
Asahi Chemical Industry Co., Ltd.
Fuji, Samejima 2—1, 416 (Japan)

Progr. Colloid & Polymer Sci. **58**, 114—120 (1975)

Ecole Supérieure de Chimie de Mulhouse (France)

Emulsions eau-huile préparées à l'aide de copolymères séquencés poly (styrène-b-oxyde d'éthylène)

Etude de l'inversion de phase

Salvatore Marti, Jacques Nervo et *Gérard Riess*

Avec 3 figures et 3 tableaux

(Reçu p. p. le 11 juillet 1974)

Dans la littérature il apparaît que les émulsions inverses, c'est-à-dire les émulsions eau dans huile stabilisées par des polymères sont moins bien étudiées que les émulsions classiques du type huile dans eau.

Comme le témoignent une série de brevets japonais récents (1—3), ces émulsions inverses présentent cependant un grand intérêt dans la fabrication de polymères ignifuges, obtenus par exemple par copolymérisation d'un monomère vinylique en présence d'un polyester non saturé comportant une fine dispersion d'eau. Un système analogue a également été décrit par *Bartl* et *von Bonin* (4—6) qui ont utilisé des copolymères greffés polystyrène — polyoxyéthylène à faible teneur en oxyde d'éthylène pour la préparation d'émulsions eau dans styrène. Il est en effet bien connu que les copolymères séquencés et greffés peuvent jouer le rôle d'émulsifiant (7—8).

Dans un précédent article (9), nous avons commencé une recherche systématique sur les émulsions inverses eau-toluène en étudiant leur stabilité en fonction du rapport eau/toluène et en fonction des caractéristiques moléculaires du copolymère polystyrène-polyoxyéthylène (Cop PS-POE), notamment de sa teneur en oxyde d'éthylène. Un tel Cop PS-POE joue en effet le rôle d'émulsifiant non ionique du fait que l'eau est un solvant sélectif de la séquence POE et le toluène un solvant sélectif de la séquence PS.

Dans le présent travail nous nous proposons d'étudier plus particulièrement le domaine d'inversion de phase, c'est-à-dire le rapport des phases eau/toluène pour lequel on passe d'une émulsion huile dans eau à une émulsion eau dans huile.

Ce phénomène d'inversion de phase sera examiné par différentes techniques et notamment par viscosimétrie et conductimétrie en fonction de la concentration en Cop PS-POE, de sa masse moléculaire et de sa composition.

Cette étude permettra ainsi également de mettre au point les techniques de détection de l'inversion de phase et de comparer leurs performances dans le cas d'émulsions obtenues à l'aide de copolymères séquencés.

I. Partie expérimentale

1. Préparation et caractéristiques des copolymères

Les copolymères séquencés polystyrène-polyoxyéthylène (Cop PS-POE) sont préparés par voie anionique selon les méthodes désormais bien connues (10, 11, 12).

Les copolymères biséquencés ont ainsi été préparés à l'aide d'un catalyseur monofonctionnel du type phényl-isopropyl-potassium en milieu tétrahydrofuranne.

En vue de purifier les copolymères, on extrait l'homopolystyrène par traitement au cyclohexane à 40°. Le polyoxyéthylène homopolymère, qui peut provenir de la polymérisation de l'oxyde d'éthylène sur les impuretés basiques provenant de la désactivation du catalyseur, est éliminé par extraction à l'eau distillée.

La masse moléculaire de la séquence polystyrène a été déterminée par les voies classiques, osmométrie et chromatographie par perméation de gel, sur un échantillon prélevé de polymère vivant.

La composition déterminée par analyse élémentaire sur le copolymère purifié permet de remonter à sa masse moléculaire. Les valeurs ainsi trouvées correspondent très bien à celles calculées d'après le $\overline{M}_n$ de la séquence PS et l'augmentation de masse déterminée après la formation de la séquence POE. Le tableau 1 donne les caractéristiques des différents copolymères.

Tableau 1. Caractéristiques des copolymères PS-POE

Copolymère	$\overline{M}_n$ séquence PS	Fraction molaire PS	$\overline{M}_n$ copolymère	Solubilité	
				eau	toluène
Cop 3	28.600	45	39.000	~ 2,5%	S
Cop 4	36.000	25	86.000	50%	40%
Cop 5	10.000	18	28.000	SM (41%)	SM (35%)
Cop 6	8.800	5,9	68.000	SM (60%)	SM (1%)
Cop 9	138.000	49	203.000	17%	SM (83%)

S: copolymère «soluble» à raison d'au moins 20 mg/cm³.

SM: solution micellaire. La valeur entre parenthèses indique un taux approximatif de polymère «soluble» dans le solvant après centrifugation de la solution micellaire.

En vue de caractériser plus en détail ces copolymères, nous avons également fait figurer dans le tableau 1 leur «solubilité» dans l'eau et le toluène. Cette «solubilité» du copolymère est déterminée en agitant pendant 24 heures 1%, 2% et 3% de Cop PS-POE dans l'eau ou le toluène. Après centrifugation on détermine la teneur en copolymère dans la phase surnageante. Comme il s'agit de copolymères séquencés, il est évident que cette «solubilité» ne correspond pas forcément à une dispersion moléculaire, mais il peut exister des agrégats de faibles dimensions.

Dans ce tableau, un copolymère est désigné comme étant «soluble» (S) s'il donne des solutions limpides à l'oeil nu pour des concentrations au moins égales à 20 mg par cm³.

On observe ainsi que l'augmentation du taux en PS confère au copolymère sa solubilité dans le toluène, tandis que l'augmentation du taux en POE n'affecte pratiquement pas la solubilité du copolymère dans l'eau.

2. Préparation des émulsions

Pour les différents essais nous avons observé les mêmes règles de préparation: le copolymère est tout d'abord mis en contact pendant 24 heures avec son meilleur solvant c'est-à-dire, suivant le cas, le toluène bidistillé ou l'eau permutée additionnée de KCl (solution N/1000 de KCl) en vue des déterminations de conductivité. On ajoute ensuite la quantité nécessaire du deuxième solvant soit la solution de KCl N/1000 ou le toluène.

L'émulsification est réalisée à l'aide d'un mixer de labo et l'homogénéisation peut être considérée comme atteinte, si les caractéristiques de l'émulsion et notamment leur viscosité, leur conductivité et la taille des particules dispersées n'évoluent plus par une agitation supplémentaire. Ce stade est généralement atteint après 3 minutes d'agitation dans un mixer du type *Brookfield*.

Les déterminations de viscosité et de conductivité électrique sont effectuées directement sur l'émulsion ainsi préparée. Ces mesures ont cependant été faites après une dizaine de jours, ce qui permet de déterminer parallèlement la stabilité de ces émulsions, mais en ayant eu soin dans ce cas de soumettre l'émulsion plus ou moins décantée, à une nouvelle homogénéisation au mixer. Aux erreurs expérimentales près, les deux techniques conduisent aux mêmes résultats.

Il faut noter à ce stade qu'il est important dans une étude systématique de se conformer à un même mode de préparation des émulsions puisqu'il est bien connu que des phénomènes d'hystérésis peuvent intervenir dans le phénomène d'inversion des phases d'une émulsion (13). Pour cette raison nous avons préféré adopter la méthode de préparation des émulsions, qui consiste à faire gonfler au préalable le copolymère dans son meilleur solvant. Des déplacements notables du point d'inversion de phase sont en effet constatés en mettant en contact le copolymère directement avec l'eau et le toluène tel que nous l'avons décrit dans une précédente publication (9). Ces différences semblent d'autant plus accentuées que le copolymère est plus riche en POE et de ce fait les valeurs indiquées dans cette précédente publication ne constituent qu'une première approche de ce problème concernant l'inversion de phases.

3. Techniques d'étude

3.1. Mesures de conductivité électrique

Différents auteurs (14—15) ont préconisé cette technique pour déterminer l'inversion de phase dans les émulsions classiques. Comme la mesure est très rapide, cette technique constitue une méthode de choix pour l'étude des émulsions.

La résistivité des émulsions est ainsi déterminée à 25° à l'aide d'un pont de conductimétrie (type *Tacussel*). Les caractéristiques des électrodes platine-platinées sont les suivantes: surface 25 mm², distance entre électrodes 5 mm, constante de la cellule $\theta = 0,84$. Cette constante de la cellule est obtenue par étalonnage du pont de mesure à l'aide de solutions de KCl $N/100$ et $N/1000$. La résistivité de la solution de KCl $N/1000$, utilisée pour la préparation des émulsions s'établit ainsi à $6,8 \cdot 10^3$ Ω/cm. La valeur expérimentale obtenue pour le toluène est de $3,6 \cdot 10^7$ Ω/cm compte tenu du fait qu'aucune compensation de la capacité n'est opérée. L'inversion de phase pour une émulsion doit donc se traduire par un saut de résistivité de l'ordre de 10^3 Ω/cm.

La reproductibilité des essais s'est avérée très satisfaisante, notamment pour les émulsions du type huile/eau où l'erreur relative peut être estimée à 5%.

3.2. *Viscosimétrie*

La viscosité est une caractéristique importante d'une émulsion. Nous avons pu montrer précédemment pour les émulsions huile-huile que la viscosité passe par un maximum à l'inversion de phase (16).

Dans une première approche de ce problème, nous avons effectué dans la précédente publication (9) les déterminations de viscosité à l'aide d'un viscosimètre à chute de bille du type *Höppler* (17). Cette méthode étant cependant peu précise dans le cas de nos émulsions, nous nous sommes tournés vers d'autres techniques de détermination des viscosités.

Le viscosimètre *Ubbelohde* donnant des résultats peu reproductibles, notamment dans le cas des émulsions huile dans eau, nous avons ainsi utilisé de préférence un viscosimètre à rotation. Signalons qu'une constatation similaire a été faite récemment par *Y. Gallot* (18).

Les viscosités absolues ont ainsi été déterminées à 25° dans un rhéomètre du type *Epprecht* (19). Un des avantages de cet appareil réside dans la possibilité d'effectuer des mesures sur des émulsions de moindre stabilité du fait que la décantation est évitée par le mouvement du rotor.

Les émulsions étudiées ont montré un comportement *Newton*ien dans un domaine de gradients de vitesse de 28 à 1760 s^{-1}.

L'erreur relative sur les mesures de viscosité aussi bien pour les émulsions huile/eau que eau/huile peut être estimée à 5%, elle peut cependant atteindre 10% dans les cas les plus défavorables, c'est-à-dire près de l'inversion de phase où une faible variation du rapport toluène/eau entraîne une variation importante de la viscosité. Nous avons également pu vérifier qu'une agitation supplémentaire au mixer n'affecte pratiquement pas les valeurs des viscosités.

3.3. *Détermination de l'inversion de phase à l'aide de colorants sélectifs ou par méthode de dilution*

Selon *E. Manegold* (20) l'inversion de phase d'une émulsion peut être déterminée en première approximation en utilisant un colorant soluble dans le toluène (rouge Soudan) et un autre soluble dans l'eau (vert Malachite). En pulvérisant ainsi un mélange des deux colorants à sec sur un papier filtre et en déposant une goutte de l'émulsion, on peut observer pour les émulsions huile/eau une tache rouge au centre entourée d'une auréole verte. Ces colorations sont inversées pour les émulsions eau/huile.

Un test qualitatif de même nature est donné par la «méthode des dilutions» qui consiste à mettre une goutte de l'émulsion dans le toluène puis dans l'eau. Une émulsion huile/eau se disperse ainsi instantanément dans l'eau mais non pas dans le toluène où elle tombe au fond du tube.

Ces méthodes sont simples à appliquer, mais l'inversion de phase ne pouvait pas être détectée avec une précision meilleure que 15—20%.

II. Résultats — Discussion

Pour déterminer l'inversion de phase, les émulsions obtenues pour différents rapports eau/toluène ont été étudiées. Comme la méthode des colorants sélectifs et la méthode de dilution n'ont pas permis de situer avec précision l'inversion de phase, nous nous sommes tournés préférentiellement vers la viscosimétrie et la mesure de la résistivité. Les caractéristiques des émulsions et notamment l'inversion de phase ont ainsi été étudiées en fonction de la concentration, de la masse moléculaire et de la composition du copolymère.

1. *Influence de la concentration en copolymère*

La Fig. 1 représente à titre d'illustration la variation de la résistivité et de la viscosité en fonction des volumes respectifs eau et toluène, en utilisant le Cop 5 à une concentration de 0,2% par rapport au volume total de l'émulsion.

L'allure de ces courbes est typique de tous les essais et on remarque essentiellement une variation brusque de la résistivité et de la viscosité à l'inversion de phase.

Le tableau 2 donne le taux de toluène à l'inversion de phase en fonction de la concentration en Cop 5, ce qui correspond au pourcentage en volume maximum de toluène dispersible dans l'eau.

Il apparaît ainsi qu'en augmentant la concentration totale de copolymères Cop 5, on peut déplacer le point d'inversion de phase, ce qui est conforme aux résultats indiqués dans la littérature (21—22). Entre 1 et 1,5% de copolymère nous obtenons des émulsions à phase continue eau comportant jusqu'à 74% de toluène sous forme de phase dispersée. Il est à noter que cette limite de 74% correspond à celle calculée par *Ostwald* (23). Les domaines «eau phase continue» et «toluène phase continue»

Tableau 2

Concentration Cop % vol. total	Inversion de phase (% toluène)			Viscosité à l'invers. de phase c P
	Viscosimétrie	Conductimétrie	Moyenne	
0,1	57—59	59	58,5	23
0,2	63—65	64—66	64,5	70
0,75	70—71	70—71	70,5	82
1	67—72	68—73	70	92
1,5	75	75	75	91
2	77—79	78—80	78,5	$\simeq$ 450

Inversion de phase en fonction de la concentration en copolymère Cop 5.

Fig. 1. Variation de la résistivité et de la viscosité en fonction de différents rapports eau-toluène. Cop 5 à une concentration de 0,2% par rapport au volume total de l'émulsion

--o -- variation de la résistivité,

——— variation de la viscosité

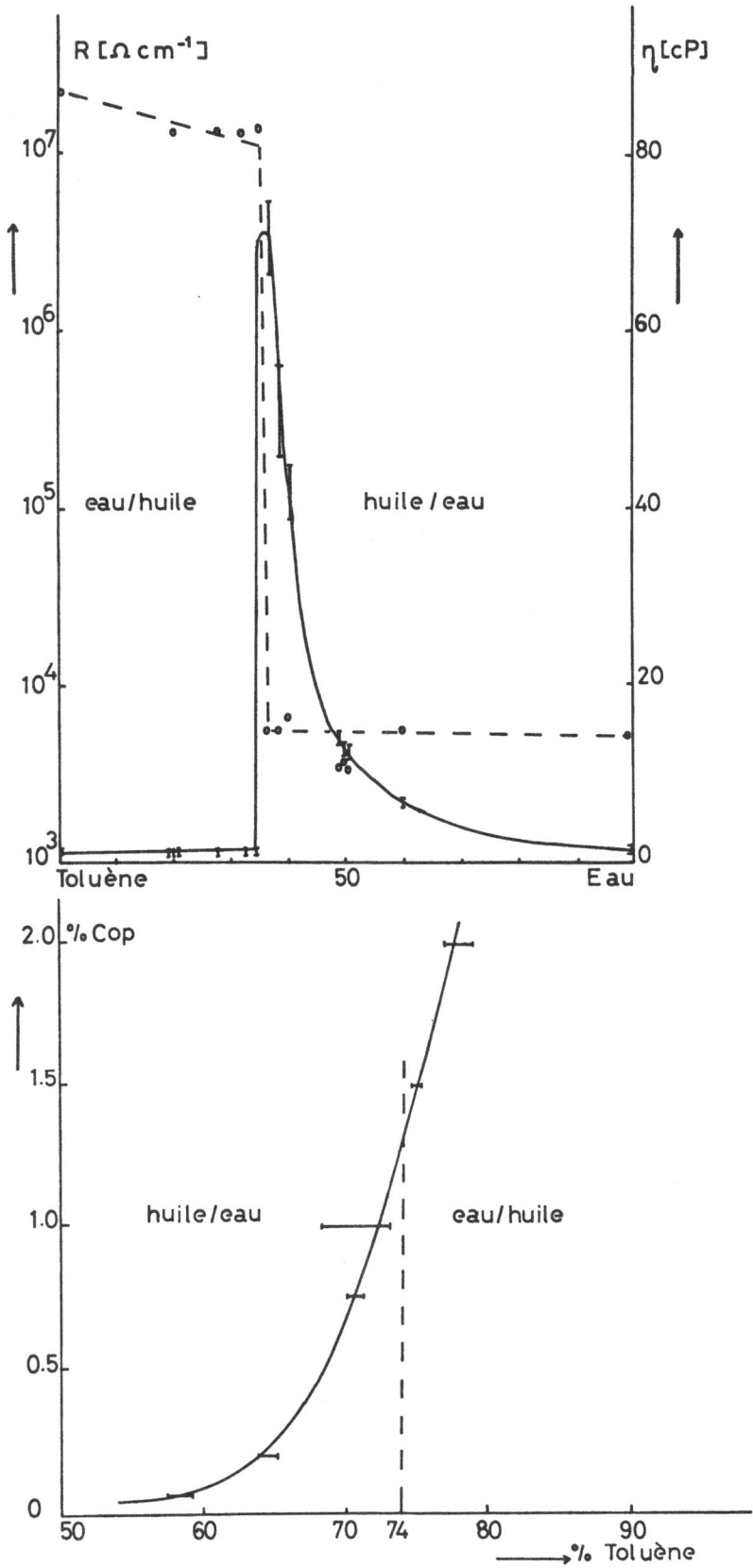

Fig. 2. Domaines «eau phase continue» et «toluène phase continue» en fonction de la concentration en copolymère Cop 5

en fonction de la concentration en copolymère sont indiqués par la Fig. 2.

Dans ce tableau figure également l'évolution des viscosités, exprimées en centipoises, des émulsions à l'inversion de phase. On peut noter tout d'abord une augmentation importante de la viscosité en passant de 0,1 à 0,2% de copolymère, suivie d'un accroissement plus faible entre 0,2 et 1,5%. La viscosité très élevée pour 2% de copolymère peut être attribuée à la présence de phase liquide cristalline dans l'émulsion (24—25). Par ailleurs, il est intéressant de remarquer que la viscosité de l'émulsion augmente avec le volume de toluène dispersé.

2. Influence de la masse moléculaire du copolymère

Les Cop 3 et Cop 9 ont des compositions sensiblement égales, mais des masses moléculaires variant dans un rapport 1 : 5. Il apparaît dans le Tableau 3 que l'inversion de phase est peu affectée par la masse moléculaire, du fait qu'elle se produit en moyenne entre 46—52% de toluène pour le Cop 3 et entre 45—51% pour le Cop 9. Les viscosités des émulsions à l'inversion de phase sont également similaires pour les deux copolymères, soit de l'ordre de 12—15 cp. On note enfin que les stabilités des émulsions obtenues avec les deux copolymères sont similaires.

3. Influence de la composition

Ayant montré que la masse moléculaire des copolymères intervient dans une moindre mesure sur l'inversion de phase, nous avons pu étudier l'influence de la composition du copolymère sur cette caractéristique.

Le tableau 3 donne ainsi le taux de toluène à l'inversion de phase pour les différents copoly-

mères. Dans ce tableau, qui montre les domaines d'inversion de phases déterminés par viscosimétrie et par conductimétrie, on peut noter une bonne concordance entre les valeurs obtenues par ces deux techniques. Dans ce tableau nous avons également indiqué les valeurs des viscosités pour les émulsions à l'inversion de phase.

Il apparaît par ailleurs que la capacité du copolymère pour stabiliser les émulsions toluène dans eau diminue si la teneur en polystyrène (PS) du copolymère augmente. En effet la solubilité du copolymère s'accroît avec la séquence PS, ce qui entraîne le toluène à devenir phase continue selon la règle de *Bancroft* (26).

Ainsi en augmentant la teneur en PS du copolymère, on favorise sa solubilité dans le toluène, c'est-à-dire on accroît le domaine où le toluène forme la phase continue des émulsions. En effet pour les Cop 5 et Cop 6, les émulsions toluène phase continue sont obtenues entre approximativement 70 et 100% de toluène, tandis que pour les Cop 3 et Cop 9, à plus forte teneur en PS, donc de meilleure solubilité dans le toluène, on peut obtenir des émulsions «toluène phase continue» entre 50 et 100% de toluène.

Il en résulte qu'inversement pour les copolymères riches en polyoxyéthylène (POE), c'est-à-dire pour les Cop 5 et Cop 6, le domaine «eau phase continue» est important et il est possible d'obtenir des émulsions comportant jusqu'à 70—72% de toluène dispersé. Ces domaines sont indiqués par la Fig. 3.

Il est à noter de plus qu'en augmentant la concentration en copolymère par rapport au volume total à émulsifier on s'approche de la limite théorique de 74% de phase dispersée définie précédemment.

Tableau 3

Cop	PS % molaire	Inversion de phase (% toluène)			Viscosité à l'inversion de phase c P
		Viscosimétrie	Conductimétrie	Moyenne	
Cop 6	5,9	63—67	66—68	66	55
Cop 5	18	67—72	68—73	70	92
Cop 4	25	58—60	50—62	59	39
Cop 3	45	44—48	50—54	49	13
Cop 9	49	43—46	50—52	48	12

Inversion de phase en fonction de la composition du copolymère; concentration en copolymère par rapport au volume total : 1%.

Fig. 3. Domaines «eau phase continue» et «toluène phase continue» en fonction de la composition du copolymère PS-POE

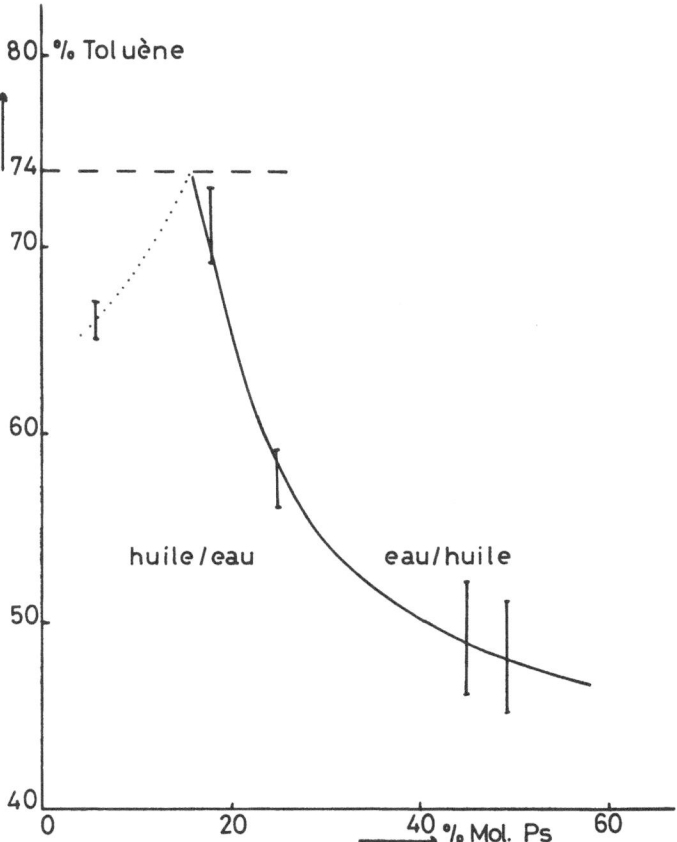

Cette limite ne semble cependant plus atteinte si le copolymère contient une proportion trop faible de PS comme dans le Cop 6 qui conduit également à des émulsions de stabilité moindre (9). Une proportion d'au moins 15% molaire en POE dans le copolymère semble donc nécessaire pour obtenir des émulsions de bonne stabilité et permettant d'atteindre la limite de 74% de toluène comme phase dispersée.

On peut également mentionner que le Cop 5, présentant des «solubilités» équivalentes dans l'eau et le toluène, conduit à des émulsions multiples de viscosité élevée ce qui est conforme aux observations déjà faites pour les émulsions huile-huile étudiées précédemment (27).

En ce qui concerne la viscosité des émulsions il apparaît une relation directe entre cette valeur et le volume de toluène qui constitue la phase dispersée. Cette observation explique également que des viscosités plus faibles ont été obtenues pour les émulsions décrites dans la précédente publication (9), où la méthode de préparation utilisée n'a pas permis de dépasser un volume de 40 à 50% de toluène comme phase dispersée.

C'est donc en perfectionnant les méthodes de mesure et de préparation des émulsions qu'il a été possible d'établir une corrélation entre l'inversion des phases et la composition des copolymères séquencés PS-POE.

Remerciements

Ce travail a pu être réalisé grâce à l'appui matériel de la Délégation Générale à la Recherche Scientifique et Technique (Contrat n° 72.7.0767).

Les auteurs tiennent à remercier également Monsieur le Professeur *R. A. Schutz* pour ses conseils et pour avoir mis à leur disposition le Rhéomètre.

Resumé

On a étudié l'inversion de phase des émulsions huile/eau et eau/huile obtenues à l'aide de copolymères bi-séquencés polystyrène PS-polyoxyéthylène POE.

Ces copolymères jouent pour le système eau-toluène le rôle d'émulsifiant non ionique.

Le phénomène d'inversion de phase a été examiné par viscosimétrie et conductimétrie en fonction de la concentration en copolymère PS-POE, de sa masse moléculaire et de sa composition.

Il apparaît que le point d'inversion de phase peut être déplacé par augmentation de la concentration en copolymère. En plus, en augmentant la teneur en PS du copolymère, on favorise sa solubilité dans le toluène et on accroît ainsi le domaine où le toluène forme la phase continue des émulsions.

Zusammenfassung

Die Phasenumkehr der durch Polystyrol (PS)-Polyäthylenoxid (POE) Blockcopolymere hergestellte Öl/Wasser- und Wasser/Öl-Emulsionen wurde untersucht.

Solche Copolymere wirken bei dem Wasser-Toluol-System als nichtionische Emulgatoren.

Mit Hilfe viskosimetrischer und konduktometrischer Messungen wurde die Phasenumkehr in Abhängigkeit der Konzentration, des Molekulargewichtes und von der Zusammensetzung des Copolymers PS-POE ermittelt.

Durch die Zunahme der Konzentration an Copolymeren ist eine Verschiebung der Phasenumkehr zu beobachten: Wird der PS-Anteil im Copolymer gesteigert, dann nimmt seine Löslichkeit im Toluol zu, und damit wird der Bereich des Toluols als durchgehende Phase der Emulsionen begünstigt.

Summary

Phase inversion for oil-water and water-oil emulsions, stabilized by polystyrene polyethylene oxide block copolymers has been studied.

These copolymers are non ionic emulsifiers for the toluene-water system.

Phase inversion has been studied by viscosimetry and conductiometry as a function of copolymer PS-POE concentration, its molecular weight and its composition.

The point of phase inversion can be displaced by increasing the copolymer concentration. The more by increasing the PS content of the copolymer, its solubility in toluene is enhanced, so the domain where toluene is the continuous phase is increased.

Littérature

1) *Negishi, S., Y. Toi* et *Y. Shishido,* Japan 72 43, 309, 01 Nov. 1972 (CA **80**, 15652 q).
2) *Ide, F., T. Kodama, H. Furutake* et *H. Mohri,* Japan 72 50, 988, 21 Dec. 1972 (CA **79**, 147162 u).
3) *Murai, K., G. Okazome, M. Yamagami, T. Sabagava* et *Y. Oka,* Japan Kokai 73 65, 239, 8 Sept. 1973 (CA **79**, 137667 y).
4) *Bartl, H.* et *W. von Bonin,* Makromol. Chem. **57**, 74 (1962).
5) *Bartl, H.* et *W. von Bonin,* Makromol. Chem. **66**, 151 (1963).
6) *Bartl, H.* et *W. von Bonin,* Ger. 1494,024, 19 Nov. 1970 (CA **74**, 32450 r).
7) *Riess, G., J. Periard* et *A. Banderet,* Colloïdal and Morphological Behavior of Block and Graft Copolymers **173** (1971). Edition G. E. Molan. Plenum Press (1971).
8) *Periard, J.* et *G. Riess,* Colloid & Polymer Sci. **251**, 97 (1973).
9) *Marti, S., J. Nervo, J. Periard* et *G. Riess,* Colloid & Polymer Sci. **253**, 220 (1975)
10) *Richards, D. H.* et *M. Szwarc,* Trans. Far. Soc. **55**, 1644 (1959).
11) *Finaz, G., P. Rempp* et *J. Parrod,* Bull. Soc. Chim. France 262, (1962).
12) *O'Malley J. J., R. G. Cristal* et *P. F. Erhardt,* Block Polymer, S. L. Aggarwal **163** (1970), Plenum Press.
13) *Becher, P.,* J. Soc. Cosmetic Chem. **9**, 141 (1958).
14) *Bhatnagar, S. S.,* J. Chem. Soc. **117**, 542 (1920).
15) *Nielsen, L. E.,* Ind. Eng. Chem. Fund. **13**, 17 (1974).
16) *Periard, J., A. Banderet* et *G. Riess,* Polymer Letters **8**, 109 (1970).
17) *Höppler, F.,* Chemiker Zeitung **57**, 62 (1933).
18) *Gallot, Y.,* CRM Strasbourg (communication privée).
19) *Epprecht, A. G.,* Kolloid-Z. **145**, 116 (1956).
20) *Manegold, E.,* "Emulsionen", Verlag Straßenbau — Chemie — Technik (Heidelberg, 1952).
21) *Sherman, P.,* Emulsion Sci. (Academic Press London/New York, 1968).
22) *Periard, J.* et *G. Riess,* Colloid & Polymer Sci. **251**, 97 (1973).
23) *Ostwald, W.,* Kolloid-Z. **6**, 103 (1910); **7**, 46 (1910).
24) *Friberg, S., L. Mandell* et *M. Larson,* J. Coll. a. Interface Sci. **29**, 155 (1969).
25) *Friberg, S.* et *P. Solyom,* Colloid & Polymer Sci. **236**, 173 (1970).
26) *Bancroft, W. D.,* J. Phys. Chem. **17**, 501 (1913); **19**, 275 (1915).
27) *Periard, J.* et *G. Riess,* Colloid & Polymer Sci. **253**, 362 (1975).

Adresse des auteurs

Salvatore Marti, Jacques Nervo, Prof. Dr. G. Riess
Ecole Supérieure de Chimie de Mulhouse
Mulhouse
(France)

Progr. Colloid & Polymer Sci. **58**, 121–135 (1975)

Kunststofflaboratorium (WK) der BASF, Ludwigshafen/Rhein

Gleichgewichtsdrücke, Löslichkeit und Mischbarkeit des Systems. Unvernetztes Polystyrol (PS)/Kohlenwasserstoffe (KW)

2. Teil: Der Einfluß der Ringgröße (c-Propan, c-Pentan, c-Hexan) *)

H. Horacek

Mit 22 Abbildungen und 8 Tabellen

(Eingegangen 23. März 1974)

1. Einleitung

c-Propan, *c*-Pentan und *c*-Hexan sind mit Monostyrol in jedem Verhältnis mischbar. Ersetzt man das Monomere durch das Polymere, so treten im untersuchten Temperaturbereich von 20 bis 250 °C obere und untere Mischungslücken auf. Die folgenden Messungen sind durchgeführt worden, um den Einfluß der Ringgröße auf die Löslichkeit zu untersuchen.

2. Ausgangsmaterialien

2.1. Polystyrol

In den Experimenten wurde das BASF Polystyrol 168 N verwendet (1).

2.2. Kohlenwasserstoffe

Es wurden folgende Kohlenwasserstoffe eingesetzt:

c-Propan der Fa. Baker,
c-Pentan der Fa. Schuchardt,
c-Hexan der Fa. Merck.

Alle hatten eine gaschromatographisch geprüfte Reinheit von größer als 99%.
Die Meßapparatur und der Arbeitsvorgang sind bereits im 1. Teil beschrieben worden (1).

3. Ergebnisse

3.1. Spezifische Volumina von Kohlenwasserstoff-Polystyrolgemischen

Die Abb. 1 und 2 zeigen das spezifische Volumen als Funktion der Konzentration. Die strich-

lierte Linie wurde additiv aus den Volumina der reinen Komponenten berechnet; die ausgezogene Linie verbindet die experimentellen Punkte (2,3). Liest man bei 50 Gew.-% die Differenz zwischen gemessenen und berechneten Werten V^E ab, so ist sie laut Tabelle 1 negativ und nimmt mit steigender Temperatur und sinkendem Molekulargewicht des KW zu.

Geht man von glasigem PS aus und bestimmt bei 20 °C das spezifische Volumen von PS mit einem KW-Gehalt kleiner gleich der Konzentration, bei der die Glastemperatur gerade 20 °C beträgt, so erhält man nach Abb. 3 Geraden.

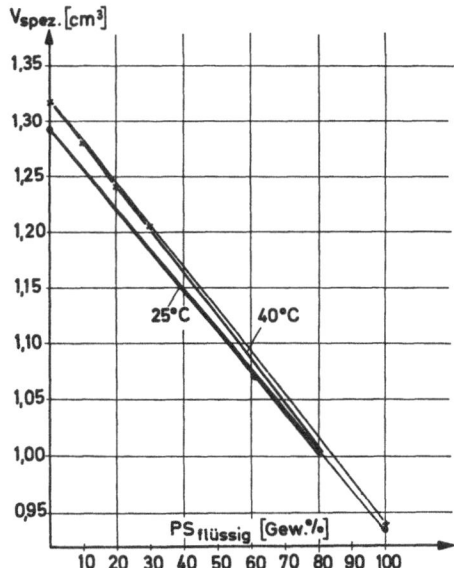

Abb. 1. Spezifische Volumina von *c*-Hexan/Polystyrol-Gemischen bei 40 °C und 25 °C (3)

*) *Herrn Professor Johann Wolfgang Breitenbach zum 65. Geburtstag nachträglich gewidmet*

Tabelle 1. Differenz zwischen experimentellen und berechneten spezifischen Volumina V^E (cm³/g) von Lösungen mit 50 Gew.-% an c-Pentan, c-Hexan und von gesättigten Lösungen mit 12 Gew.% n-Pentan, 16 Gew.-% n-Hexan

Temperatur (°C)	20	25	40	100	110	120	130	140	150
— V^E c-Pentan	0,007	—	0,011	0,020	0,029	0,033	0,037	0,043	0,051
— V^E c-Hexan	—	0,002	0,004	—	—	—	—	—	—
— V^E n-Pentan	0,020	—	—	—	—	—	—	—	—
— V^E n-Hexan	0,018	—	—	—	—	—	—	—	—

An der Glaskonzentration sollten die spezifischen Volumina der flüssigen und glasigen PS-Lösung gleich sein. Nach Tabelle 2 ist dies auch annähernd der Fall.

3.2. Dampfdrücke zwischen 20—250 °C und 0– 19 atm.

In den Abb. 4—6 sind die Ergebnisse der Dampfdruckmessungen zusammengefaßt. Die spezifischen Gasvolumina, die zur Berechnung der im Gasraum befindlichen Kohlenwasserstoffmenge benötigt werden, sind der Literatur (4, 5) entnommen worden.

3.3. Sättigungslöslichkeit und Mischungslücke

c-Pentan und c-Propan wurden mit PS in Glasbombenrohre in verschiedenen Mischungsverhältnissen eingeschmolzen und im Temperaturbereich von Raumtemperatur bis zur kritischen Temperatur des Kohlenwasserstoffes im Hinblick auf eine auftretende Trübung und Entmischung beobachtet.

Die Glastemperaturen von vernetzten kohlenwasserstoffhaltigen Polystyrol wurden differentialkalorimetrisch bestimmt. Aus Abb. 7 ist er-

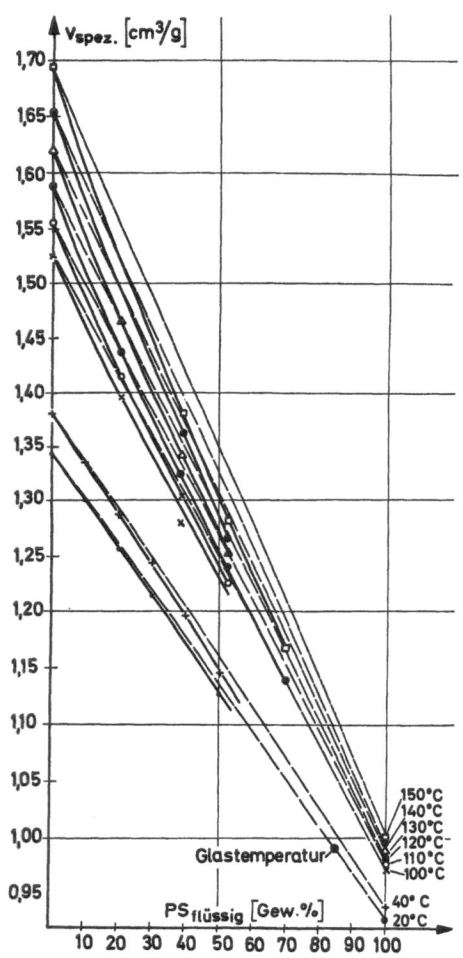

Abb. 2. Spezifische Volumina von c-Pentan/Polystyrol-Gemischen bei Temperaturen zwischen 20 und 150 °C Meßwerte von *E. Brunner*

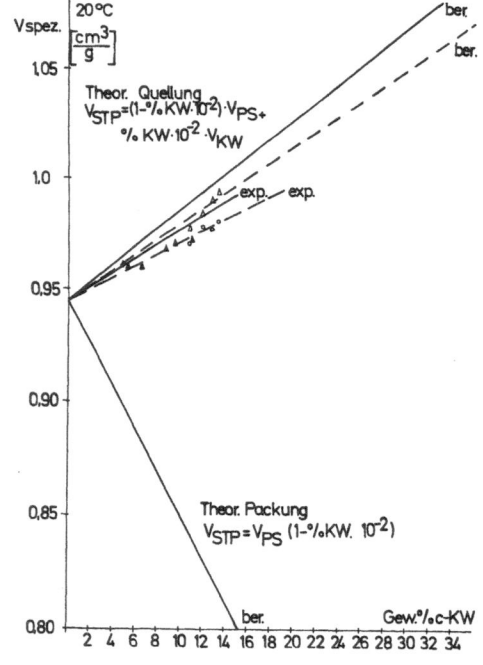

Abb. 3. Spezifische Volumina von glasigem Polystyrol, das c-Hexan (-- $V_{spez} = 1{,}29$ cm³/g) und c-Pentan (——— $V_{spez} = 1{,}35$ cm³/g) enthält. Meßwerte von *O. Riedel*. [▲ △ Methanol, ○ Petroleum]

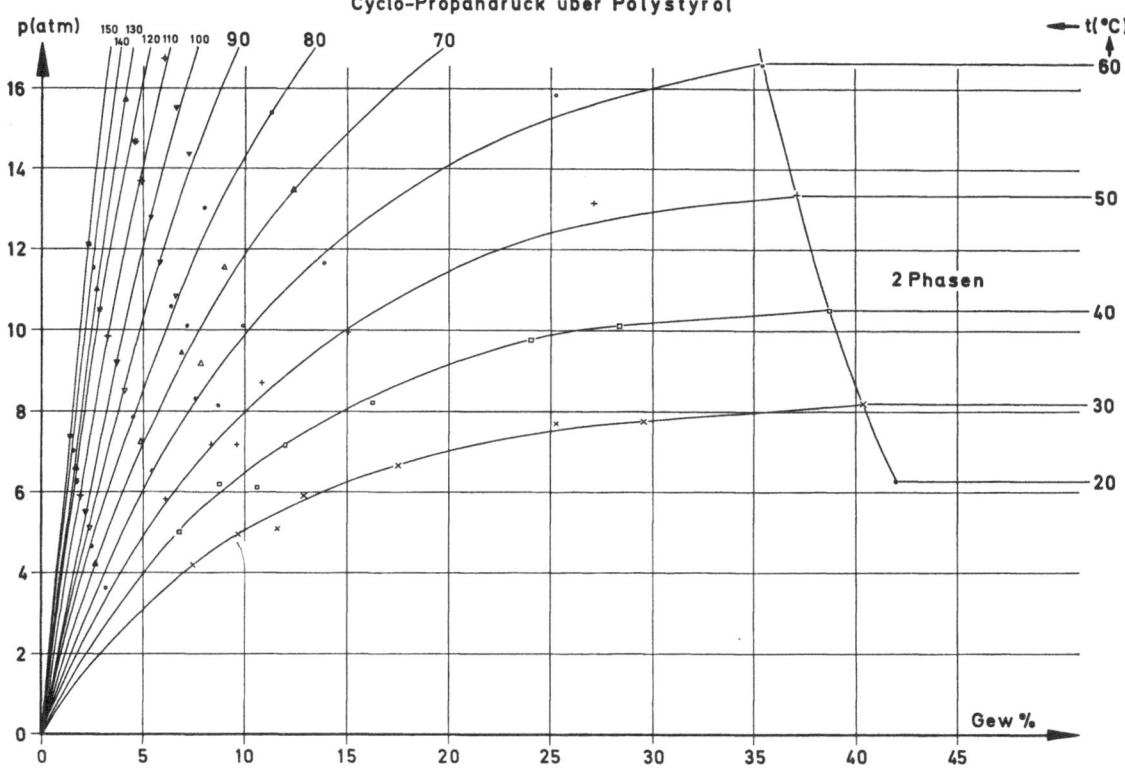

Abb. 4. *c*-Propandruck über Polystyrol

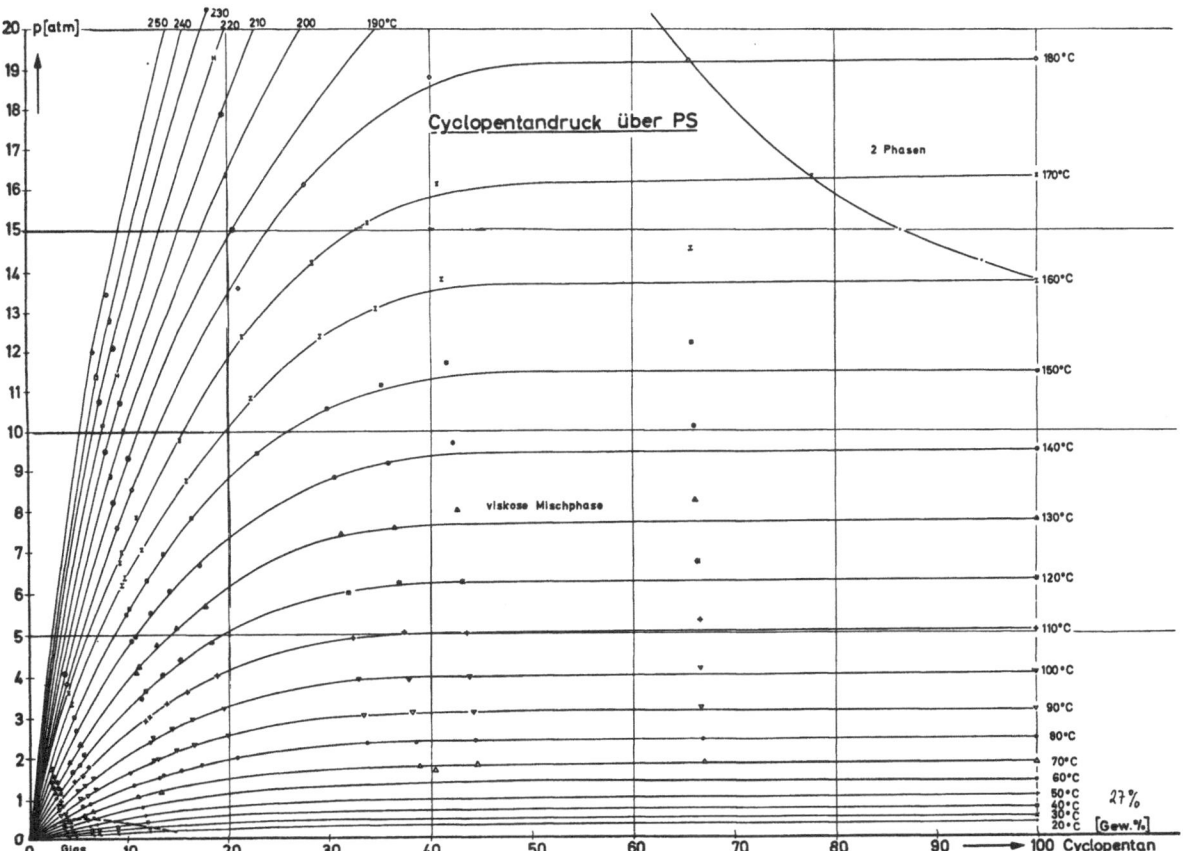

Abb. 5. *c*-Pentandruck über Polystyrol

Tabelle 2. Spezifische Volumina bei der Glaskonzentration und bei 20 °C

Kohlen-wasserstoff	Glaskonzen-tration (Gew.-%)	V_{spez} (flüssig) (cm³/g)	V_{spez} (glasig) (cm³/g)	Sättigungs-konzentration (Gew.-%)
c-Hexan	19	0,997 (25 °C)	0,995	95
c-Pentan	15	0,994	0,997	100
c-Propan	11	—	—	42
n-Hexan	12	—	0,999	16
n-Pentan	11	—	1,000	12
Propan	9 (extr.)	—	1,010	8,5

Tabelle 3. Sättigungslöslichkeit an KW bei Temperaturen höher als LCST

Tempe-ratur (°C)	c-Propan Gew.-% KW	x_2^{S+}	c-Pentan Gew.-% KW	x_2^{S+}	c-Hexan Gew.-% KW
20	42,0	0,35	100	1	95
40	39	0,385	100	1	100
60	35,5	0,42	100	1	100
80	34	0,44	100	1	100
100	29	0,49	100	1	100
120	26	0,535	100	1	100
140	—	—	100	1	100
160	—	—	95	—	100
180	—	—	59	0,315	100
200	—	—	53	0,375	100
220	—	—	47	0,430	83
$\Delta x_1^{S+}/\Delta T$	$-19 \cdot 10^{-4}$		$-34,5 \cdot 10^{-4}$		—

sichtlich, daß mit wachsender Ringgröße die unteren kritischen Entmischungstemperaturen (LCST) zu höheren Temperaturen verschoben werden. Die Sättigungslöslichkeit nimmt oberhalb der LCST mit der Temperatur zu, wobei der Anstieg ebenfalls mit der Temperatur wächst. Aus Abb. 8 erkennt man, daß er streckenweise nahezu konstant ist und durch eine Zahl angegeben werden kann. $\Delta x_1^{S+}/\Delta T$ beträgt $-34,5 \cdot 10^{-4}$ für c-Pentan und $-19 \cdot 10^{-4}$ für c-Propan.

Die kritischen Entmischungstemperaturen sind vom Molekulargewicht des Polymeren abhängig: mit steigendem Molekulargewicht steigt die UCST und fällt die LCST, d. h. das Einphasengebiet wird mit steigendem Molekulargewicht kleiner (15, 16, 18, 21, 22).

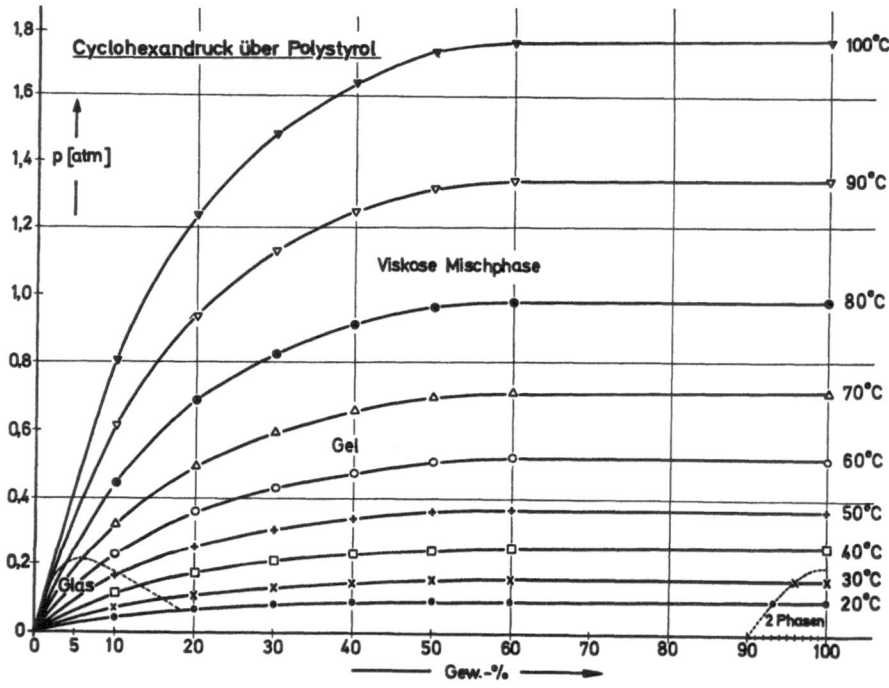

Abb. 6. c-Hexandruck über Polystyrol

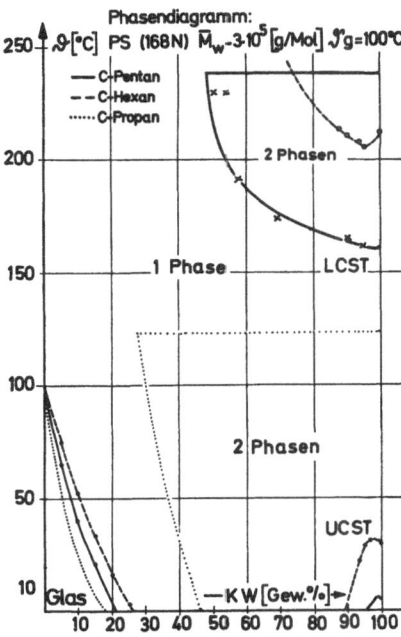

Abb. 7. Phasendiagramm von Polystyrol und —— c-Pentan, -- c-Hexan, ... c-Propan. Die Glastemperaturen sind Meßwerte von *K. H. Illers*.

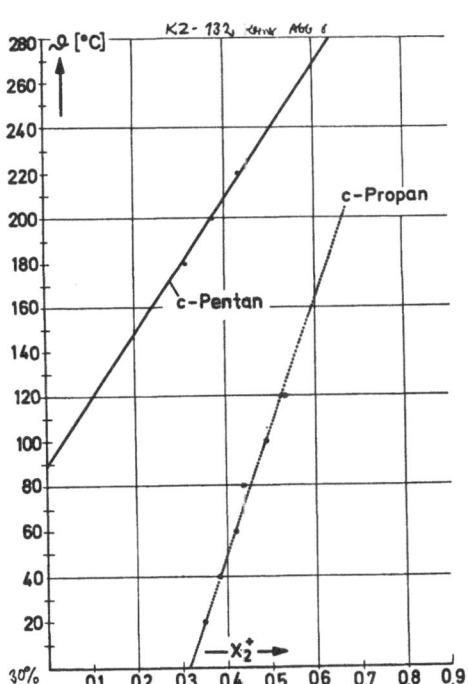

Abb. 8. Temperaturabhängigkeit der Sättigungslöslichkeit (obere Mischungslücke); —— c-Pentan, ... c-Propan

Tabelle 4. ΔH Enthalpiewerte (kcal/mol).

Kohlenwasserstoffe	ΔH_L aus σ	ΔH_L^+	ΔH_1 aus $\dfrac{p_1}{p_{01}}$	$\dfrac{\Delta H_1^*}{x_2^{+2}}$ aus	ΔH_K	ΔH_M^{**}	δ_1 $(\text{cal/cm}^3)^{1/2}$	V_{01} $\left(\dfrac{\text{cm}^3}{\text{Mol}}\right)$	V_{spez} $\left(\dfrac{\text{cm}^3}{\text{g}}\right)$ 20 °C	MG $\left(\dfrac{\text{g}}{\text{Mol}}\right)$
c-Hexan	$-8{,}2$	$-7{,}0 + \dfrac{n_2^+}{n_1}(\Delta H_2^+ - 0{,}85)$	$0{,}2{-}0{,}3\ (6)/0{,}14\ (9)$	$0{,}008$	$-7{,}2$	-1	$8{,}2\ (27)$	109	$1{,}29$	$84{,}16$
c-Pentan	$-7{,}6$	$-6{,}0 + \dfrac{n_2^+}{n_1}(\Delta H_2^+ - 0{,}85)$	$0{,}5{-}1{,}5$	$0{,}00047$	$-6{,}5$	$-1{,}1$	$8{,}7\ (27)$	$94{,}6$	$1{,}35$	$70{,}13$
c-Propan	$-3{,}8$	$-4{,}5 + \dfrac{n_2^+}{n_1}(\Delta H_2^+ - 0{,}85)$	$0{,}3{-}0{,}5$	$0{,}0126$	$-4{,}8$	$+1$	$9{,}2^{***}$	$69{,}8$	$1{,}66$	$42{,}08$

*) $\dfrac{\Delta H_1}{x_2^{+2}} = \beta = \dfrac{(\delta_1 - 8{,}6)^2 \cdot V_{01}}{R}$.   **) $\Delta H_M = \Delta H_L - \Delta H_K$ ***) $\delta_1 = 1{,}25\ \sqrt{p_{\text{krit.}}} = 9{,}2\ (p_{\text{krit.}} = 54{,}2\ \text{atm})$.

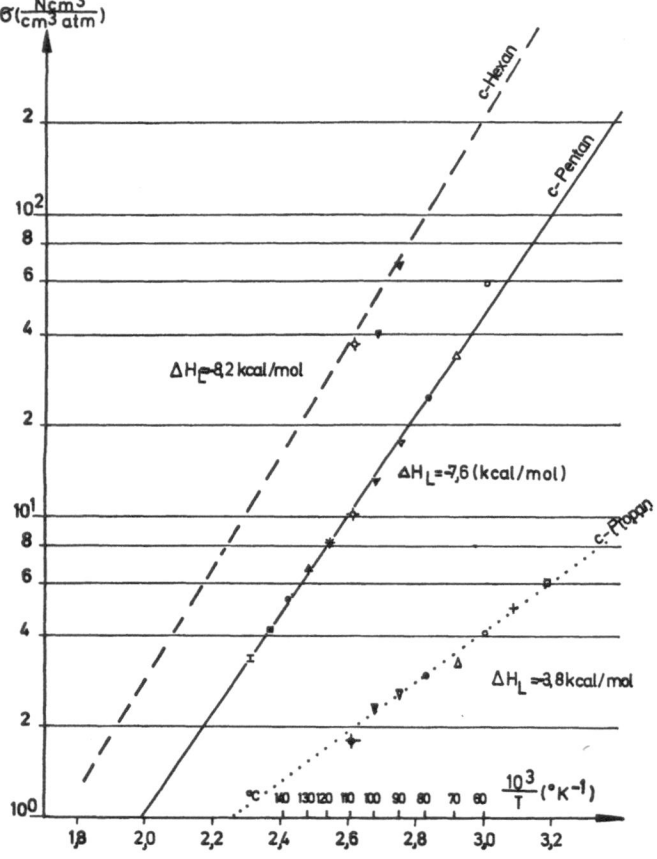

Temperaturabhängigkeit der Bunsenschen
Löslichkeit von Cyclo-Pentan in PS

Abb. 9. Temperaturabhängigkeit der *Bunsen*-schen Löslichkeit von *c*-Hexan – – –, *c*-Pentan ——, und *c*-Propan . . . . in Polystyrol

3.4. Wärmeeffekte

3.4.1. Bunsenscher Löslichkeitskoeffizient und integrale Lösungswärme

Aus der *Bunsen*schen Löslichkeit σ, die sich aus den bei 1 atm gelösten KW-Mengen der Abb. 4—6 errechnet, wird in Abb. 9 nach einer *van't Hoff*-Auftragung die integrale Lösungswärme ΔH_L bestimmt. Wie aus Tabelle 4 ersichtlich, liegt sie in der Größenordnung der Kondensationswärme.

3.4.2. Differentielle, molare Verdünnungswärme

Entnimmt man für jeweils konstante Konzentrationen die Aktivitäten p_1/p_{01} aus Abb. 4—6 und trägt sie gegen die reziproke Temperatur auf, so erhält man die differentiellen molaren Verdünnungswärmen ΔH_1. Nach Abb. 10 und 11 sind sie klein und positiv. Mit steigender Konzentration werden die Verdünnungswärmen kleiner, die letzten Verdünnungswärmen sind wegen des negativen Wertes von $\Delta x_1^s{}^+/\Delta T$ ebenfalls negativ.

3.5. Wechselwirkungsparameter

Versucht man die Meßwerte nach der Quasigittertheorie vermittels der *Flory-Huggins*-Gleichung auszuwerten, so stößt man auf Schwierigkeiten.

a) Der Wechselwirkungsparameter, der nach *Flory* und *Huggins* berechnet worden ist, ist sowohl temperatur- als auch konzentrationsabhängig. Tabelle 5, Abb. 12—14.

b) Die differentielle molare Verdünnungswärme ΔH_1 ist schwächer als gefordert konzentrationsabhängig (Abb. 10, 11).

c) Beim Mischen erfolgt eine Kontraktion der Volumina. *Flory* und *Huggins* gehen von der Additivität der Volumina aus (Abb. 1—3).

d) Es tritt eine LCST auf, obwohl von der Quasigittertheorie ein monotoner Anstieg der Lösungsmittelgüte mit steigender Temperatur gefordert wird (Abb. 7).

Die obere Mischungslücke ist eine Besonderheit von Mischungen aus Polymeren und nieder-

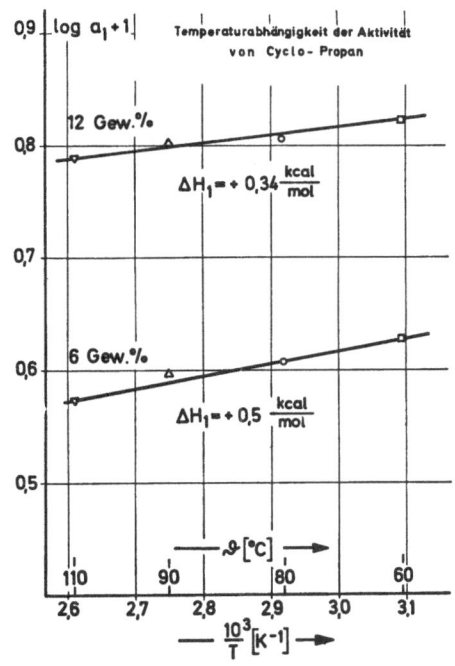

Abb. 10. Temperaturabhängigkeit der Aktivität von *c*-Propan (6 und 12 Gew.-%) in Polystyrol

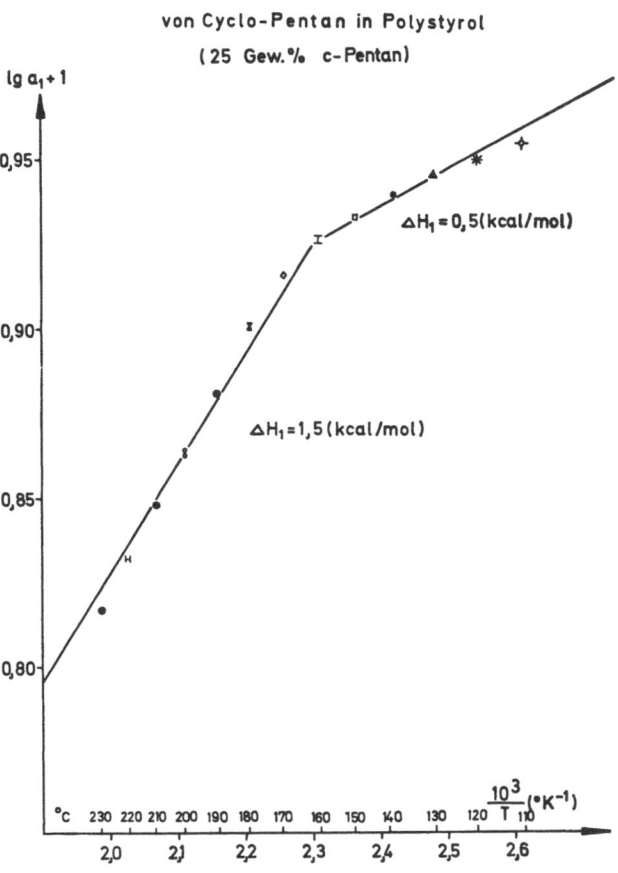

Abb. 11. Temperaturabhängigkeit der Aktivität von *c*-Pentan (25 Gew.-%) in Polystyrol

molekularen Verbindungen und beruht auf der Ungleichheit der freien Volumina von Polymeren und Lösungsmitteln aufgrund der Größenunterschiede der beiden Spezies. Das Lösungsmittel ist viel weiter ausgedehnt als das Polymere, so daß das Mischen der Kondensation eines Gases (Lösungsmittel) in einem dichten Medium (Polymeres) nahe kommt (12, 13, 14, 17).

Eine obere Mischungslücke (LCST) wurde auch in anderen Lösungsmitteln wie Benzol (19) und Methylazetat (20) beobachtet und tritt prinzipiell

bei Polymer-Lösungsmittel-Systemen unterhalb der kritischen Temperatur des Lösungsmittels auf. Der Wechselwirkungsparameter setzt sich

Tabelle 5. Mittlerer Wechselwirkungsparameter χ nach *Flory-Huggins.* $\chi = \alpha + \beta/T$

Kohlen-wasserstoff	ϑ (°C)	χ	α	β (grad)	Literatur
c-Hexan	20	0,89			(26)
	20	0,95			(7—10)
	23	0,60			(23)
	37	0,499			(24)
	49	0,489			(24)
	60	0,485			(24)
	6	0,62			(25)
	20	0,70	0,575 (20—120 °C)	50	
c-Pentan	20	0,74	−3,34 (120—250 °C)	1700	
c-Propan	20	0,53	−0,451	287	
n-Hexan	20	1,49	−0,487	580	(46)
n-Pentan	20	1,43	−0,11	450	
Propan	20	0,94	2,575	−480	

aus zwei Beiträgen zusammen, einmal aus dem,
der auf der Ungleichheit der Volumina der beiden
Komponenten beruht, andermal aus dem, der in
der Ungleichheit der Kontaktenergien zwischen
Lösungsmittel und Polymeren seine Ursache hat.

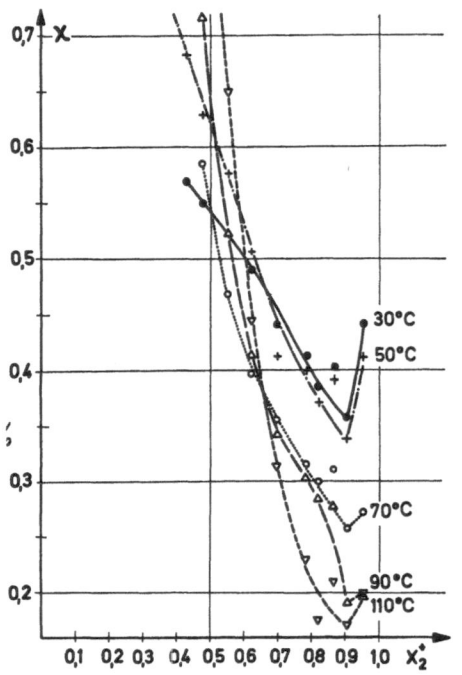

Abb. 12. Wechselwirkungsparameter χ von c-Propan in
PS

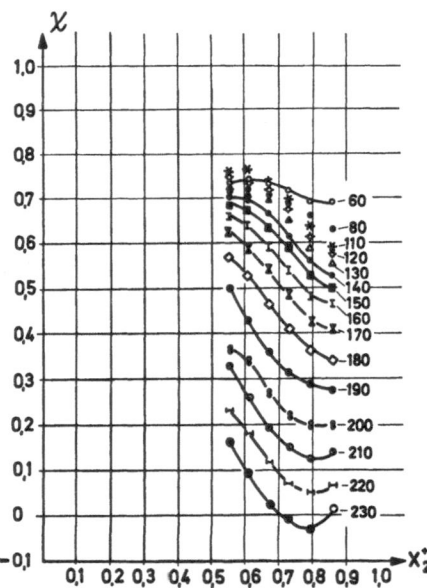

Abb. 13. Wechselwirkungsparameter χ von c-Pentan in
PS

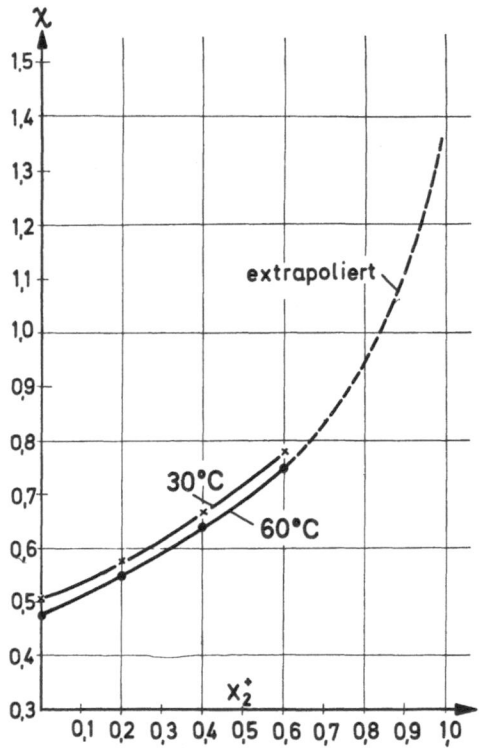

Abb. 14. Wechselwirkungsparameter χ von c-Hexan in
PS (6—9)

4. Versuch einer Auswertung der Meßergebnisse nach einer reduzierten Zustandsgleichung von Prigogine und Flory

Es werden die Eigenschaften der Mischung auf
die charakteristischen Größen der reinen Flüssigkeiten, die sich in den Parametern der Zustandsgleichung manifestieren, zurückgeführt. Ähnlich
ist das Vorgehen von *Patterson* (32), (35), der mit
der Theorie der korrespondierenden Zustände
arbeitet. Die reduzierte Zustandsgleichung nach
Prigogine und *Flory* hat die Form (33, 34, 36, 39)

$$\tilde{p}/(\tilde{v} \cdot \tilde{T}) = \tilde{v}^{1/3}/(\tilde{v}^{1/3} - 1) - 1/(\tilde{v} \cdot \tilde{T}). \qquad [1]$$

Damit lassen sich die thermodynamischen Eigenschaften der Mischungen beschreiben.

Die reduzierten Zustandsgrößen aus Gl. [1]
werden als p/p^*, V/V^* und T/T^* definiert. Die
Symbole mit Stern sind die charakteristischen
Größen des Reinstoffes. Sie lassen sich aus dem
spezifischen Volumen, dem kubischen Ausdehnungskoeffizienten und dem Druckkoeffizienten
berechnen und sind für PS, n-Pentan, n-Hexan,
c-Pentan und c-Hexan in den Abb. 15—22 in
Abhängigkeit von der Temperatur dargestellt.

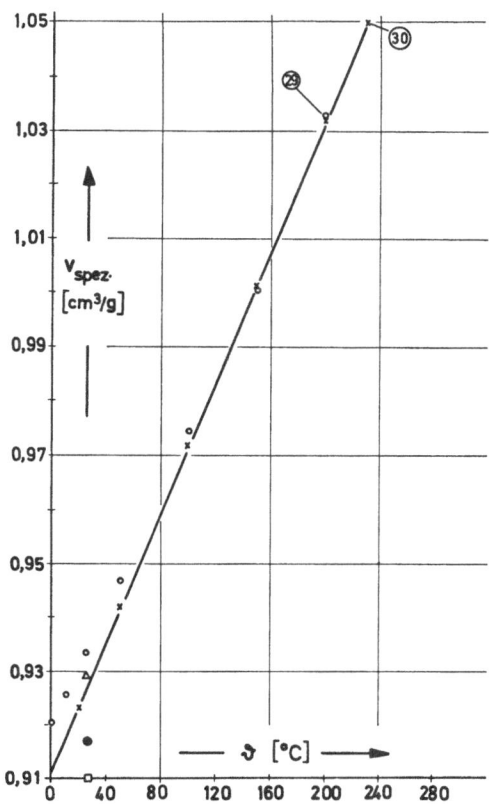

Abb. 15. Spezifisches Volumen von flüssigem Polystyrol Partielles molares Volumen in c-Hexan △, Benzol □ und Toluol ⊕

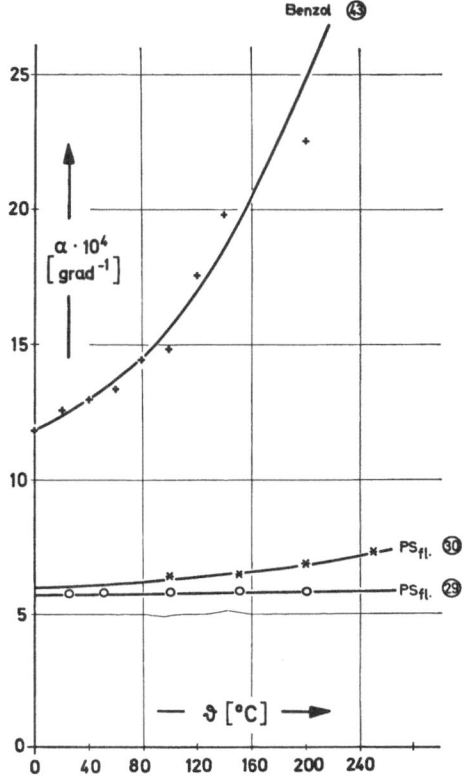

Abb. 16. Kubischer Ausdehnungskoeffizient von flüssigem Polystyrol PS_{fl} und Benzol

Die Daten wurden der Literatur entnommen und weisen aus folgenden Gründen große Schwankungen auf:

a) Bei PS resultieren die größten Abweichungen aus dem Umstand, daß die Stoffkonstanten einmal aus konzentrierten Lösungen (29), ein andermal von höheren zu tieferen Temperaturen extrapoliert werden (30). Bei beiden Verfahren trifft man willkürliche Annahmen, bei der Extrapolation über die Konzentration setzt man die partiellen Größen gleich den Größen selbst, bei der Extrapolation über die Temperatur nimmt man Linearität an.

b) Die Werte für den Ausdehnungskoeffizienten und der Kompressibilität der Kohlenwasserstoffe leiden darunter, daß manchmal die Angabe des Volumens, durch das die entsprechenden Anstiege $(\mathrm{d}V/\mathrm{d}T)_p$ und $(\mathrm{d}V/\mathrm{d}p)_T$ dividiert werden, fehlen. In der vorliegenden Arbeit diente das Volumen bei 0 °C und 760 Torr als Bezugsvolumen.

Das Exzeßvolumen beim Mischen, die kritische Temperatur, bei der Entmischung einsetzt, die Mischungswärmen und das reduzierte chemische Restpotential lassen sich mit Hilfe der folgenden Gleichungen berechnen:

1. Das Exzeßvolumen V^E, das beim Mischen auftritt

$$\frac{V^E}{V_0} = \frac{\tilde{v}}{\tilde{v}_0} - 1 = \frac{\tilde{v}}{(\varphi_1 \cdot \tilde{v}_1 + \varphi_2 \cdot \tilde{v}_2)} - 1 \qquad [2]$$

und $\tilde{v}$ erhält man aus Gl. [3]

$$\tilde{v}^{4/3}/(\tilde{v}^{1/3} - 1) = \frac{p_1^* + \varphi_2 \cdot p_2^*/\varphi_1 - \theta_2 \cdot X_{1,2}}{p_1^* \cdot \tilde{T}_1 + \tilde{T}_2 \cdot p_2^* \cdot \varphi_2/\varphi_1} . \qquad [3]$$

2. Die kritische Temperatur T_{krit}, bei der Entmischung auftritt,

$$T_{krit} = X_{1,2} \cdot \left\{ \left[\frac{\left(\frac{S_1}{S_2}\right)^2 \cdot R}{2 \cdot V_1^*} + Q_{1,2} \right] \right.$$

$$\left. \cdot \tilde{v}_1 - \frac{A^2 \cdot \alpha_1 \cdot p_1^* \cdot \left(\frac{S_1}{S_2}\right)^2}{2} \right\}^{-1} , \qquad [4]$$

$$A = \left(1 - \frac{T_1^*}{T_2^*}\right) \frac{p_2^*}{p_1^*} - \left[X_{1,2}/p_1^* \cdot \left(\frac{S_1}{S_2}\right) \right] . \qquad [5]$$

9

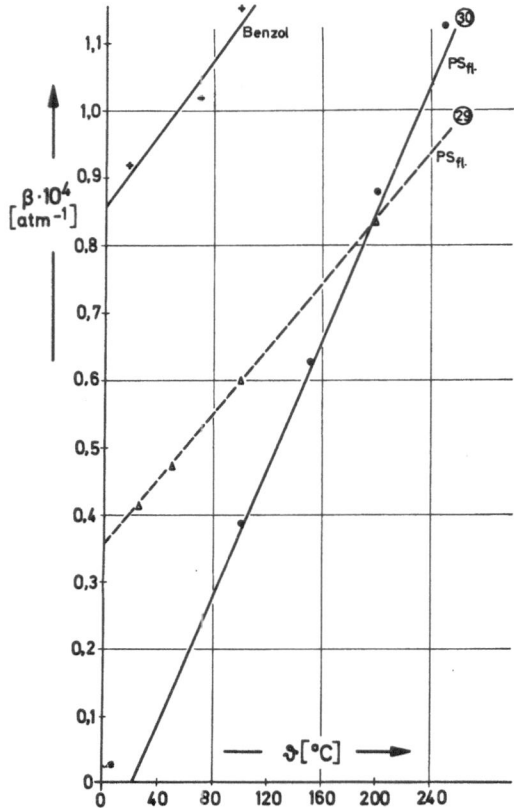

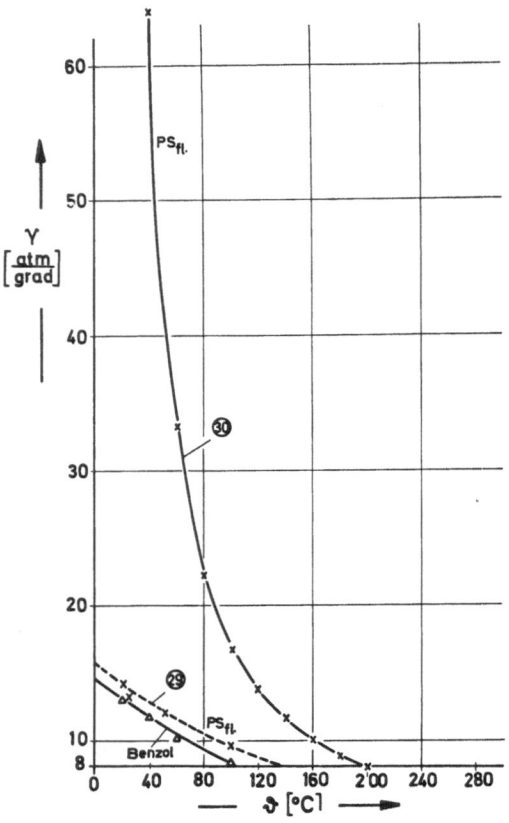

Abb. 17. Kompressibilität von flüssigem Polystyrol PS$_{fl}$ und Benzol

Abb. 18. Thermischer Druckkoeffizient von flüssigem Polystyrol PS$_{fl}$ und Benzol

3. *Die Mischungswärme* ΔH_M, *die auftritt, wenn man* N_1 *Mole Lösungsmittel mit* N_2 *Molen gelösten Stoff mischt,*

$$\Delta H_M = N_1 \cdot p_1^* \cdot V_1^* \cdot \left[\frac{1}{\tilde{v}_1} - \frac{1}{\tilde{v}} \right] + N_2 \cdot p_2^* \cdot V_2^*$$

$$\cdot \left[\frac{1}{\tilde{v}_2} - \frac{1}{\tilde{v}} \right] + X_{1,2} \cdot N_1 \cdot V_1^* \cdot \theta_2/\tilde{v}. \qquad [6]$$

4. *Das reduzierte chemische Restpotential* χ

$$\chi = \frac{p_1^* \cdot V_1^*}{R \cdot T \cdot \varphi_2^2} \cdot \{3 \cdot \tilde{T}_1 \cdot \ln[(\tilde{v}_1^{1/3} - 1)/(\tilde{v}^{1/3} - 1)]$$

$$+ (\tilde{v}_1^{-1} - \tilde{v}^{-1})\} + [V_1^* \cdot X_{1,2} \cdot \theta_2^2/(\tilde{v} \cdot R \cdot T \cdot \varphi_2^2)]$$

$$- [V_1^* \cdot Q_{1,2} \cdot \theta_2^2/(R \cdot \varphi_2^2)], \qquad [7]$$

$$R = 82{,}05 \left[\frac{\text{Ncm}^3 \cdot \text{atm}}{\text{grad} \cdot \text{mol}} \right],$$

$$\theta_2 = \varphi_2 \cdot S_2/(\varphi_1 \cdot S_1 + \varphi_2 \cdot S_2),$$

$$V_1^* = M_1 \cdot v_{\text{spez}_1}/\tilde{v}_1 \quad [\text{cm}^3/\text{mol}].$$

M_1 = Molekulargewicht [g/mol].

S_1/S_2 = Segmentoberflächenverhältnis, das aus molekularen Abmessungen abgeschätzt werden kann.

$Q_{1,2}$ = Anpaßbarer Parameter, der der Entropie pro Volumeneinheit Rechnung trägt, die sich aus der Wechselwirkung benachbarter Moleküle ableitet [atm/grad].

$X_{1,2}$ = Anpaßbarer Parameter, der dem Unterschied in der Wechselwirkung zwischen gleichen und ungleichen Segmentpaaren Rechnung trägt [atm].

Das Exzeßvolumen und die Mischungswärme sind nur eine Funktion von $X_{1,2}$; hingegen sind die kritische Entmischungstemperatur und das reduzierte chemische Restpotential sowohl von $X_{1,2}$ als auch von $Q_{1,2}$ abhängig.

Deshalb wurde versucht, den anpaßbaren Parameter $X_{1,2}$ aus den Messungen des spezifischen Volumens von PS-KW-Mischungen zu berechnen. Aus Tabelle 6 erkennt man, daß bei Verwendung der PS-Daten, die durch Extrapolation von höheren Temperaturen gewonnen worden sind (30), $X_{1,2}$ mit steigender Temperatur kleiner wird. Benützt man die Zahlenwerte, die aus PS-Lösungen erhalten worden sind (29), so sind die $X_{1,2}$-Werte annähernd konstant und be-

Tabelle 6. Der anpaßbare Parameter $X_{1,2}$ berechnet aus den Abweichungen der Volumenadditivität nach *Prigogine* und *Flory*

Temperatur (°C)	$\dfrac{V^E}{V_0} \cdot 10^3$	φ_2 [a]	$V^*_{spez_1}$ [b] (cm³/g)	$V^*_{spez_2}$ [b] (cm³/g)	$\tilde{v}_0$ [c]	$\tilde{v}_1$ [d]	$\tilde{v}_2$ [d]
50 Gew.-% c-Hexan in PS ($S_1/S_2 = 2$)							
25	− 1,8	0,45	0,992	0,810	1,235	1,302	1,153
		0,447	0,992	0,800	1,238	1,302	1,159
40	− 3,6	0,45	0,993	0,812	1,251	1,330	1,160
		0,446	0,993	0,800	1,257	1,330	1,170
50 Gew.-% c-Pentan in PS ($S_1/S_2 = 2$)							
20	− 6,2	0,448	1,025	0,8093	1,243	1,320	1,150
40	− 9,6	0,440	1,031	0,812	1,266	1,350	1,160
		0,437	1,031	0,800	1,270	1,350	1,170
100	− 16,3	0,447	1,018	0,821	1,363	1,505	1,188
		0,442	1,018	0,808	1,335	1,505	1,202
150	− 37,7	0,453	0,996	0,829	1,476	1,695	1,210
		0,448	0,996	0,810	1,510	1,695	1,230
16 Gew.-% n-Hexan in PS ($S_1/S_2 = 1,75$)							
20	− 17	0,786	1,159	0,809	1,184	1,314	1,150
12 Gew.-% n-Pentan in PS ($S_1/S_2 = 1,75$)							
20	− 20	0,834	1,181	0,809	1,177	1,320	1,150

[a] $\varphi_2 = \dfrac{w_2 \cdot v^*_{spez_2}}{w_2 \cdot v^*_{spez_2} + w_1 \cdot v^*_{spez_1}}$

[b] $V^*_{spez_i} = \dfrac{v_{spez_i}}{\tilde{v}_i}$

[c] $\tilde{v}_0 = \varphi_1 \cdot \tilde{v}_1 + \varphi_2 \cdot \tilde{v}_2$

[d] $\tilde{v}_1 = \left[1 + \dfrac{\alpha_i \cdot T}{3(\alpha_i \cdot T + 1)} \right]^3$

$\tilde{v}$	p_1^* [f] (atm)	p_2^* [f] (atm)	$\tilde{T}_1 \cdot 10^2$ [g]	$\tilde{T}_2 \cdot 10^2$ [g]	$\tilde{T} \cdot 10^2$ [g]	θ_2 [h]	$X_{1,2}$ (atm)	Lit.
1,222	5 750	5 470	6,4	4,14	5,44	0,290	200	(29)
1,237	5 750	121 000	6,4	4,12	5,72	0,288	92 600	(30)
1,245	5 610	5 380	6,83	4,20	5,62	0,290	250	(29)
1,250	5 610	24 800	6,83	4,30	5,92	0,287	16 500	(30)
1,235	6 430	5 500	6,9	4,12	5,51	0,289	− 1 670	(29)
1,263	5 200	5 380	7,05	4,2	5,84	0,282	400	(29)
1,256	5 200	24 800	7,0	4,3	6,2	0,34	15 340	(30)
1,340	7 350	5 060	8,6	4,78	7,06	0,288	− 950	(29)
1,360	7 350	7 430	8,6	4,93	7,33	0,284	6 400	(30)
1,420	10 500	5 045	9,5	5,35	7,8	0,293	− 3 350	(29)
1,422	10 500	5 650	9,5	5,5	7,8	0,288	1 310	(30)
1,163	4 300	5 500	6,6	4,12	4,27	0,678	− 4 500	(29)
1,152	3 600	5 500	7,08	4,12	4,05	0,742	− 3 000	(29)

[e] $\tilde{v} = (1 + V^E/V_0) \cdot \tilde{v}_0$

[f] $p_i^* = \gamma_i \cdot T \cdot \tilde{v}_i^2$

[g] $\tilde{T}_1 = \dfrac{\tilde{v}_i^{1/3} - 1}{\tilde{v}_i^{4/3}}$

[h] $\theta_2 = \left(\dfrac{\varphi_1 \cdot S_1}{\varphi_2 \cdot S_2} + 1 \right)^{-1}$

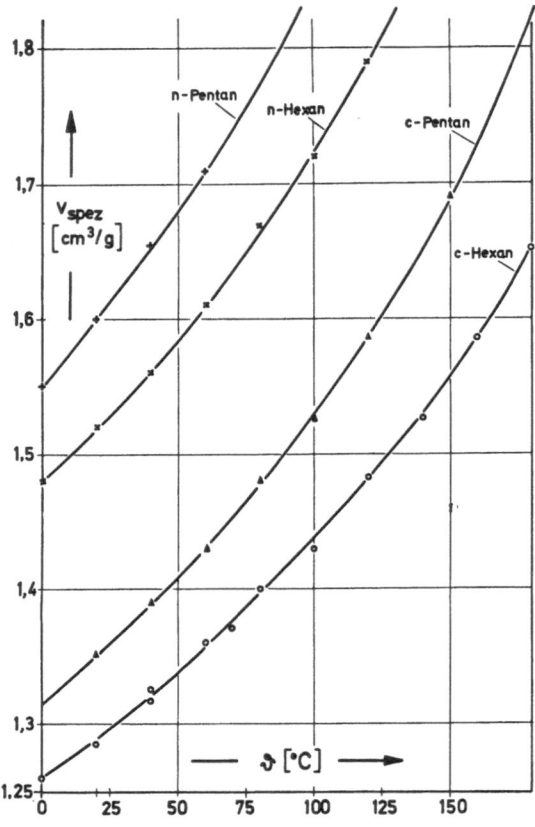

Abb. 19. Spezifisches Volumen von Kohlenwasserstoffen

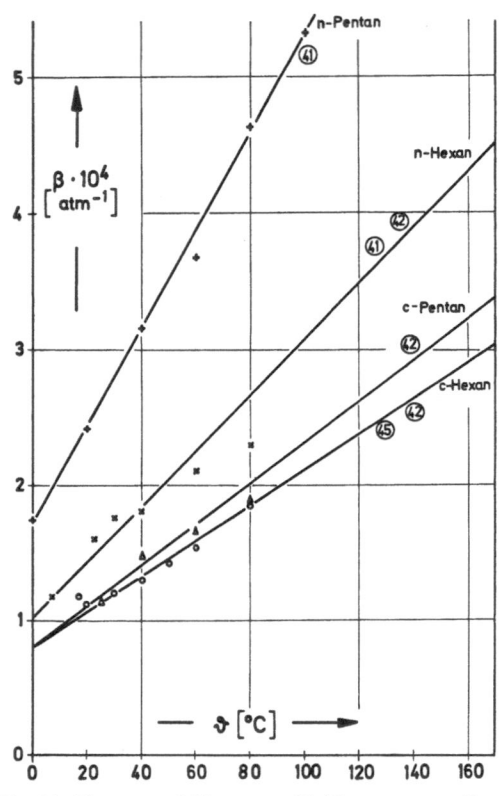

Abb. 21. Kompressibilität von Kohlenwasserstoffen

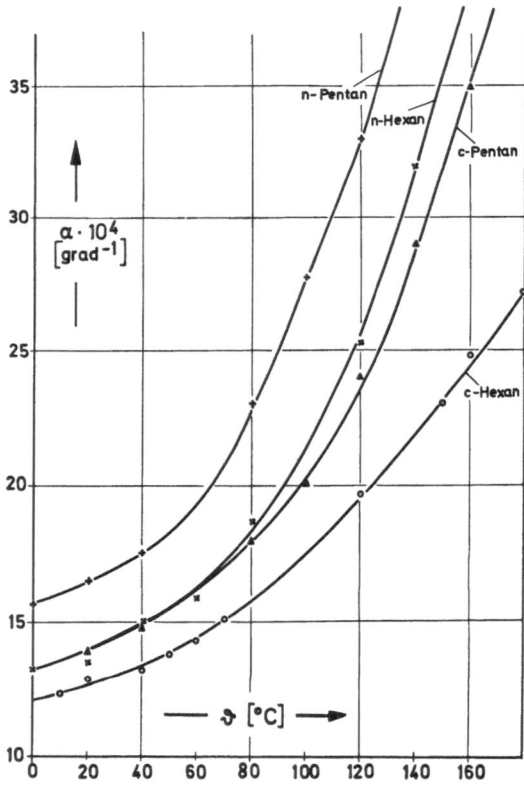

Abb. 20. Kubischer Ausdehnungskoeffizient von Kohlenwasserstoffen

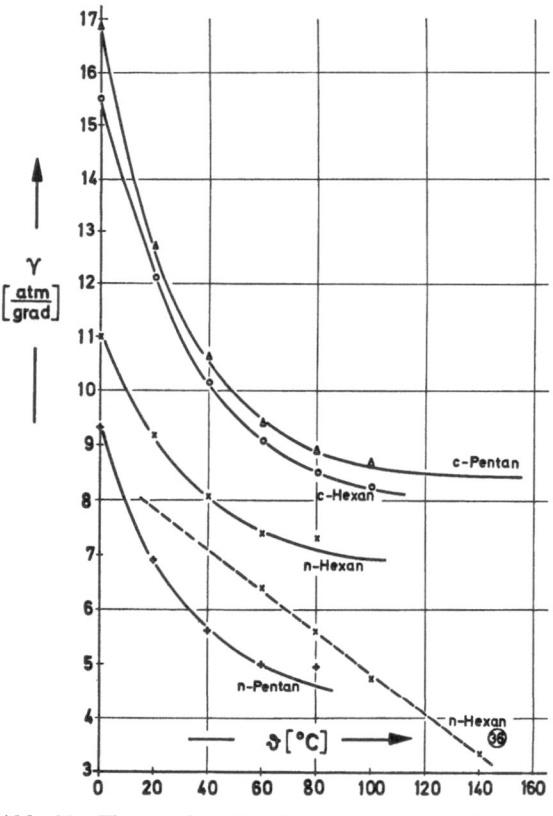

Abb. 22. Thermischer Druckkoeffizient von Kohlenwasserstoffen

tragen für PS-*c*-Hexan 250, für PS-*c*-Pentan — 1250, für PS-*n*-Hexan — 4500, und für PS-*n*-Pentan — 3000 atm. (Tabelle 6).

5. Diskussion der Ergebnisse

Zyklische Kohlenwasserstoffe unterscheiden sich von linearen vor allem dadurch, daß sie zum Schwingen befähigt sind: *c*-Pentan schwingt zwischen der Briefumschlag- und der Halbsessel-form, *c*-Hexan zwischen der Wannen- und Sessel-form. Die Unfähigkeit der linearen Alkane zum Schwingen äußert sich darin, daß sie schwieriger in ein Kristallgitter eingepaßt werden können, was makroskopisch einen tieferen Schmelzpunkt bedeutet, und daß sie zu Nachbarn des gleichen Typs weniger gastfreundlich sind, was makro-skopisch zu tieferen Siedepunkten, niedrigeren kritischen Temperaturen und niedrigeren Dich-ten führt (Tabelle 7).

Im Verein mit Polystyrol zeigt die Betrach-tung der Mischungslücken, der Sättigungs-löslichkeit bei 20 °C, der Wechselwirkungspara-meter bei 20 °C, der spezifischen Weichmacher-wirkung und der Löslichkeitsparameter bei 20 °C, daß die zyklischen besser als die entsprechenden aliphatischen KW in PS löslich sind. Die Lös-lichkeitsparameter der zyklischen KW liegen näher an dem Wert von 8,6 für PS. Die Lage der Mischungslücken ist dadurch gekennzeich-net, daß bei den zyklischen KW die oberen und unteren kritischen Löslichkeitstemperaturen unterhalb der kritischen Temperaturen, bei den linearen KW hingegen oberhalb der kritischen Temperaturen liegen. Der Wechselwirkungspara-meter und die spezifische Weichmacherwirkung weisen bei den ringförmigen Molekülen kleinere Werte auf (Tabelle 8). Die Regel, daß Ringe Lösungsmittel und Ketten Nichtlösungsmittel für PS darstellen, ist nur beschränkt auf Hetero-ringe und -ketten übertragbar. Sie gilt uneinge-schränkt für Äther, für Ketone und Amine nur bei dreigliedrigen Verbindungen, die höherglied-rigen sind alle Lösungsmittel, Alkohole sind unterschiedlos Nichtlösungsmittel.

Tabelle 7. Materialkonstanten der reinen Komponenten

Kohlenwasserstoff	Schmelzpkt. (°C)	Siedepkt. (°C)	krit. Temp. (°C)	Dichte (20 °C) (g/cm³)
n-C$_2$	— 172	— 88,3	32,3	(0,561/— 100 °C)
c-C$_2$ (Doppelbindung)	— 169	— 103,9	9,25	(0,566/— 100 °C)
n-C$_3$	— 187	— 42	96,8	0,539
c-C$_3$	— 127	— 33	124,4	0,605
n-C$_5$	— 130	— 36,5	196,6	0,623
c-C$_5$	— 94	49,4	238,6	0,746
n-C$_6$	— 95	69	234,2	0,659
c-C$_6$	7	81	281	0,778
n-C$_{12}$	— 10	215	386	0,751
c-C$_{12}$	61	213	—	0,800

Tabelle 8. Vergleich von Daten, die an Lösungen aus PS in zyklischen und linearen KW bestimmt worden sind.

Kohlen-wasserstoff	Sättigungs-löslichkeit/20 °C (Gew.-%)	Wechsel-wirkungs-parameter χ	Spez. Weich-macherwirkung (grad/Gew.-%)	Löslichkeitsparameter $\left(\dfrac{cal}{cm^3}\right)^{1/2} \delta_1$	UCST (°C)	LCST (°C)
n-C$_3$	8,5	0,94	11	5,3	—	—
c-C$_3$	42	0,53	12	6,4	—	— 155 (extr.)
n-C$_5$	12	1,4	9	7,0	—	—
c-C$_5$	100	0,7	7	7,2	5	160
n-C$_6$	16	1,49	8,5	7,3	—	—
c-C$_6$	100	0,7	5	8,2	30	205
n-C$_8$	14	2,9	9	7,6	—	—
c-C$_8$	100	1,6	—	7,9	15	355
n-C$_{12}$	10	—	5	7,8	—	—
c-C$_{12}$	100 (65 °C)	—	—	7,6	65	460 (extr,)

Vergleicht man die zyklischen Kohlenwasserstoffe untereinander, so sieht man, daß mit wachsender Ringgröße die Abweichungen von der Volumenadditivität abnehmen (Tabelle 1) und der Temperaturbereich, in dem unbegrenzte Mischbarkeit vorliegt, immer größer wird. Denn es liegen die oberen kritischen Löslichkeitstemperaturen nahezu unabhängig von der Kohlenstoffzahl n der Cycloalkane zwischen 15 und 65 °C, hingegen steigen die unteren kritischen Löslichkeitstemperaturen in °K mit wachsender Ringgröße nach folgender Gleichung:

$$T_{krit} = 1043 - 3700/(n + 1) \qquad [8]$$

[8] ist eine Lösung von [4] für die Abhängigkeit der LCST von der Kohlenstoffzahl n.

Die Unterschiede zwischen Ringen und Ketten verringern sich einmal mit sinkendem Molekulargewicht, weil die Möglichkeit zu schwingen abnimmt, z. B. Äthan/Äthylen, ein andermal mit steigendem Molekulargewicht, weil dann die Konformationen von Ringen und Ketten einander immer ähnlicher werden, z. B. n-C_{12}/c-C_{12}.

Ich möchte es nicht versäumen, an dieser Stelle den Herren Dr. *K. H. Illers*, Dr. *O. Riedel* und Dr. *E. Brunner* für die Überlassung ihrer Meßwerte, Herrn *St. Lück* für die experimentelle Unterstützung und der BASF für die Erlaubnis, diese Arbeit veröffentlichen zu dürfen, *zu danken*.

Zusammenfassung

Es wurden spezifische Volumina, Gleichgewichtsdampfdrücke und Phasendiagramme von Mischungen aus Polystyrol und Cycloalkanen ermittelt und diese mit den entsprechenden Größen von Mischungen aus Polystyrol und Alkanen verglichen. Dabei zeigte sich, daß sich Ketten und Ringe in ihren Löseeigenschaften für Polystyrol stark unterscheiden: Cycloalkane sind Lösungsmittel, Alkane sind Nichtlösungsmittel. Innerhalb der homologen Reihe der Cycloalkane steigen die unteren kritischen Löslichkeitstemperaturen mit der Zahl der Kohlenstoffatome. während die oberen kritischen Löslichkeitstemperaturen nahezu konstant bleiben. Zyklische und lineare Äther befolgen die Regel, lineare Ketone und Amine sind als dreigliedrige Verbindungen Nichtlösungsmittel, die höhergliedrigen sind alle Lösungsmittel, Alkohole sind unterschiedslos Nichtlösungsmittel. Die thermodynamische Beschreibung dieses Sachverhaltes durch eine Zustandsgleichung, die den Zusammenhang zwischen Daten der reinen Komponenten und den Mischungserscheinungen herstellt, bedarf noch weiterer Meßwerte.

Summary

Specific volumes, vapour pressures and phase diagrams were measured for mixtures of polystyrene and cycloalkanes. The data was compared with that obtained for polystyrene-alkane-mixtures. It was found that acyclic and cyclic hydrocarbons strongly differ in their ability to dissolve polystyrene: cycloalkanes are solvents and acyclic alkanes are non-solvents. Within the homologous series of cycloalkanes the lower critical solubility temperature increases with the number of carbon atoms whereas the upper critical solubility temperature remains practically constant. Cyclic and acyclic ethers behave like hydrocarbons, but only three-membered ketons and amines follow this pattern, the higher members being solvents. All alcohols exhibit non-solvent activity. A thermodynamic relationship as an equation of state which correlates the data of the pure compounds with mixing phenomena is still uncertain and requires further work.

Literatur

1) *Horacek, H.*, Kolloid-Z. u. Z. Polymere **250**, 863 (1972).
2) *Brunner, E.*, Bericht No. 71.81, 17. 12. 1971 (BASF intern).
3) *Höcker, H.* und *P. J. Flory*, Trans. Farad. Soc. **67**, 2271 (1971).
4) *Lin, D. C. K.*, *J. H. Silberberg* und *J. J. Mcketta*, J. of Chem. and Eng. Data, Vol. **15**, 483 (1970).
5) *McCullough, I. P.*, *R. E. Pennington*, *J. C. Smith*, *J. A. Hossenlopp* und *Waddington*, J. Americ. Soc. **81**, 5880 [$V = RT/(P + B)$] (1959).
6) *Jenckel, E.* und *K. Gorke*, Z. f. Elektrochem. **60** 573 (1956).
7) *Krigbaum, W. R.* und *D. O. Geymer*, J. Americ. Soc. **81**, 1859 (1959).
8) *Koningsveld, R.*, *L. A. Kleintjens* und *A. R. Schultz*, J. Pol. Sci. A – **2**, 8, 1261 (1970).
9) *Höcker, H.*, *H. Shih* und *P. J. Flory*, Trans. Farad. Soc. **67**, 2275 (1971).
10) *Schmoll, K.* und *E. Jenckel*, Z. Elektrochem. **60**, 756 (1956).
11) *Baughan, E. C.*, Trans. Farad. Soc. **44**, 495 (1948).
12) *Billmeyer* jr., *F. W.*, Textbook of Polymer Science, 2nd Ed. (1971).
13) *Patterson, D.*, Rubber Chem. and Techn. **40**, 2 (1967).
14) *Wolf, B. A.*, Adv. Sci. **10**, 110 (1972).
15) *Baker, C. H.*, *C. S. Clemson* und *G. Allen*, Pol. **7**, 525 (1966).
16) *Allen, G.* und *C. H. Baker*, Pol. **6**, 181 (1965).
17) *Patterson, D.* Macromol. Vol. 2, No. 6, 672 (1969).
18) *Tager, A. A.*, *A. A. Anikeyeva*, *V. M. Andreyeva*, *T. Ya. Gumarova* und *L. A. Chemoskutova*, Pol. Sci. (USSR) **10**, 1926 (1968).
19) *Freeman, P. J.* und *J. S. Rowlinson*, Pol. **1**, 20 (1960).
20) *Myrat, C. D.* und *J. S. Rowlinson*, Pol. **6**, 645 (1965).
21) *Bataille, P.* und *D. Patterson*, J. of Pol. Sci. A 1, 3265 (1963).
22) *Delmas, G.* und *D. Patterson*, Pol. **7**, 513 (1966).
23) *Horth, A.*, *D. Patterson* und *M. Rinfret*, J. Pol. Sci. **39**, 189 (1959).
24) *Schick, M. J.*, *P. Doty* und *B. H. Zimm*, J. Americ. Soc. **72**, 530 (1950).

25) *Huggins, M. L.*, Ann. N.Y. Acad. Sci. **44**, 442 (1943).

26) *Baughan, E. C.* und *J. N. Brønsted*, Trans. Farad. Soc. **42 B**, 48 (1946).

27) *Mellan, J.*, Compatibility and Solubility (1968). ndc

28) *Flory, P. J.*, J. Americ. Soc. **87**, 9, 1833 (1965).

29) *Höcker, H., G. J. Blake* und *P. J. Flory*, Trans. Farad. Soc. **67**, 2251 (1971).

30) *Breuer, H.* und *G. Rehage*, Kolloid-Z. u. Z. Polymere **216, 217**, 159 (1967).

31) *Patterson, D.* und *A. A. Tager*, J. Pol. Sci. (USSR) **9**, 2051 (1967).

32) *Houwink* und *Staverman*, Chemie und Technologie der Kunststoffe, Bd. **1**, 4. Aufl., 513 (1962).

33) *Flory, P. J., J. E. Ellenson* und *B. E. Eichinger*, Macromolecules, **1**, No. 3, 279 (1968).

34) *Flory, P. J., R. A. Orwoll* und *A. Vrij*, J. Americ. Soc. **86**, 3507, 3515 (1964).

35) *Delmas, G., D. Patterson* und *T. Somcynsky*, J. Pol. Sci. **57**, 79 (1962).

36) *Orwoll, R. A.* und *P. J. Flory*, J. Americ. Soc. **89**, 6814 (1967).

37) *Patterson, D.* und *J. M. Bardin*, Trans. Farad. Soc. **66**, 321 (1970).

38) *Flory, P. J.*, Disc. Farad. Soc. **49**, 7 (1970).

39) *Hammers, W. E., C. L. De Ligny* und *L. A. Vraas*, J. Pol. Sci. Vol. **11**, No. 3, 499 (1973).

40) *Dupp, G.*, Kunststofftechnik, Bd. **8**, 271 (1969).

41) *Landolt-Börnstein*, Hauptwerk, Bd. I, 94 (Berlin, 1923).

42) *Kuss, E.* u. a., Chem. Ing. Techn. **42**, 1073 (1970).

43) *Timmermans, J.*, Physico-Chemical Constants of Pure Organic Compounds (1950).

44) *Landolt-Börnstein*, Bd. II, 633 (Berlin-Heidelberg-New-York, 1971).

45) *Moelwyn, Hughes* e. a. und *P. L. Thorpe*, Proc. Roy. Soc. (A) **278**, 574 (1964).

46) *Rehage, G.*, Kolloid-Z. u. Z. Polymere **196**, 97 (1964).

47) *Small, P. A.*, J. Appl. Chem. **3**, 71 (1953).

Adresse des Autors:

H. Horacek
Kunststofflaboratorium
der BASF
Ludwigshafen/Rhein

Progr. Colloid & Polymer Sci. **58**, 136—140 (1975)

*Faculty of Technology, Tokyo University of Agriculture and Technology, Koganei-shi, Tokyo (Japan)
and Research Institute for Polymers and Textiles, Yokohama (Japan)*

Dye aggregations or crystals of acridine orange within natural fibres

T. Ohtsu, K. Nishida and *K. Tsuda*

With 10 figures

(Received June 18, 1974)

1. Introduction

It is very important to understand the state of dyes within fibres not only theoretically but practically accounting for the light fading or washing fastness of dyeings.

In our previous papers (1, 2, 3), the authors suggested that the visible absorption spectra of the cross-sections of the synthetic-polymer fibres dyed with Acridine Orange free base were measured by means of a microspectrophotometer and showed three peaks at about 440 nm (crystal band), about 470 nm (aggregate band) and about 490 nm (monomer band). It is assessed from these results that the dye is in the monomer, aggregate, and crystal state within the synthetic-polymer fibres.

In the present paper, the visible absorption spectra of the cross-sections of the various *natural* fibres dyed with Acridine Orange hydrochloride or its free base are measured in order to obtain further informations on the state of the dye within fibres.

2. Experimental

Materials. The hydrochloride or free base of Acridine Orange (abbreviated as AO) was purified by the column chromatographic method (4). The following various natural fibres were used; American cotton, jute (Vietnam), silk (Taihei × Choan, Japan, 1972), wool (Australian-Merino) and viscose rayon (Mitsubishi Rayon Co., Ltd., Shinko, 3D, bright). The cotton and the jute were purified by the method of standard cellulose preparation, and were treated by washing with boiling alcohol and ether for 6 h respectively. The silk was used after the scouring by common method. The wool was purified with boiling methylenechloride. The viscose rayon was washed with ethanol for 8 h. The technique of the dyeing, the measurement of absorption spectra using

microspectrophotometer (Olimpus model MSP-AlV) (abbreviated as MSP) and the scanning-electron microscopy (Nippon Denshi model JMS-2) have been described in our previous papers (1, 2).

3. Results and Discussion

Fig. 1 is the absorption spectrum of the cross-section of the cotton fibre dyed with AO free base. The absorption spectrum shows a high maximum at 440 nm, which will be referred to as the crystal band. And two slight hamps are observed. The peak at 470 nm will be referred to as aggregate band (β-band) and the peak at 490 nm will be referred to as the monomer band (α-band). Therefore, AO free base mostly exists as crystals. AO hydrochloride did not dye the cotton fibre, and the spectra measurement could not be carried out.

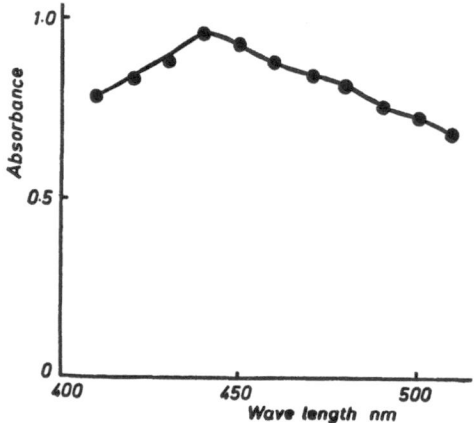

Fig. 1. Absorption spectrum of the cross-section of cotton fibre dyed with Acridine Orange free base. Amount of dye; 1.19×10^{-2} mole/kg of fibres. Thickness of section: 8 μ

Fig. 2 is the scanning microscopic photograph of the cross-section of the dyed cotton. The presence of the fibril structure and the channel are very clear. The fibres were buried in methacrylic resin and then frozen in liquid nitrogen. The cross-section of the fibre were prepared by the cracking of the frozen fibres.

Fig. 3 is the absorption spectrum of the cross-section of the jute fibre dyed with AO free base or hydrochloride. The spectrum for the free base shows three peaks. The highest maximum of 470 nm will be referred to as the aggregate band.

The second highest peak at 450 nm will be referred to as the crystal band and the slight peak at 490 nm will be referred to as the monomer band. These results show that AO free base within the jute as with the dyed cotton fibre, mostly exists as crystals. In the case of AO hydrochloride, the spectrum shows the monomer band (500 nm) and aggregate band (480 nm).

Fig. 4 is the scanning-electron microscopic photograph of the dyed cross-section of the jute. As with the cotton fibre, the presence of the fibril structure and channel on the jute are very clear.

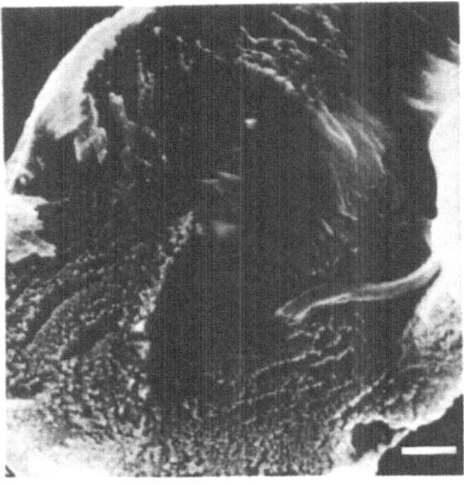

Fig. 2. The electron microscopic photograph of the cross-section of cotton dyed with Acridine Orange free base. The white bar indicates 1 μ

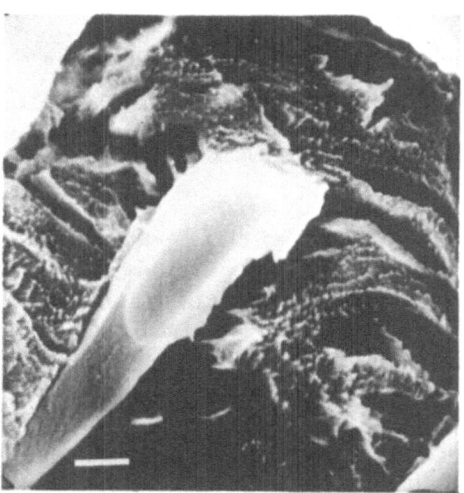

Fig. 4. The electron microscopic photograph of the cross-section of jute dyed with Acridine Orange free base. The white bar indicates 1 μ

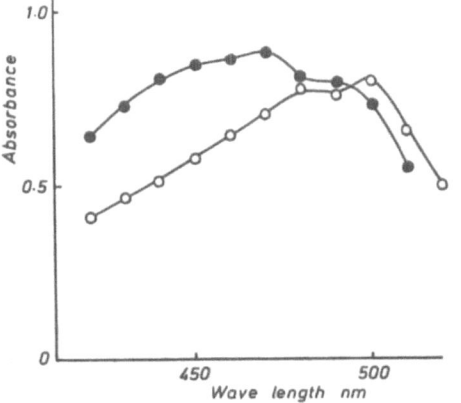

Fig. 3. Absorption spectra of the cross-section of jute dyed with Acridine Orange. ○ Hydrochloride: Amount of dye: 0.7×10^{-2} mole/kg of fibres. Thickness of section: 6 μ. ● Free base. Amount of dye: 1.28×10^{-2} mole/kg of fibres. Thickness of section 8 μ

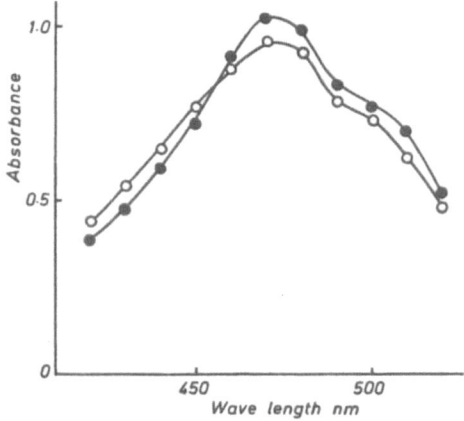

Fig. 5. Absorption spectra of the cross-section of viscose rayon dyed with Acridine Orange. ○ Hydrochloride: Amount of dye: 2.82×10^{-2} mole/kg of fibres. Thickness of section: 8 μ. ● Free base. Amount of dye: 5.48×10^{-2} mole/kg of fibres. Thickness of section: 8 μ

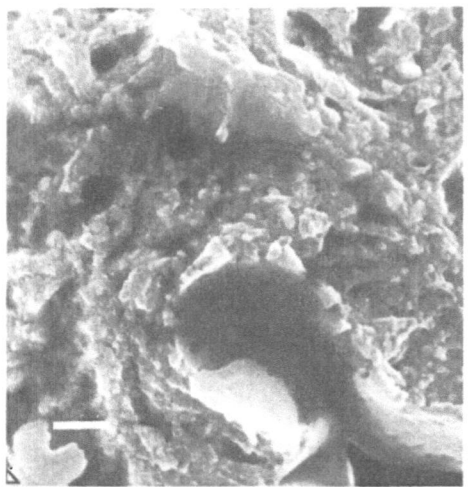

Fig. 6. The electron microscopic photograph of the cross-section of viscose rayon dyed with Acridine Orange free base. The white bar indicates 1 μ

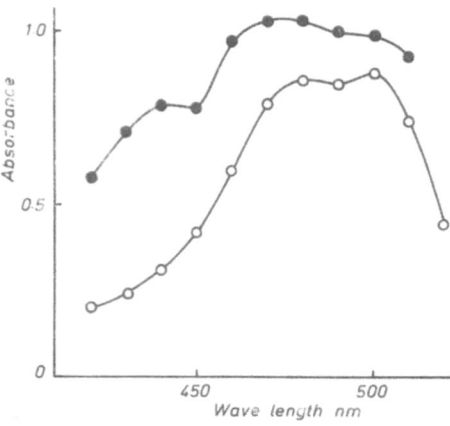

Fig. 7. Absorption spectra of the cross-section of silk dyed with Acridine Orange. ○ Hydrochloride: Amount of dye: 2.48×10^{-2} mole/kg of fibres. Thickness of section: 6 μ. ● Free base. Amount of dye: 1.43×10^{-1} mole/kg of fibres. Thickness of section: 8 μ

Fig. 5 is the absorption spectrum of the cross-section of the viscose rayon dyed with AO free base or hydrochloride. Both these spectra show two peaks, that is, monomer band (495 nm) and aggregate band (470 nm).

Fig. 6 is the scanning-electron microscopic photograph of the dyed cross-section of the viscose rayon. Many void places are observed in the fibre cross-section.

Fig. 7 is the absorption spectrum of the cross-section of the silk dyed with AO free base or hydrochloride. The spectrum of AO free base shows three peaks. The high maximum of 495 nm and 475 nm, will be referred to as the monomer and aggregate band, respectively. The slight peak of 440 nm will be referred to as the crystal-band. In the case of hydrochloride, the spectrum shows monomer band (500 nm) and aggregate band (480 nm).

Fig. 8 is the scanning-electron microscopic photograph of the cross-section of the dyed silk. Many void places are observed and the structure is very rough.

Fig. 9 is the absorption spectrum of the cross-section of the wool fibre dyed with AO free base and hydrochloride. Both these spectra show monomer band (495 nm) and aggregate band (475 nm).

Fig. 10 is the scanning-electron microscopic photograph of the cross-section of the wool fibre dyed with AO free base or hydrochloride. Many void places are observed and the structure is relatively regular.

From above results, the following summary is obtained:

AO hydrochloride shows only monomer and aggregate band in any case, but AO free base shows monomer and aggregate band in the fibres of the wool and the viscose rayon, and monomer, aggregate and the crystal band in the fibres of the cotton, the jute and the silk fibres.

These results may be explained as follows. The formation of the dye crystals within the fibres occurs in the correlation of the dye affinity for the fibres, that is, the fibre which has strong

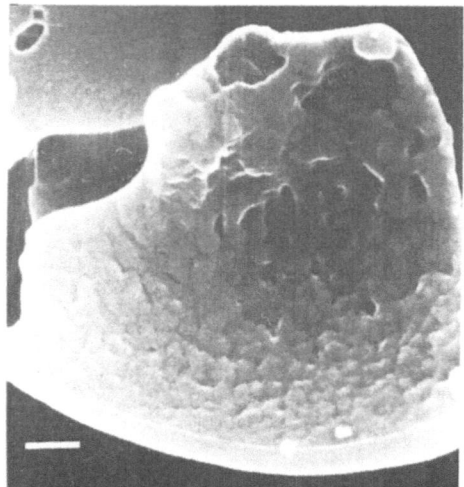

Fig. 8. The electron microscopic photograph of the cross-section of silk dyed with Acridine Orange free base. The white bar indicates 1 μ

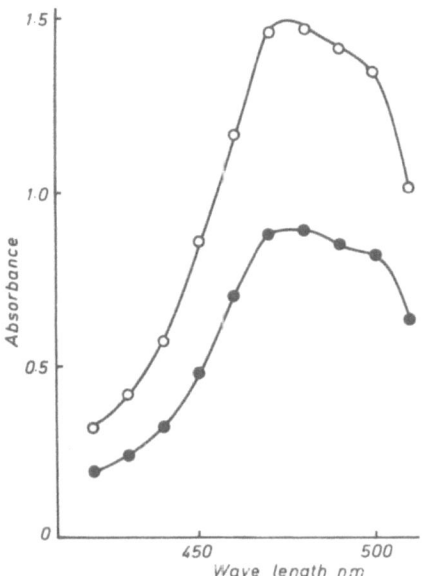

Fig. 9. Absorption spectra of the cross-section of wool dyed with Acridine Orange. ○ Hydrochloride: Amount of dye: 2.44×10^{-2} mole/kg of fibres. Thickness of section: 10 μ. ● Free base. Amount of dye: 1.02×10^{-1} mole/kg of fibres. Thickness of section: 6 μ

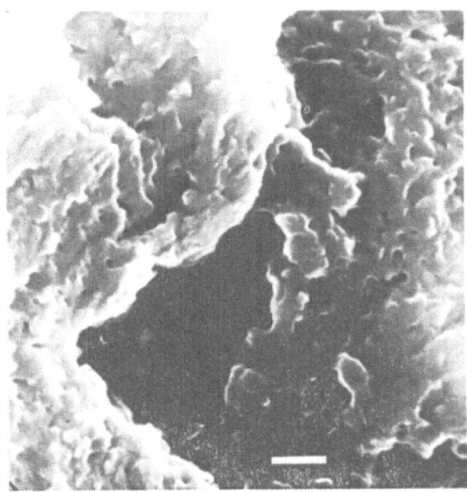

Fig. 10. The electron microscopic photograph of the cross-section of wool dyed with Acridine Orange free base. The white bar indicates 1 μ

affinity for the dye retards the formation of the dye crystals. However, the fibre which has weak affinity for the dye does not retard the crystal formation. The cotton and the jute are in the former cases, and the wool and the viscose rayon are the latter case. The viscose rayon contains

the oxycellulose and the fibre is dyed with strong affinity (5). The silk has the strong affinity for AO. But the fibre structure is very rough and very large void places are observed. Thus, the formation of the crystals is not retarded relatively.

In our previous paper (3), it is reported that the spectra of AO within the acrylic fibres show monomer and aggregate band but do not show the crystal band. This is the former case above. It may be that the fibres which have clear absorption site for the dye do not cause the formation of the crystals.

The spectral curve of the AO hydrochloride on fibres, compared with those in alcohol solution, suggested that in the natural fibre, about 30% of total dye is at a dimeric state, while, in synthetic-polymer fibres the ratio is close to about 50% (2, 3). These results correspond to the fibre structure. In calculation, the assumption has been made that aggregate is at dimeric state.

The authors acknowledge the help of Mr. *T. Ozawa* (Toppan Insatsu Co., Ltd.) for experimental works.

Summary

The visible absorption spectra of the cross-section of natural fibres (cotton, jute, wool and silk) dyed with Acridine Orange hydrochloride or free base were measured by means of a microspectrophotometer. As in the case of synthetic-polymer fibres, the spectra showed monomer, aggregate and crystal band. Many void places or channels were observed in these natural fibres by *scanning*-electron microscopy. In general, the fibres which have weak affinity for Acridine Orange free base show crystal band. This may be because the dye affinity for fibre retards the formation of crystals.

Zusammenfassung

Mit Hilfe des Mikrospektrophotometers wurden die Absorptionsspektren im Sichtbaren von Querschnitten der mit Acridine Orange gefärbten natürlichen Fasern (Baumwolle, Hanf, Wolle und Seide) gemessen. Wie im Fall von synthetischen Polymerfasern zeigen die Spektren der Querschnitte der gefärbten Faser Monomeren-dimeren- und Kristallbanden. Mit dem *Scanning*-Elektronenmikrospektroskop wurden verschiedene Hohlräume und Kanäle in den natürlichen Fasern beobachtet. Im allgemeinen zeigen die Fasern, die eine schwache Affinität für Acridine Orange haben, eine Kristallbande der freien Base. Dies läßt vermuten, daß die Farb-Affinität von Fasern die Formation der Kristalle zögert.

References

1) *Ohtsu, T., K. Nishida, K. Nagumo* and *K. Tsuda,* Kolloid-Z. u. Z. Polymere **249,** 1077 (1971).
2) *Ohtsu, T., K. Nishida, K. Nagumo* and *K. Tsuda,* ibid., **250,** 860 (1972).
3) *Ohtsu, T., K. Nishida, K. Nagumo* and *K. Tsuda,* ibid., **252,** 377 (1974).
4) *Zanker, V.,* Z. physik. Chem. **199,** 225 (1952).
5) *Doree, C.,* The methods of cellulose chemistry **112,** (London, 1933).

Authors' addresses:

T. Ohtsu and Dr. *K. Nishida*
Faculty of Technology
Tokyo University of Agriculture
and Technology, (Tokyo No-Ko Daigaku)
Koganei-shi, Tokyo (Japan)
Dr. *K. Tsuda*
Research Institute for Polymers
and Textiles, Sawatari, Kanagawa
Yokohama (Japan)

Progr. Colloid & Polymer Sci. **58**, 141—144 (1975)

Toppan Printing Co. Ltd., Taitoh-ku, Tokyo, Japan and Faculty of Technology, Tokyo University of Agriculture and Technology, Koganei-shi, Tokyo (Japan)

Interaction between acridine orange and sulphated polysaccharides

H. Watanabe and *K. Nishida*

With 5 figures and 1 table

(Received June 29, 1974)

1. Introduction

In our previous papers, from the results of interaction between the dyes and some polymers; amylose (1, 2), synthetic sucrose polymer *Ficoll* (3), sodium carboxymethylcellulose (4, 5), sodium polyacrylate (6) and poly-α, L-glutamate (7), it was suggested that the positive entropy change ($\Delta S°$) for binding of the dyes to the polymers were principally caused by changes in the conformation of the polymer chain. Thus, the interaction was thought to be dependent on the flexibility of the polymer chain. But the effect of the destruction of water structure (8) on the positive entropy changes is unlikely to be negligible. In the present paper, we describe the interaction between Acridine Orange and the sodium salts of α- and β-cyclodextrin sulphates, amylose sulphates and amylopectin sulphate. The structure of α- and β-cyclodextrin sulphates is rigid. But amylose and amylopectin sulphates have flexible polymer structures. From the studies of the interaction between the dyes and the

polymers, we conclude that the positive $\Delta S°$ results from the conformation change of polymers caused by the dye binding and also from the formation of hydrophobic bonding along with the destruction of the water structures about polymers.

2. Experimental

Materials. Acridine Orange sulphate used (subsequently referred to as AO) was purified by column chromatographic method (9, 10).

The sulphate of α- and β-cyclodextrin (Tokyo Kasei Co., Ltd.), amylose (Hayashibara Seibutsukagaku laboratory) and of amylopectin (Tokyo Kasei Co., Ltd.) were subsequently referred to as α-CDS, β-CDS, AS and APS, respectively. The preparation of the polymer sulphates was done by known method (11). The degree of the sulphation (from sulphur content) of the polymers is shown in Table 1. The values are nearly equal. The sulphates of polymers were purified by repeated fractionations using sodium iodide-alcohol solution and pure alcohol. The purity was assessed with gel filtration method (Sephadex G-25) and conductivity method.

Table 1. Thermodynamics of the binding of Acridine Orange and polysaccharides

		α-CDS	β-CDS	A$_{350}$S	A$_{700}$S	APS
$k_1 (\times 10^{-7})$	16 °C	5.48	5.31	6.00	6.22	5.94
	24 °C	5.24	5.11	6.39	6.61	6.36
	30 °C	5.50	5.39	7.08	7.03	7.50
$-\Delta F°$ (Kcal/mol)		10.5	10.5	10.6	10.6	10.6
$\Delta H°$ (Kcal/mol)		0.0	0.0	1.47	1.11	2.60
$\Delta S°$ (e. u.)		35.4	35.4	40.7	39.6	44.6
Degree of sulphation (%)		18.0	18.0	18.0	18.5	18.3
n	16 °C	1.6	1.7	1.6	2.0	1.9
	24 °C	1.6	1.7	1.6	2.0	1.9
	30 °C	1.6	1.7	1.6	1.9	1.8

Spectral method. The spectral technique and the representation of binding data have been described in our previous paper (1). In our notation, r is the number of moles of bound dye per unit mole of polymers, $[A]$ is the free dye concentration, n is the number of binding sites on the polymer and k_1 is the binding constant. No attempt was maid to adjust the pH.

3. Results and Discussion

The effect of β-CDS on the spectral curve in the aqueous AO solution at 1×10^{-5} mole/l is shown in Fig. 1 at 24 °C. That of $A_{700}S$ (the suffix shows the degree of polymerisation) in the same dye solution is shown in Fig. 2. The polymer concentration is 1×10^{-6} to 1×10^{-7} unit mole of polymer per liter. Additional polymers caused enhancement about 470 nm and depression at 492 nm. The equilibrium between only two species (free and bound dye) is supported by the existence of isosbestic points. Same tendencies are also observed in the cases of α-CDS, $A_{350}S$ and APS. The spectral change may be taken as an evidence of binding. The binding constant is figured out from *Scatchard* plot ($r/[A]$ against $[A]$) and the n value is also obtained from *Klotz* plot ($1/r$ against $1/[A]$). Fig. 3 is the example of *Scatchard* plot of β-CDS-AO complex. These results are shown in Table 1. The molar changes in enthalpy (subsequently referred to as $\Delta H°$) and $\Delta S°$ for the binding of

AO on polymers can be calculated from the temperature dependence of binding constants. The values are also listed in Table 1. Both $\Delta H°$ and $\Delta S°$ are positive. $\Delta H°$ is very small in the cases of α-CDS and β-CDS and $\Delta S°$ of α-CDS and β-CDS is nearly equal. On the other hand,

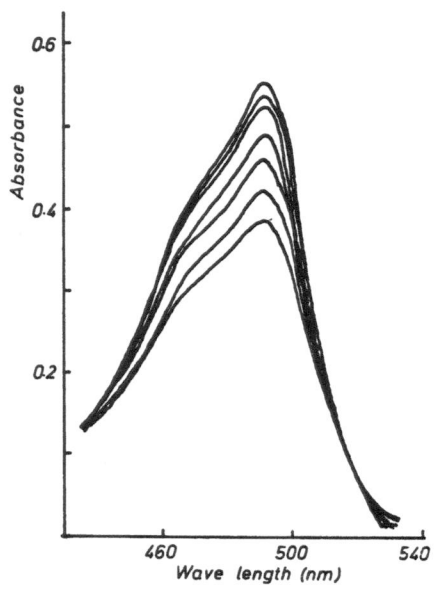

Fig. 2. Effect of the presence of the sodium salt of amylose sulphate ($A_{700}S$) on the spectral curve of Acridine Orange at 24 °C

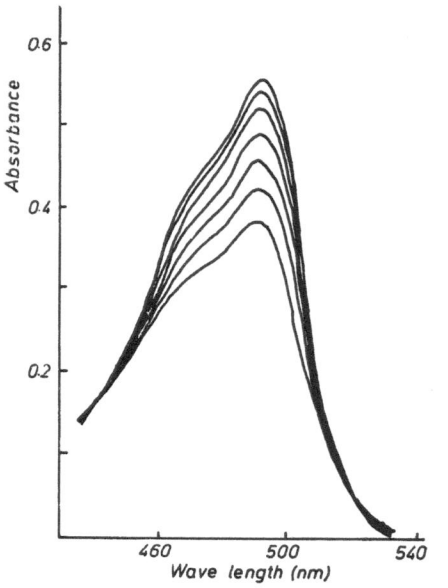

Fig. 1. Effect of the presence of the sodium salt of β-cyclodextrin sulphate on the spectral curve of Acridine Orange at 24 °C

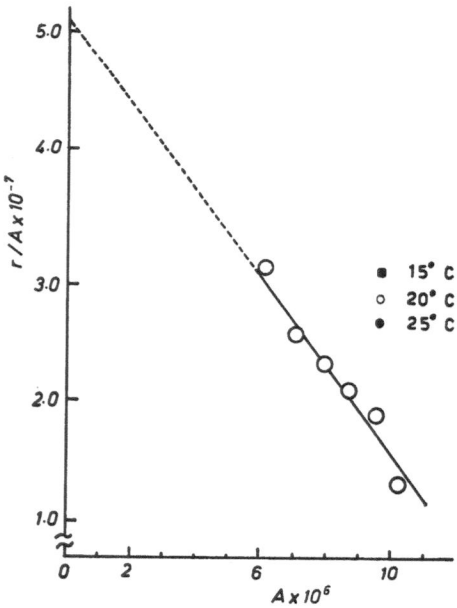

Fig. 3. Extrapolation to determine the binding constant for binding of Acridine Orange and the sodium salt of β-cyclodextrin sulphate

the values for $A_{350}S$ and $A_{700}S$ are higher by 5 e.u. and the value of APS is still higher by 5 e.u. These results are very interesting.

$\Delta H°$ of the dye binding may be defined as the heat of binding per mole of dye when a small quantity of dye is transfered from a very large volume of standard state dye solution to a very large quantity of standard state polymer phase. Consequently, the heat of binding may be regarded as the sum of the heat of formation of the various bonds existing between dye and polymers. So, the zero enthalpy of α- or β-CDS shows that the sum of the heat of various bonds formation in dye-polymer complex is zero. $\Delta H°$ contains the heat of the dye-polymer bond formation, hydration, dehydration, the destruction of iceberg etc. in dye binding.

The entropy change difference of 5 e.u. between CDS and AS corresponds to the structure difference between rigid polymer and relatively flexible polymer. That is, rigid polymer does not change the conformation on dye binding, while flexible polymer does. So, the value of 5 e.u. is attributed to the conformation changes on the dye binding.

The polymer structures of α- and β-CDS are very rigid in the molecular structure model (CPK model) (Fig. 4).

The entropy change difference of 5 e.u. between AS and APS is attributed to the network structure of APS. Since α- and β-CDS do not change the conformation, the positive $\Delta S°$ of α- or β-CDS is not attributable to the conformation

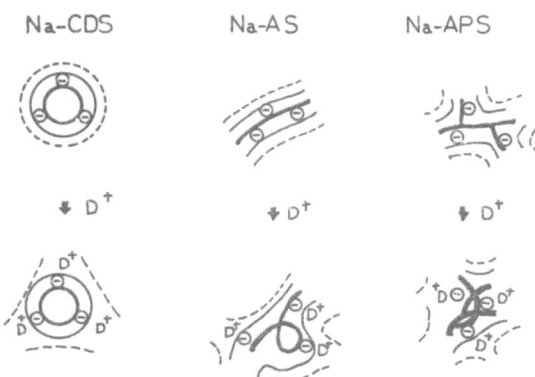

Fig. 5. The schematic representation of Acridine Orange — the sodium salt of polymer sulphates complex in aqueous solution. —— and – – – indicate hydration layer

change in the binding. The positive $\Delta S°$ must be caused by dehydration and by hydrophobic bonding along with the destruction of the water structure.

Although in our cases, the electrostatic interaction is a very important factor, the positive $\Delta S°$ of α-CDS or β-CDS must be principally caused by the hydrophobic bonding along with the destruction of the water structure about polymers and dye.

Fig. 5 is a schematic representation of AO-polymer complexes in aqueous solution. Above explanation is evident from the Fig. 5.

The n value is $1.6 \sim 2.0$ in Table 1. It is interesting that the degree of sulphation corresponds to n values. The dye is bound on the sulphate site of the polymers.

Summary

This paper described the interaction between Acridine Orange and the sodium polysaccharides sulphates; the sodium salts of α- and β-cyclodextrin sulphate, amylose or amylopectin sulphate. α- and β-cyclodextrin sulphate have a rigid polymer structure. But amylose sulphate and amylopectin sulphate have flexible polymer structures in the aqueous solutions. The binding data show that the positive entropy difference of 5 e.u. between cyclodextrin sulphate and amylose sulphate correspond to the conformation changes of the polymers on dye binding. And the positive entropy change of α- and β-cyclodextrin must be caused by the hydrophobic bonding with the destruction of the water structure about polymers.

Also, the saturated binding sites are corresponding to the degree of the sulphated sites. These results show that the dye binding is principally caused by the electrostatic interaction.

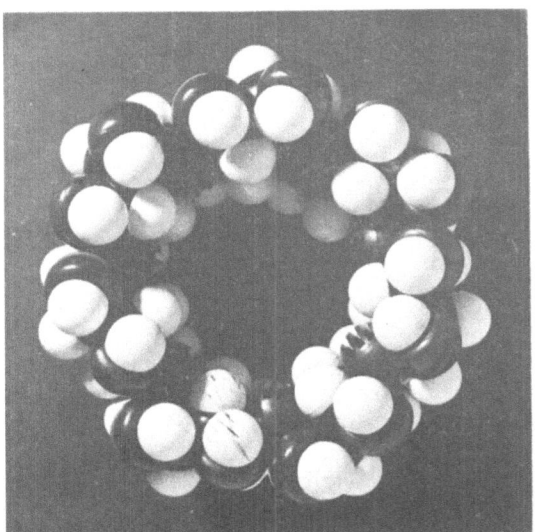

Fig. 4. The molecular structure model of β-cyclodextrin

Zusammenfassung

In dieser Mitteilung wird die Wechselwirkung zwischen Acridine Orange und Natrium-polysaccharidsulfaten (α- und β-Cyclodextrin, Amylose und Amylopektin-sulfat) beschrieben. α- und β-Cyclodextrin-sulfat sind steife Polymere, Amylosesulfat und Amylopektinsulfat sind dagegen flexible Polymere. Den Bindungswerten entspricht der Entropieunterschied 5 e.u. zwischen Cyclodextrin und Amylose der Konformationsänderung der Amylose.

Und die positive Entropieänderung der α- und β-Cyclodextrine muß durch die Zerstörung der Wasserstruktur mit der hydrophoben Bindung bewirkt werden. Auch die Zahl der Bindungsplätze korrespondiert mit der der Sulfatplätze. Dieses Resultat zeigt, daß die Hauptursache der Bindung die elektrostatische Wechselwirkung ist.

References

1) *Nishida, K., T. Akimoto* and *H. Uedaira*, Kolloid-Z. u. Z. Polymere **233**, 896 (1969).
2) *Nishida, K.*, Colloid & Polymer Sci. **252**, 107 (1974).
3) *Nishida, K., S. Yoshida, K. Ishige* and *H. Uedaira*, Kolloid-Z. u. Z. Polymere **238**, 423 (1970).
4) *Nishida, K., T. Akimoto, H. Shindate* and *H. Uedaira*, ibid., **243**, 97 (1971).
5) *Nishida, K.* and *H. Watanabe*, ibid., **244**, 346 (1971).
6) *Nishida, K.* and *H. Watanabe*, Colloid & Polymer Sci. **252**, 392 (1974).
7) *Nishida, K.* and *N. Takahashi*, unpublished data.
8) *Eisenberg, D.* and *W. Kauzmann*, The structure and properties of water (Oxford, 1969).
9) *Tsuda, K., K. Nishida, H. Watanabe* and *T. Hirata*, Kolloid-Z. u. Z. Polymere **240**, 827 (1970).
10) *Zanker, V.*, Physik. Chem. **199**, 225 (1952).
11) *Tanabe, R.*, Biochem. Z. **141**, 274 (1923). Starch: Chemistry and Technology (New York, 1965). p. 451, Ar. Biochem. & Biophys. **95**, 36 (1961).

Authors' address:

H. Watanabe
Toppan Printing Co. Ltd.
(Cyuoh-laboratory),
Taitoh-ku, Taitoh, 1—5, Tokyo, Japan
Dr. *K. Nishida*
Faculty of Technology, Tokyo University
of Agriculture and Technology
Koganei-shi, Tokyo, Japan

Progr. Colloid & Polymer Sci. **58**, 145–151 (1975)

Department of Textile Industrial Chemistry, Faculty of Textile Science and Technology,
Shinshu University, Ueda (Japan)

Effect of syndiotacticity on the aqueous poly (vinyl alcohol) gel

K. Ogasawara, T. Nakajima, K. Yamaura and *S. Matsuzawa*

With 8 figures in 10 details and 1 table

(Received May 14, 1974)

1. Introduction

An aqueous solution of poly(vinyl alcohol) (PVA) has been well known to form a thermally reversible gel upon standing at low temperatures. *Mandelkern* (1) has pointed out that gelation is attributed to the copolymeric nature as a result of polymerization of two or more monomers, stereoregularities, head to head polymerization, etc.. The factors favoring gelation are just those that do not favor solubility such as lowering the temperature, adding a nonsolvent, or using a poor solvent. The melting point of gel is the temperature at which the tie points are dissolved. By a thermodynamic treatment, *Eldridge* and *Ferry* (2) have derived the following relationship for thermally reversible gel;

$$\log C = (\text{constant})_1 - \Delta H / 2.303\, RT_m \qquad [1]$$

where C is the concentration of polymer in grams per liter, T_m, the absolute melting point of the gel and ΔH, the heat of cross-linking expressed in calories per mole of cross-links. *Maeda, Kawai* and *Kashiwagi* (3) have reported that ΔH for PVA derived from poly(vinyl acetate) (VAc-PVA) is estimated to be 8.8 kcal/mole which corresponds to one or two hydrogen bonds. In the previous paper (4), we have reported that the rates of gelation for aqueous solutions of PVAs derived from poly(vinyl formate) (VF-PVA) and poly(vinyl trifluoroacetate) (VTFA-PVA) were more rapid than that for VAc-PVA and that the values of ΔH are estimated to be 10–12 kcal/mole for VF-PVA and 20 kcal/mole for VTFA-PVA, each of which corresponds to two or four hydrogen bonds.

The appearance of X-ray crystalline diffraction pattern has been shown for a filament gel spun from a very concentrated solution by *Sone, Hirabayashi* and *Sakurada* (5). *Shibatani* (6) and *Takahashi* and *Hiramitsu* (7) reported that the cross-linking loci in PVA gel are syndiotactic sequences. We estimated ΔH according to *Ferry*'s eq. [1] for syndiotacticity-rich VTFA-PVA polymerized at lower temperature and obtained wide angle X-ray diffraction patterns of these gels. Further, a change of concentration or stress with the enhancement of syneresis was observed. From these results, we tried to elucidate the mechanism of gelation of PVA-water systems and the structure of junctions in gel networks.

2. Experimental

2.1. Materials

Vinyl trifluoroacetate was polymerized in *n*-pentane or ethylidene chloride at the temperatures of 0 °C and −78 °C. The polymerization was initiated by photodecomposition of 2,2'-azobis-(2,4-dimethyl valeronitrile) by light generated from a 400 W mercury lamp. Poly(vinyl trifluoroacetate) was converted to PVA by ammonolysis with diethylene triamine. Table 1

Table 1. Samples used

Sample	Polymerization conditions		DP[a]	s-(diad) %[b]
No.	Solvent	Temp.		
1	CH_3CHCl_2	0	2800	60.9
2	CH_3CHCl_2	−78	1300	66.2
3	n-C_5H_{12}	0	5400	58.4
4	n-C_5H_{12}	−78	840	64.3

[a] Degree of polymerization from intrinsic viscosity.
[b] Content of syndiotactic diads.

shows the methods of preparation and the pertinent physical properties of these PVAs. PVAs were acetylated in an acetic anhydride/pyridine mixture. Then these poly(vinyl acetate)s were fractionated into 10—13 parts by adding water to the acetone solution. These fractions were hydrolyzed before use.

2.2. The Melting Temperature of Gel

The apparent melting point of gel was determined according to *Ferry*'s procedure. Aqueous solutions were obtained by dissolving a known amount of PVA about 140 °C in a sealed test-tube (11 × 120 mm) with 2.5 ml water. Then the aqueous solution in a sealed test-tube was held for 24 hours at the temperature of 0 °C, 30 °C, 45 °C, or 60 °C. Finally it was placed upside down in a well stirred polyethylene glycol-bath at each temperature and the bath was warmed at the rate of about 12 °C per hour. The temperature at which the gel fell to the bottom of the test-tube was taken to be the melting point.

2.3. X-ray Diffraction of Gel

10% aqueous solution of PVA was injected into a glass fine capillary (inner diameter: 1 mm) and gelation was conducted at 0 °C or 60 °C for 40 hours in the capillary. The capillary was mounted on a Rigaku X-ray diffraction camera and wide angle X-ray diffraction pattern was obtained using CuKα point beam for 6 hours. The X-ray diffraction patterns of PVA films were also obtained by the same procedure for 4 hours.

2.4. Syneresis of Gel

The degree of syneresis was determined by the change of concentration of gel with rising temperature at the rate of about 12 °C per hour. The change of stress of rod-like gel (2.5 × 5 × 70 mm) under the constant length (5 cm) accompanying the rise of temperature was also measured using the tensilon (UTM-11).

3. Results

3.1. Melting Point of Gel

The gelation of aqueous solution of more syndiotacticity-rich PVA used in this experiment was a very rapid process as compared with that of commercial PVA and it occurred even at 100 °C. The melting point for 10% aqueous gel

of commercial PVA ($DP = 1700$) was about 20 °C, nevertheless for more syndiotacticity-rich PVA gel it was higher than that for former. For example, it is about 100 °C for the gel of PVA having the content of syndiotactic diads (s-(diad) %) of 60.9 (No. 1—8, $DP = 1500$). Fig. 1 shows the plots of syndiotactic diads against reciprocals of melting point ($1/T_m$) for gels chilled at 30 °C. As might be expected, the melting point rised with increasing concentration of PVA and syndiotactic diads. The degree of the melt-

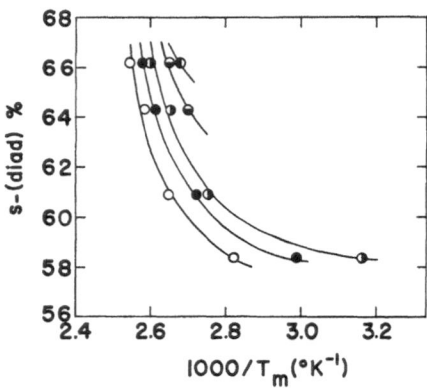

Fig. 1. Plots of s-(diad) % against $1/T_m$ for the gels chilled at 30 °C of PVAs of $DP = 1200—1400$. Concentration of PVA: (○) = 5 wt%; (●) = 3 wt%; (◑) = 2 wt%; (◒) = 1 wt%; (◐) = 0.75 wt%

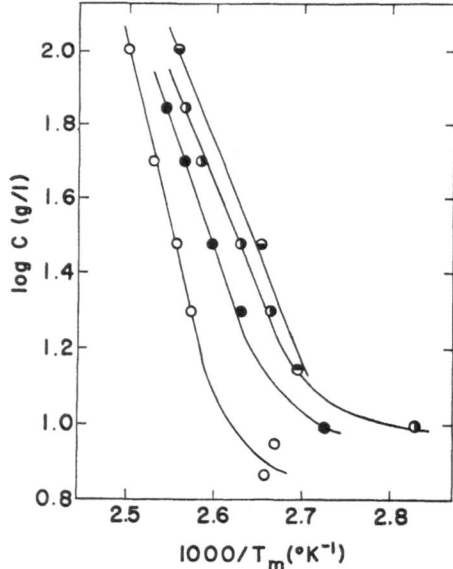

Fig. 2. Plots of log C against $1/T_m$ for gels chilled at 0 °C of PVA No. 4.
Degree of polymerization (DP): (○) = 2000; (●) = 1100; (◐) = 600; (◒) = 300

ing point depression of gel is remarkable with decreasing content of syndiotactic diads.

For gel of PVA having the value of *s*-(diad) % above 60.9, the relationship between the melting point of gel and the concentration of PVA satisfied eq. [1]. In Fig. 2, the plots of log *C* against $1/T_m$ are shown for gel of PVA No. 4. The ΔH estimated from the slopes of this plot was about 40 kcal/mole. It is very high as compared with 8.8 kcal/mole for atactic PVA gel. For atactic PVA gel, the ΔH has been assumed to correspond to the heat of fusion of a cross-link which consists of one or two hydrogen bonds (3) assuming an energy of hydrogen bond to be 5 kcal/mole. A cross-link for the gel of PVA No. 4 consists of about 8 hydrogen bonds.

Figs. 3a and 3b show the plots of log *C* against $1/T_m$ for the gels chilled at 0 °C and 30 °C of PVA having the value of *s*-(diad) % of 58.4 (No. 3). The relationships between the melting point of gel and the concentration of PVA deviate from a linear line for the gel chilled at 0 °C of PVA No. 3 and two straight lines intersected each other for that chilled at 30 °C in the region of lower concentration of PVA as we have reported in the previous paper (4). Fig. 4 shows the plots of ΔH against gelling temperature for different syndiotactic diads. The ΔH increased with the rise of gelling temperature for gel of PVA No. 3. It has been reported in the case of gelatin gel that the cross-links for gels chilled at 15 °C is more stable than that chilled at 0 °C, as the ΔH estimated from the melting points for the former is higher than that for the latter (2). It is also reported that the junction for the commercial PVA gel chilled at 25 °C is more stable than that chilled at 0 °C (3). However, the value of ΔH revealed minimum in the temperature region of 30 °C to 45 °C for the gels of PVAs having the value of *s*-(diad) % above 60.9. In Fig. 5, the plots of ΔH against syndiotactic diads are shown for different gelling temperatures. As might be expected, the ΔHs increased with increasing syndiotactic diads at each gelling temperature. This fact seems to support that the cross-links are formed by intermolecular hydrogen bonds between the syndiotactic sequences.

Ferry et al. (2) have also derived the relationship between the T_m and the molecular weight (*M*) of the polymer for thermally reversible gels as

$$\log M = (\text{constant})_2 - \Delta H/16\,RT_m \qquad [2]$$

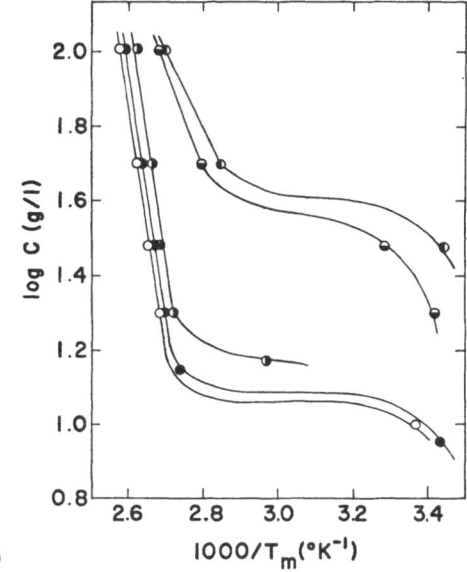

(a)

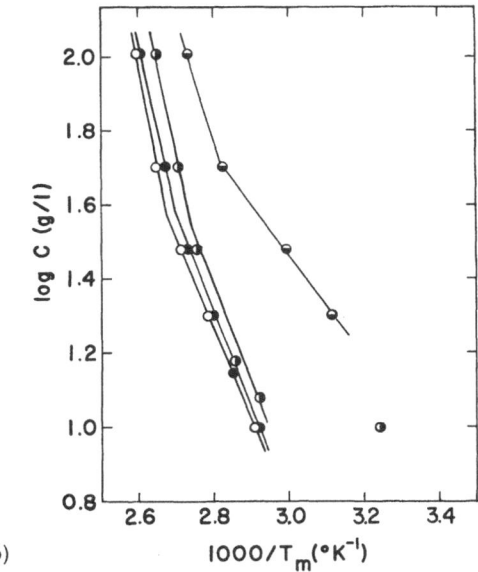

(b)

Fig. 3. Plots of log *C* against $1/T_m$ for
(a) gels chilled at 0 °C of PVA No. 3,
(b) gels chilled at 30 °C of PVA No. 3.
DP: (○) = 5000; (●) = 3100; (◑) = 2000;
(◒) = 1200; (◔) = 200
In a from left /; A — B, — ; B — C, /; C — D.

and have described that the ΔH estimated from eq. [2] do not closely agree with that from eq. [1].

Although we plotted the relationship between the *M* of PVA and the T_m of gel according to the eq. [2], the linear relation was only recognized in the region of higher concentration of PVA.

The ΔHs estimated from those slopes of straight lines are higher than that according to eq. [1]. Fig. 6 shows the plots of DP (degree of polymerization) against C_{crit} for different gelling temperature, where C_{crit} is critical gelling concentration of PVA expressed in weight percentage (wt-%). The C_{crit} for PVA gel chilled at 45 °C was lower than those at the other temperatures.

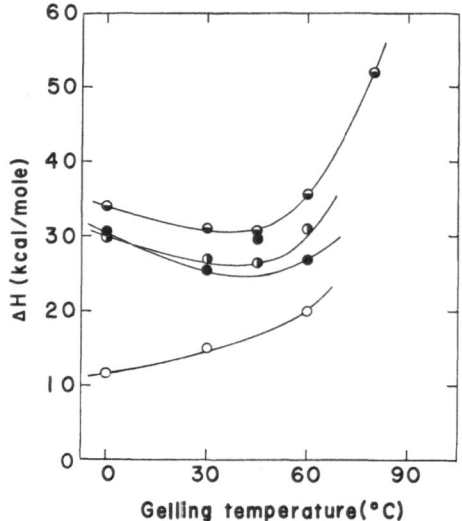

Fig. 4. Plots of ΔH against gelling temperature for PVA gels.
s-(diad) %: ($\circ$) = 58.4 (DP = 12200);
($\bullet$) = 60.9 (1400); (◑) = 64.3 (1300);
(◓) = 66.2 (1300)

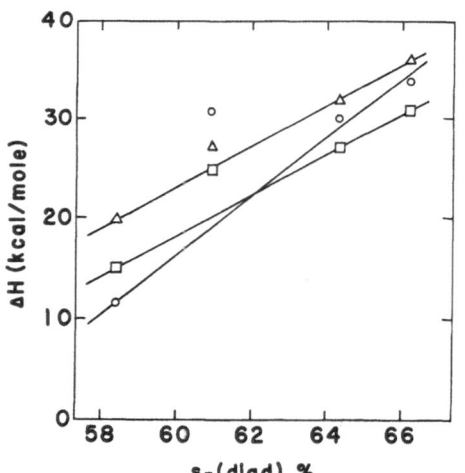

Fig. 5. Plots of ΔH against s-(diad) % for PVA gels.
Gelling temperature: (◠) = 0 °C; ($\square$) = 30 °C;
($\triangle$) = 60 °C

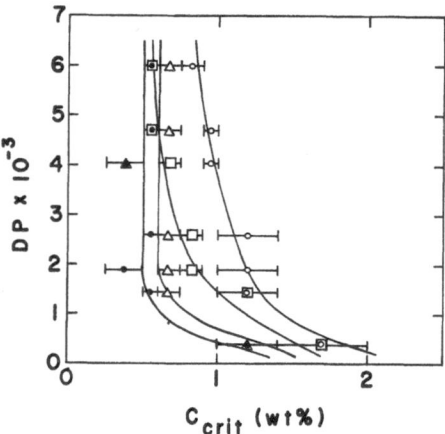

Fig. 6. Plots of DP against critical concentration for gels of PVA No. 1.
Gelling temperature: ($\circ$) = 0 °C; ($\square$) = 30 °C;
($\bullet$) = 45 °C; ($\triangle$) = 60 °C

3.2. X-Ray Diffraction of Gel

The polymer concentration in the gel was 10 wt-%. Fig. 7a shows a X-ray diffraction pattern for a film of PVA No. 4. Obviously, the PVA film is crystalline, and the d-spacings are in good agreement with those from the X-ray diffraction pattern of PVA fiber (8). The clearer X-ray diffraction pattern was observed for gel chilled at 60 °C than that chilled at 0 °C and for that of PVA having the value of s-(diad) % of 64.3 than that of 60.9. As is seen in Fig. 7b for gel of PVA No. 4, three diffraction rings were observed, the inner ring had moderate diffraction intensity, the middle ring very strong and the outer ring strong. These diffraction rings corresponding to $2\theta = 11°18'$, $19°39'$ and $23°28'$ or $d = 7.82$ Å, 4.51 Å and 3.78 Å, respectively. These values are nearly equal to that of PVA film and their $(h\,k\,l)$'s correspond to $(1\,0\,0)$, $(1\,0\,1)$ and $(2\,0\,0)$.

The gel network of PVA appears to be formed by microcrystallites and the amount of crystallites which exists in the gel increase with increasing content of syndiotacticity and the rise of gelling temperature.

3.3. Syneresis of Gel

On measuring the melting point of gel, gels showed syneresis, i.e., they tend to contract with separation of pure solvent on prolonged standing (9). It was remarkable in gel chilled at 0 °C.

In Fig. 8, the change of the concentration with the rise of temperature for 10% aqueous gel

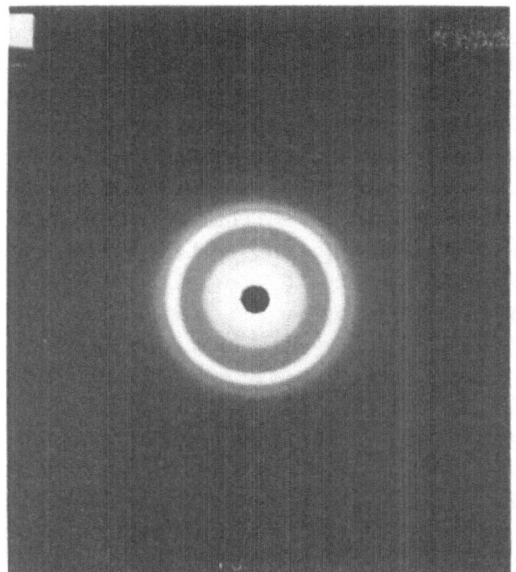 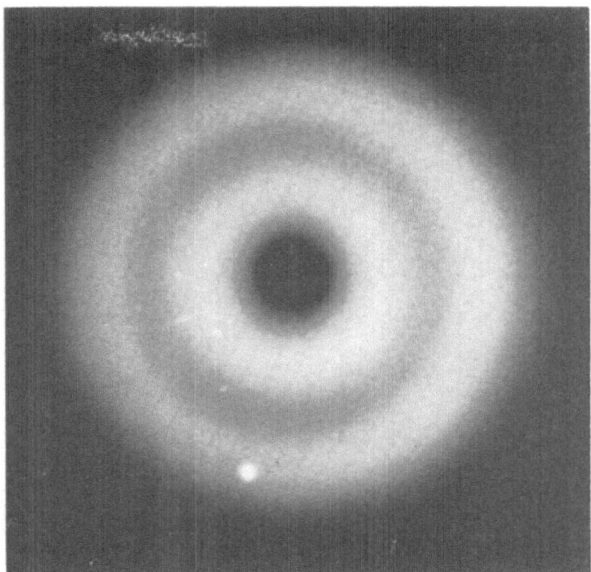

Fig. 7. X-ray diffraction patterns of   (b) 10% aqueous gel chilled at 60 °C of PVA No. 4
(a) film of PVA No. 4,

chilled at 0°C for 24 hours and the change of stress under the constant length (5 cm) accompanying the rise of temperature for 1% aqueous gel chilled at 0°C for 24 hours of PVA No. 1—2 are shown. Obviously the shrinkage of the volume enhances rapidly from about 30 °C accompanying the rise of temperature. As the melting point for 1% aqueous gel chilled at 0 °C of PVA No. 1—2 was less than 100 °C, the decrease of stress at the temperatures of higher than 75 °C would be ascribed to the initiation of the fusion of PVA gel.

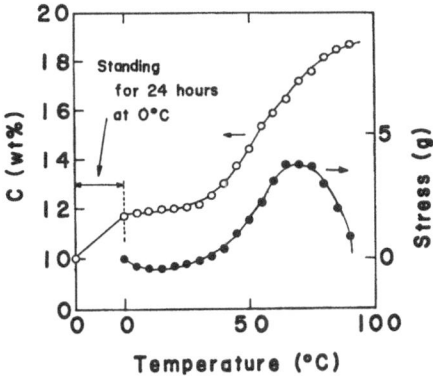

Fig. 8. The change of the concentration with rise of temperature for 10% aqueous gel chilled at 0 °C for 24 hours (○) and the change of stress under the constant length (50 mm) accompanying the rise of temperature for 1% aqueous gel ($2.5 \times 5 \times 50$ mm) chilled at 0 °C for 24 hours (●) of PVA No. 1 ($DP = 6000$)

4. Discussion

The results show that the gelling property is highly promoted by slight increase in syndiotacticity. Isotactic diad parts in PVA are inclined to form intramolecular hydrogen bonds (10, 11) and isotactic PVA is known to dissolve easily in cold water, while syndiotactic samples would not dissolve even in boiling water. Syndiotactic parts in PVA chains would be inclined to form intermolecular junctions in water. These facts seem to support the idea that the cross-linking loci in PVA solution are syndiotactic sequences.

The critical gelling concentration for PVA gel chilled at 45 °C was lower than at the other temperatures. It would be due to the difficulty of formation of junctions at higher temperature than 45 °C because of the vigorous thermal motion of molecules and at lower temperature than 45 °C because of formation of individual cluster or aggregate. For PVA solution chilled at about 45 °C, it would be easy to form gel network because of moderate thermal motion or tension of PVA molecules in aqueous solution.

As the concentrations corresponding to the part of B—C in Fig. 3 a is near the closest packing one of hydrodynamical equivalent sphere (12), gel networks were expected not to form in the lower concentration than that. Practically gel network, however, was formed in the lower concentration for the aqueous solution of PVA No. 3.

The PVAs used in this experiment would form intermolecular hydrogen bonds rather than intra-molecular ones because of higher syndiotacticity. We have assumed the types of intermolecular hydrogen bond formed in gel networks as follows;

(a) $\begin{array}{c} |-O-H:O-H:O-| \\ |\quad\quad|\quad\quad|\quad\quad| \\ |\quad\quad H\quad\quad H\quad\quad| \end{array}$

(b) $\begin{array}{c} |-O-H:O-H:O-H:O-| \\ |\quad\quad|\quad\quad|\quad\quad|\quad\quad| \\ |\quad\quad H\quad\quad H\quad\quad H\quad\quad| \end{array}$

(c) $\begin{array}{c} |-O-H:O-| \\ |\quad\quad|\quad\quad| \\ |\quad\quad H\quad\quad| \end{array}$

(a) or (b) type seems to be formed in the region of lower concentration of PVA below the C point shown in Fig. 3a and (c) type seems to be formed in the region of higher concentration beyond the B point (4). The ΔH estimated from the slope of line C—D is lower than that of the line A—B. From Fig. 3, the gel formed in region between B and C is expected to be formed by unstable hydrogen bonds in the transition region from (c) to (a) or (b), or from (a) or (b) to (c).

It has been reported (6) that the formation of nucleus is the rate-determining step in the gela-tion process, and the activation energy of the gelation is estimated as 13 kcal/mole using the *Arrhenius* equation.

2 (free cross-linking loci) $\xrightarrow{k}$ nucleus → cross-link

The ΔH increased with the rise of gelling tem-perature for gels of PVA No. 3 having the value of s-(diad) % of 58.4. This reason is considered as follows: the rate of gelation at the lower tem-perature is more rapid and the structures of these gels is more unstable than those chilled at higher temperature. However, the values of ΔH of gels chilled at 0 °C of PVA having the values of s-(diad) % above 60.9 were higher than those chilled at 30 °C or 45 °C. This reason is considered as follows. The gels chilled at 0 °C of the PVA having the value of s-(diad) % above 60.9 seem to have larger nuclei than those of the PVA having the value of s-(diad) % of 58.4 and these nuclei are inclined to grow and recrystallize rather than to melt accompanying the rise of temperature during the melting point measure-ment. Therefore, these gels showed syneresis. This fact is well known in terms of annealing effect. Gel is known to show syneresis accom-panying the annealing (4, 5). However, this

phenomena has been observed for a few of thermally reversible gels of synthetic polymer (13). *Sone* et al. (5) have indicated existence of crystallites in a filament gel spun from about 50% aqueous solution of atactic PVA, while we recognized the existence of fairly large crystal-lites even in the 10% aqueous gel of more syn-diotacticity-rich PVA. Therefore, the structure of network for aqueous gel of PVA must consist of crystallite formed by the syndiotactic sequence parts and the number and size of this crystallite seems to increase with increasing syndiotactic sequence. Therefore, the effect of annealing on the melting point for the gels chilled at lower temperature of the PVA having the value of s-(diad) % above 60.9 is remarkable. The syn-diotactic sequence length in the syndiotacticity-rich PVA seems to be sufficiently long to form the crystallites which serve as cross-links. The syneresis accompanying the rise of temperature seems to be a result of the breakdown of the unstable junction point and further growth of the crystallites. The detailed experiment of syn-eresis will be reported in future.

Summary

The mechanism of gelation of aqueous solution of syndiotacticity-rich poly(vinyl alcohol) (PVA) and the structure of junctions in gel were studied. Melting points (T_m) of the gels were measured according to *Ferry*'s procedure. Gelling ability was highly promoted by slight increase in syndiotacticity. The relation be-tween the concentration of PVA and T_m was not linear in the region of lower concentration of PVA. Three types of intermolecular hydrogen bond were assumed following the concentration of PVA. The gels chilled at lower temperature of PVA having the content of syndiotactic diads above 60.9% showed remarkable syneresis accompanying the rise of temperature. This phenomena seems to be a result of the breakdown of the unstable junctions and further growth of the crys-tallites. The X-ray diffraction pattern of the gel chilled at higher temperature was clearer than that at lower temperature and that pattern became progressively clearer with increasing content of syndiotacticity. These results lead to the conclusion that the structure of junc-tion in network of PVA-water gel must consist of crystallites formed by the syndiotactic sequence parts.

Zusammenfassung

Der Mechanismus der Gelierung einer wäßrigen Lö-sung von Polyvinylalkohol (PVA) und die Struktur des Bindungspunktes in dem Gel wurden studiert. Die Schmelzpunkte (T_m) des Gels wurden nach dem *Ferry*schen Verfahren gemessen.

Die Gelierungsmöglichkeit wurde durch die geringe Zunahme der Syndiotaktizität sehr verstärkt. Die Abhängigkeit zwischen der Konzentration des PVA und T_m war im Bereich der dünnen Konzentration nicht linear. Nach der Konzentration des PVA wurden die drei Arten intermolekularer Wasserstoffbrücken vorgeschlagen. Das bei tiefer Temperatur abgekühlte Gel vom PVA, dessen Syndiotaktizitätsgehalt in Diad über 60,9% beträgt, zeigt mit der Zunahme der Temperatur eine auffallende Syneresis. Dieses Phänomen scheint ein Ergebnis des Zusammenbruchs der unstabilen Bindung und des ziemlichen Wachstums des Kristallits zu sein. Die Röntgenbeugungsbilder des bei höherer Temperatur abgekühlten Gels waren klarer als diejenigen bei tieferer Temperatur. Das Bild wurde mit dem zunehmenden Gehalt von Syndiotaktizität allmählich klarer.

Es folgt daraus, daß die Struktur der Bindung im Netzwerk des aus wäßriger Lösung von PVA bestehenden Gels aus den aus syndiotaktischen Sequenzteilen gebildeten Kristalliten bestehen muß.

References

1) *Mandelkern, L.*, Crystallization of Polymer, p. 112 (New York, 1964).
2) *Eldridge, J. E.* and *J. D. Ferry*, J. Phys. Chem. 58, 992 (1954).
3) *Maeda, H., T. Kawai* and *R. Kashiwagi*, Kobunshi Kagaku 13, 193 (1956).
4) *Go, Y., S. Matsuzawa* and *K. Nakamura*, Kobunshi Kagaku 25, 62 (1968).
5) *Sone, Y., K. Hirabayashi* and *I. Sakurada*, Kobunshi Kagaku 10, 1 (1953).
6) *Shibatani, K.*, Polymer J. 1, 348 (1970).
7) *Takahashi, A.* and *S. Hiramitsu*, Polymer J. 6, 103 (1974).
8) *Mochizuki, T.*, Nippon Kagaku Zashi 81, 15 (1960).
9) *Morawets, H.*, Macromolecules in Solution, p. 77 (New York, 1966).
10) *Nagai, E., S. Kuribayashi, M. Shiraki* and *M. Ukida*, J. Polymer Sci. 35, 295 (1959).
11) *Shimanouchi, T.* and *M. Oka*, Preprint (Physics) of 12th Symposium on Polymer Chemistry, p. 323 (Nagoya, 1963).
12) *Go, Y., S. Matsuzawa, K. Nakamura, I. Saito, T. Hayashi* and *T. Ina*, Kobunshi Kagaku 24, 715 (1967).
13) *Pines, E.* and *W. Prins*, Macromolecules 6, 888 (1973).

Authors' address:

K. Ogasawara, T. Nakajima, K. Yamaura
and *S. Matsuzawa*
Department of Textile Industrial Chemistry
Faculty of Textile Science and Technology
Shinshu University
Ueda, Nagano 386 (Japan)

Progr. Colloid & Polymer Sci. **58**, 152–158 (1975)

Ingenieurbereich Angewandte Physik der Bayer AG, Leverkusen

Sedimentationskonstante und scheinbares Molekulargewicht von Gelatine in Abhängigkeit von der Rotordrehzahl bei Ultrazentrifugenversuchen

W. Scholtan und *H. Lange*

Mit 3 Abbildungen und 2 Tabellen

(Eingegangen am 17. Juli 1974)

1. Einleitung

Für die Molekulargewichtsabhängigkeit der Sedimentationskonstante und der Viskositätszahl von Gelatine wurden in einer vorangehenden Arbeit von *Scholtan, Lange, Rosenkranz* und *Moll* (1) folgende Beziehungen gefunden:

$$[\eta]_{40} = K_\eta \cdot M^a = 7{,}75 \cdot 10^{-2} \, \text{ml/g} \cdot M_w^{0{,}51}, \qquad [1]$$

$$s_{40,w}^{0,w} = K_s \cdot M^u = 0{,}61 \, \text{Svedberg} \cdot M_w^{0{,}17}. \qquad [2]$$

Bei undurchspülten und schwach durchspülten Knäuelmolekülen bestehen zwischen den Exponenten a und u bzw. dem Exponenten b der Diffusionskonstante - Molekulargewichts - Beziehung $D = K_D \cdot M^b$ nach *Volmert* (2) die Gleichungen:

$$u = \frac{2-a}{3} \quad \text{und} \quad b = -\frac{1+a}{3}. \qquad [3], [4]$$

Bei der untersuchten Gelatine ergibt sich mit $a = 0{,}51$ für den Exponenten der s-M-Beziehung $u = 0{,}50$. Im Gegensatz dazu wird experimentell nur der Wert $u_{\exp} = 0{,}17$ gefunden. Zur Erklärung dieser Diskrepanz haben wir in unserer vorangehenden Arbeit (1) angenommen, daß die in der Lösung als Knäuel vorliegenden Gelatinemoleküle bei der Sedimentation in der Ultrazentrifuge stärker durchspült werden als bei der Viskositätsmessung. Wahrscheinlich ist mit der stärkeren Durchspülung der Moleküle auch eine Formänderung in Richtung auf abgeflachte Rotationsellipsoide verbunden. Für stark durchspülte Knäuel läßt sich die Abhängigkeit der Sedimentationskonstante vom Molekulargewicht nach *Volmert* (2) folgendermaßen beschreiben:

$$s^0 \sim \frac{M^u}{K + M^u}. \qquad [5]$$

Hieraus ergibt sich, daß die Steigung der s-M-Kurve im doppelt logarithmischen Maßstab um so geringer ist, je kleiner die Konstante K ist und je größer M ist. Wenn K gegenüber M^u vernachlässigt werden kann, wird die Sedimentationskonstante sogar vom Molekulargewicht unabhängig.

In der vorliegenden Arbeit werden zusätzliche Versuchsergebnisse, die für eine stärkere Durchspülung der Gelatinemoleküle bei der Sedimentation sprechen, mitgeteilt. Zunächst haben wir die Abhängigkeit der Sedimentationskonstanten von der Rotordrehzahl und der Konzentration für eine hoch- und eine niedermolekulare Gelatine untersucht. Dann wurde die Diffusionskonstante D^0 einer Reihe von Gelatinefraktionen bestimmt und die Molekulargewichtsabhängigkeit von D^0 ermittelt. Aus dem Exponenten der D-M-Beziehung ergeben sich nämlich zusätzlich Rückschlüsse auf das Durchspülungsverhalten der Gelatinemoleküle.

Aus den Diffusions- und Sedimentationskonstanten wurden ferner nach der *Svedberg*schen Formel (Gl. [6]) die Molekulargewichte $M_{s,D}$ berechnet und mit den nach der Methode von *Archibald* und der Sedimentationsgleichgewichtsmethode bestimmten Molekulargewichten M_w verglichen.

2. Präparate und Untersuchungsmethoden

Die Untersuchungen wurden an einer durch alkalischen Aufschluß von Rinderknochen hergestellten Gelatine I (1. Abzug) und zusätzlich an einigen daraus gewonnenen Fraktionen durchgeführt. Der isoionische Punkt (pI) der Präparate lag bei 4,9. Als Lösungsmittel wurde eine wässrige Pufferlösung (0,15 m K-Acetat, pH 5,0) verwendet. Die physikalischen Daten der Gelatinepräparate (Viskositätszahl, Sedimentationskonstante, Molekulargewicht) und die Einzelheiten der Versuchsmethoden sind in der vorangehenden Arbeit (1) aufgeführt. Ferner wurden zum Vergleich die entsprechenden Untersuchungen an einem Polystyrol mit dem Molekulargewicht $M_w = 450000$ in Methyläthylketon durchgeführt.

Die Sedimentationskoeffizienten s und die Diffusionskoeffizienten D der Präparate wurden mit der analytischen Spinco-Ultrazentrifuge[1]) unter Verwendung einer 12 mm Überschichtungszelle bestimmt. Die Temperatur betrug bei der Gelatine 40 °C und beim Polystyrol 20 °C. Die Drehzahl des Rotors betrug bei den Diffusionsversuchen 2000 min^{-1} und bei den Sedimentationsversuchen 10000 bis 60000 min^{-1}. Das Diffusions- und Sedimentationsverhalten der Präparate wurde durch *Philpot-Svensson*-Aufnahmen registriert. Aus diesen Aufnahmen wurden die Diffusionskoeffizienten D_{40} der Gelatine-Präparate nach der Flächenmethode bestimmt und auf Wasser von 40 °C umgerechnet

$(D_{40,w} = 1,033 \cdot D_{40})$.

Aus jeweils drei bei den Konzentrationen $c = 3$; 5 und 10 g/l durchgeführten Versuchen wurde durch Extrapolation auf $c = 0$ die Diffusionskonstante $D_{40,w}^0$ ermittelt.

Die Berechnung der Sedimentationskoeffizienten aus der zeitlichen Wanderung des Kurvenmaximums der *Philpot-Svensson*-Aufnahmen erwies sich als zu ungenau. Die Gelatinemoleküle sedimentierten nämlich vor allem bei kleineren Drehzahlen selbst bei langer Versuchsdauer (8 h) nur wenig, und das Kurvenmaximum konnte wegen der Diffusion der Gelatinemoleküle nicht sehr genau bestimmt werden. Deshalb haben wir die *Philpot-Svensson*-Aufnahmen 10fach vergrößert, eine Parallele zur Basislinie in halber Höhe des Kurvenmaximums gezogen, die Lage des Halbierungspunktes der Halbwertsbreite des Sedimentationsdiagramms bestimmt und daraus den Abstand r dieses Punktes von der Rotationsachse berechnet. Aus der zeitlichen Veränderung der Lage dieses Punktes wurde mit Hilfe der Gleichung $s = 2,303 \, \mathrm{dlg}\, r/\omega^2 \, \mathrm{d}t$ der Sedimentationskoeffizient berechnet. Dazu wurde der Logarithmus von r (cm) gegen die Versuchszeit t auf-

getragen. (Der aus der Steigung der erhaltenen Geraden berechnete Sedimentationskoeffizient s_{40} liegt zwischen dem Sedimentationskoeffizient s_{40}^h der häufigsten Molekülart und dem Gewichtsmittel des Sedimentationskoeffizienten s_{40}^w). Die s_{40}-Werte wurden auf Wasser von 40 °C umgerechnet ($s_{40,w} = 1,043 \, s_{40}$).

Aus jeweils vier bei den Konzentrationen $c = 2,5$; 5; 7,5 und 10 g/l durchgeführten Versuchen wurde durch Extrapolation auf $c = 0$ die Sedimentationskonstante $s_{40,w}^0$ berechnet.

Aus den Sedimentations- und Diffusionskonstanten ($s_{40,w}^0$ bzw. $s_{40,w}^{0,w}$ und $D_{40,w}^0$) wurde das Molekulargewicht $M_{s,D}$ der Gelatine-Präparate nach der *Svedberg*schen Formel

$$M_{s,D} = \frac{RT}{1 - V \varrho_0} \cdot \frac{s}{D} \qquad [6]$$

bestimmt. Hierbei bedeutet:

R = Gaskonstante,
T = absolute Temperatur,
V = partielles spezifisches Volumen (für Gelatine in 0,15 m K-Acetatpuffer, pH 5,0 bei 40 °C beträgt $V = 0,695$ ml/g),
ϱ_0 = Dichte des Lösungsmittels (für 0,15 m K-Acetatpuffer, pH 5,0 bei 40 °C beträgt $\varrho_0 = 0,9974$ g/ml),
s = Sedimentationskonstante,
D = Diffusionskonstante.

3. Versuchsergebnisse und Diskussion

Die experimentell bestimmten Sedimentationskoeffizienten s_{40} der Gelatine I und der Gelatinefraktion b sind in Tabelle 1 für verschiedene Konzentrationen und Drehzahlen aufgeführt. In Abb. 1 ist die Konzentrationsabhängigkeit von $1/s_{40}$ mit der Drehzahl als Parameter graphisch dargestellt. Zusätzlich ist in Abb. 1 der reziproke Sedimentationskoeffizient $1/s_{20}$ einer Polystyrolprobe in Methyläthylketon als Funktion der Konzentration eingezeichnet.

Die Konzentrationsabhängigkeit von $1/s_{20}$ läßt sich bei Polystyrol sowohl bei der Drehzahl $N = 60000$ min^{-1} als auch bei $N = 10000$ min^{-1} durch dieselbe Gerade wiedergeben. Dies gilt auch für $1/s_{40}$ bei der Gelatine I für $N = 40000$ bis 60000 min^{-1}. Ebenso läßt sich die Konzentrationsabhängigkeit der Gelatinefraktionen bei $N = 60000$ min^{-1} durch Geraden darstellen (1). Für alle Geraden gilt die Gl. [7] (vgl. (3)):

$$1/s = 1/s_0 \cdot (1 + k_s c) = 1/s_0 \cdot (1 + \gamma [\eta] c) \cdot \qquad [7]$$

---

[1]) *Modell E, Beckman, München.*

Tabelle 1. Sedimentationskoeffizienten s_{40}, s_{40}^0 und $s_{40,w}^0$ (*Svedberg*) der Gelatine I und der Gelatinefraktion b in 0,15 m K-Acetatpuffer, pH 5,0 bei 40 °C in Abhängigkeit von Konzentration c und Drehzahl N

c [g/l] N [min⁻¹]	s_{40}											s_{40}^0 $c=0$	$s_{40,w}^0$ $c=0$
	2	2,5	3,5	4	5	5,5	7	7,5	8,0	8,5	10		
Gelatine I:													
60 000	—	4,40	—	—	4,10 4,10	—	—	3,85		—	3,60	4,75	5,0
40 000	4,65	—	—	4,15	3,85	—	3,75	—		—	3,55 3,70	4,75	5,0
30 000	5,7	4,5	—	4,0	—	4,2	—	—		3,8	3,4	5,75	6,0
20 000	7,0 7,3	—	7,1	7,0	—	5,3	4,8	—		4,1	3,8	7,1	7,4
10 000	10,1	—	—	9,4	10,5	9,6	6,6	—		4,9	4,1	10	10,4
Gelatinefraktion b:													
60 000			3,3	—					3,1		3,0	3,7	3,8
30 000			—	3,4					—		—	3,8ᵃ⁾	≈ 4,0
20 000			—	3,9					—		—	3,9ᵇ⁾	≈ 4,1

ᵃ) Konzentrationsabhängigkeit wie bei $N = 60\,000$ min⁻¹ angenommen.
ᵇ) Keine Konzentrationsabhängigkeit vorausgesetzt.

In dieser Gleichung stellt k_s eine Konstante dar, die die Steigung der Geraden bestimmt. $[\eta]$ bedeutet die Viskositätszahl und γ eine Größe, die die Durchspülbarkeit der Moleküle charakterisiert. In mehreren Arbeiten wurde gefunden, daß $\gamma = k_s/[\eta]$ für eine große Zahl von Polymeren unabhängig vom Molekulargewicht etwa 1,7 beträgt (4, 5, 6). Dies gilt speziell für Polymere mit flexiblen Ketten, deren Moleküle in guten Lösungsmitteln als undurchspülte Knäuel vorliegen. Mit zunehmender Durchspülbarkeit der Makromoleküle nimmt γ ab und beträgt bei völlig durchspülten Molekülen Null.

Aus Abb. 1 ergibt sich für γ beim Polystyrol ein Wert von 1,67 und für die Gelatine I bei $N = 40\,000$ und $60\,000$ min⁻¹ ein Wert von 0,67. Daraus folgt, daß die Polystyrolmoleküle in Methyläthylketon als undurchspülte Knäuel vorliegen, die Gelatinemoleküle in wässriger Lösung bei $N = 40\,000$ und $60\,000$ min⁻¹ jedoch teilweise durchspült werden.

Bei kleineren Drehzahlen kann die Konzentrationsabhängigkeit von $1/s_{40}$ nicht mehr durch eine Gerade dargestellt und durch Gl. [7] beschrieben werden. Die Abb. 1 zeigt, daß sich für Drehzahlen von $N = 30\,000$, $20\,000$ und $10\,000$

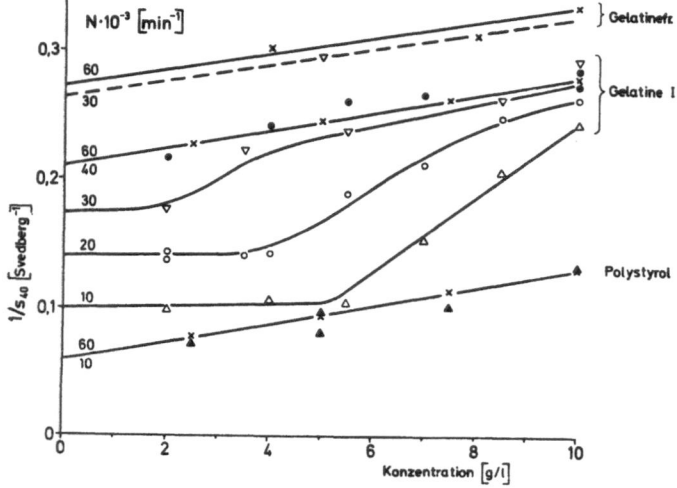

Abb. 1. Abhängigkeit des reziproken Sedimentationskoeffizienten $1/s_{40}$ von Konzentration und Drehzahl N für die Gelatine I und die Gelatinefraktion b in 0,15 m K-Acetatpuffer, pH 5,0 bei 40 °C sowie für ein Polystyrol ($M_w = 450\,000$) in Methyläthylketon bei 20 °C.

× $N = 60\,000$; ● $40\,000$; ▽ $30\,000$; ○ $20\,000$; △, ▲ $10\,000$ min⁻¹

min^{-1} Kurven ergeben, die zwar im Konzentrationsbereich $c = 5$ bis 10 g/l annähernd linear ansteigen, bei kleineren Konzentrationen aber horizontal verlaufen. Der Sedimentationskoeffizient ist hier von der Konzentration unabhängig.

Eine völlig befriedigende Erklärung für diesen anomalen Verlauf der Konzentrationsabhängigkeit von $1/s_{40}$ vermögen wir zunächst nicht anzugeben. Die zunehmende Steilheit der Kurven (Zunahme des γ-Wertes) im Konzentrationsbereich $c = 5$ bis 10 g/l bei Abnahme der Drehzahl deutet u. E. darauf hin, daß die Durchspülbarkeit der Gelatinemoleküle mit kleiner werdender Drehzahl abnimmt.

Die Abb. 1 und Tabelle 1 zeigen ferner, daß die Sedimentationskonstante s_{40}^0 der Gelatine von der Drehzahl und somit von der Sedimentationsgeschwindigkeit der Moleküle abhängig ist. Von Interesse ist die durch Extrapolation bestimmte Sedimentationskonstante bei der Drehzahl $N = 0$, da hier die Gelatinemoleküle wahrscheinlich nicht mehr durchspült werden. Eine Theorie über die Abhängigkeit der Sedimentationskonstante von der Drehzahl liegt allerdings nicht vor, so daß diese Extrapolation nur mit einer gewissen Willkür durchführbar ist. Wir haben dazu in Abb. 2 die reziproke Sedimentationskonstante gegen die Drehzahl N aufgetragen und die Extrapolation auf $N = 0$ unter der Annahme durchgeführt, daß bei Drehzahlen $N < 10000$ min^{-1} die Sedimentationskonstante von der Drehzahl unabhängig ist. Bei diesen Drehzahlen sedimentieren die Makromoleküle nämlich so langsam, daß sie wahrscheinlich vom Lösungsmittel nicht durchspült werden. (Eine experimentelle Bestimmung von $s_{40,w}^0$ bei $N < 10000$ min^{-1} ist wegen der großen Fehlerbreite bei diesen Versuchsbedingungen nicht mehr möglich.)

Aus der Abb. 2 ist ersichtlich, daß die Sedimentationskonstante mit abnehmender Drehzahl ansteigt. Dieser Effekt ist besonders stark bei den Molekülen der hochmolekularen Gelatine I, in geringerem Maße jedoch auch bei der niedermolekularen Gelatinefraktion b zu beobachten. Zur Erklärung dieser Erscheinung nehmen wir an, daß eine Drehzahl- und Molekulargewichtsabhängigkeit der Durchspülbarkeit der Gelatinemoleküle vorliegt.

Bei kleinem Molekulargewicht (Fraktion b) besitzt ein Gelatinemolekül in Lösung eine relativ gestrecktere Knäuelform als bei großem Molekulargewicht (Gelatine I). Die Durchspülbarkeit ist deshalb bei kleinem Molekulargewicht von vornherein groß. Drehzahländerungen können also in diesem Fall die Durchspülbarkeit und damit die Reibungsverhältnisse im gelösten Molekül nur wenig beeinflussen. Die Sedimentationskonstante wird bei Erniedrigung der Drehzahl nur geringfügig erhöht (vgl. Abb. 2).

Bei großem Molekulargewicht (Gelatine I) ist dagegen ein Molekül in Lösung erheblich stärker geknäuelt und damit bei geringen Drehzahlen vermutlich weniger durchspült als bei kleinem Molekulargewicht. Eine Drehzahlerhöhung kann hier die Durchspülbarkeit und damit die Reibung im Molekülknäuel stark vergrößern. Die Sedimentationskonstante nimmt dann mit steigender Drehzahl stark ab (vgl. Abb. 2). Daß dies nicht auch bei der relativ hochmolekularen Polystyrolprobe auftritt, liegt vermutlich an den andersartigen Wechselwirkungen zwischen Makromolekül und Lösungsmittel.

Die für $c = 0$ erhaltenen Sedimentations- und Diffusionskonstanten sind für verschiedene Gelatinepräparate in Tabelle 2 zusammengestellt. Aus diesen Werten wurden nach der *Svedberg-*

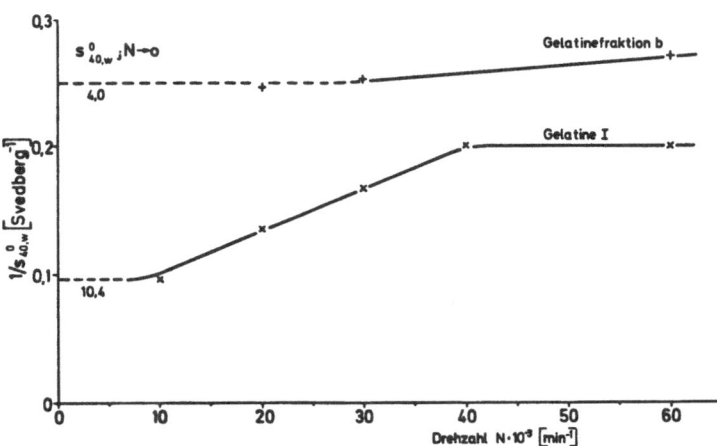

Abb. 2. Abhängigkeit des Sedimentationskoeffizienten $s_{40,w}^0$ von der Drehzahl N für die Gelatine I und die Gelatinefraktion b in 0,15 m K-Acetatpuffer, pH 5,0 bei 40 °C

Tabelle 2. Sedimentationskonstante, Diffusionskonstante und Molekulargewicht von verschiedenen Gelatinepräparaten in 0,15 m K-Acetatpuffer, pH 5,0 bei 40 °C und von Polystyrol in Methyläthylketon bei 20 °C

Probe	$s_{40,w}^{0,w}$ [*Svedberg*] $N = 60\,000$ min^{-1}	$s_{40,w}^{0}$ [*Svedberg*] $N = 0$	$D_{40,w}^{0} \cdot 10^7$ [cm²/sec]	$M_{s,D} \cdot 10^{-3}$ $N = 60\,000$ min^{-1}	$M_{s,D} \cdot 10^{-3}$ $N = 0$	$M_w \cdot 10^{-3}$
Gelatine I	5,4	10,4	3,5	132	253	220 [a] / 250 [b]
Fr. 13	5,78	—	2,6	190	—	318 [a]
Fr. d	4,90	—	4,7	89	—	98 [b]
Fr. c	—	—	5,2	—	—	84,5 [b]
Fr. b	3,85	4,0	7,5	44	45,7	31,9 [a]
Polystyrol	—	16,7 (s_{20}^{0})	3,4 (D_{20}^{0}) [c]	460	—	450

[a] Ultrazentrifuge. [b] Lichtstreuung. [c] Vergleiche (7).

schen Formel die Molekulargewichte $M_{s,D}$ berechnet und den aus Lichtstreuungs- und Sedimentationsversuchen (Sedimentationsgleichgewicht bzw. Methode *Archibald*) ermittelten Molekulargewichten M_w gegenübergestellt (vgl. Spalte 5, 6, 7 der Tabelle 2).

Die $M_{s,D}$- und M_w-Werte stimmen nur für Gelatinefraktionen $M < 100\,000$ annähernd überein. Bei der hochmolekularen Gelatine I weichen diese Werte jedoch sehr stark voneinander ab. Für M_w erhält man 220 000 bzw. 250 000, dagegen für $M_{s,D}$ nur 132 000, wenn man die Sedimentationskonstante $s_{40,w}^{0} = 5,4$ Svedberg bei $N = 60\,000$ min^{-1} zugrundelegt. Erst wenn man bei der Berechnung von $M_{s,D}$ die auf die Drehzahl 0 extrapolierte Sedimentationskonstante $s_{40,w}^{0} = 10,4$ Svedberg verwendet, erhält man für das Molekulargewicht $M_{s,D} = 253\,000$. Dieses stimmt mit dem M_w-Wert relativ gut überein.

Die beobachtete Diskrepanz zwischen den $M_{s,D}$- und den M_w-Werten läßt sich ebenfalls durch die unterschiedliche Durchspülbarkeit der Gelatinemoleküle erklären. Für die Gültigkeit der *Svedberg*schen Formel wird vorausgesetzt, daß der Reibungswiderstand der Makromoleküle bei der Sedimentation und der Diffusion (und damit auch ihre Durchspülbarkeit) gleich ist. Dies gilt z. B. dann, wenn die Gelatinemoleküle als undurchspülte Knäuel vorliegen, wie dies u. E. bei der Diffusion und bei der bei sehr kleinen Drehzahlen erfolgenden Sedimentation sowie bei der Viskositätsmessung der Fall ist. Bei steigenden Drehzahlen und zunehmender Sedimentationsgeschwindigkeit nimmt die Durchspülbarkeit der Gelatinemoleküle dagegen zu und ihre Sedimentationskonstante ab. Die *Svedberg*sche Formel ist dann nicht mehr anwendbar, da ihre

Voraussetzungen nicht erfüllt sind. Die Gelatinemoleküle besitzen offenbar bei Sedimentationsversuchen mit hohen Drehzahlen einen anderen Reibungswiderstand als bei Diffusionsversuchen, verursacht durch eine veränderte Durchspülbarkeit und/oder durch eine Formänderung der Molekülknäuel. Unter diesen Umständen erhält man fehlerhafte Molekulargewichte $M_{s,D}$.

In Abb. 3 ist die Molekulargewichtsabhängigkeit der Diffusionskonstante und der Sedimentationskonstante, die wir bei früheren Untersuchungen (1) an Gelatinefraktionen ermittelt haben, im doppelt logarithmischen Netz dargestellt. Ferner sind die bei verschiedenen Drehzahlen bestimmten Sedimentationskonstanten $s_{40,w}^{0}$ für die Gelatine I und die Gelatinefraktion b eingezeichnet. Die Molekulargewichtsabhängigkeit der Sedimentations- und Diffusionskonstanten läßt sich im doppelt logarithmischen Netz durch Geraden darstellen. Ihre Gleichungen lauten:

$$s_{40,w}^{0,w} = 0,65 \text{ Svedberg} \cdot M_w^{0,17}$$
$$(\text{für } N = 60\,000 \text{ min}^{-1}), \qquad [2]$$

$$s_{40,w}^{0} = 2,2 \cdot 10^{-2} \text{ Svedberg} \cdot M_w^{0,50}$$
$$(\text{für } N = 0), \qquad [2a]$$

$$D_{40,w}^{0} = 1,5 \cdot 10^{-4} \text{ cm}^2/\text{sec} \cdot M_w^{-0,50}. \qquad [8]$$

Der Gl. [2a] liegen allerdings nur 2 Meßpunkte und die Annahme zugrunde, daß sich die $s_{40,w}^{0}$-M-Beziehung im doppelt logarithmischen Maßstab durch eine Gerade darstellen läßt. Sie ist streng genommen nur als Näherung für den Molekulargewichtsbereich von ca. $30 \cdot 10^3$ bis $300 \cdot 10^3$ anzusehen.

Die Abhängigkeit der Sedimentationskonstante von der Drehzahl des Rotors bedingt, daß man

Abb. 3. Beziehungen zwischen der Sedimentationskonstante $s_{40,w}^{0,w}$ bzw. $s_{40,w}^{0}$, der Diffusionskonstante $D_{40,w}^{0}$ und dem Molekulargewicht M_w für Gelatinefraktionen (●) und für Gelatine I (+) in 0,15 m K-Acetatpuffer, pH 5,0 für verschiedene Drehzahlen bei 40 °C. (M_w bestimmt durch Sedimentationsversuche (●) und Streulichtmessungen (*) an Gelatinefraktionen nach *Scholtan et al.* (1).)

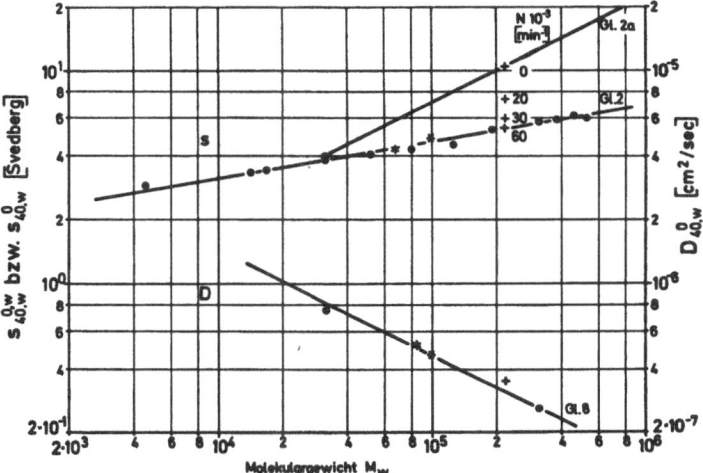

für unterschiedliche Drehzahlen auch unterschiedliche *s-M*-Beziehungen mit verschiedenen Exponenten *u* erhält (vgl. Gl. [2] und [2a]). Bei $N = 60\,000$ min^{-1} beträgt $u = 0,17$. Mit abnehmender Drehzahl steigt *u* an und erreicht schließlich für $N = 0$ den Wert 0,5, wie er auch aus der Molekulargewichtsabhängigkeit der Viskositätszahl nach Gl. [3] für undurchspülte Knäuel berechnet wird. Daraus folgern wir, daß die gelösten Gelatinemoleküle sowohl bei den Viskositäts- als auch bei den Sedimentationsmessungen mit sehr kleinen Drehzahlen als undurchspülte Knäuel vorliegen.

Eine entsprechende Schlußfolgerung ergibt sich aus der Größe des Exponenten *b* der *D-M*-Beziehung (Gl. [8]). Mit Gl. [4] erhält man nämlich aus den Exponenten *a* der [η]-*M*-Beziehung $b = -0,5$. Dieser Wert stimmt mit dem experimentell beobachteten Wert überein. Die Gelatinemoleküle müssen also auch bei den Diffusionsversuchen als undurchspülte Knäuel vorliegen.

Diese Betrachtung der Exponenten der *s-M*-, [η]-*M*- und *D-M*-Beziehungen läßt u. E. nur den Schluß zu, daß die Drehzahlabhängigkeit der Sedimentationskonstante der hochmolekularen Gelatine I durch eine mit steigender Drehzahl zunehmenden Durchspülbarkeit der Molekülknäuel verursacht wird.

Orientierende Sedimentationsversuche an einer hochmolekularen Polyvinylpyrrolidon-Probe in Wasser deuten ebenfalls darauf hin, daß die Sedimentationskonstante von der Drehzahl des Rotors abhängt. Bei einigen Stoffen wird es daher notwendig sein, zur Molekulargewichtsberechnung nach *Svedberg* die Drehzahlabhängigkeit der Sedimentationskonstanten zu berücksichtigen.

Wir danken unseren Mitarbeitern *U. Bertram, H. Hartung* und *K. Langfeld* für die sorgfältige Durchführung der Experimente.

Zusammenfassung

Die Abhängigkeit der Sedimentationskonstanten von der Drehzahl des Rotors bei Ultrazentrifugenversuchen wurde für zwei Gelatinepräparate in 0,15 m K-Acetatpuffer, pH 5,0 und für ein Polystyrolpräparat in Methyläthylketon bei verschiedenen Konzentrationen untersucht.

Beim Polystyrol ist die Sedimentationskonstante s^0 von der Drehzahl unabhängig. Bei der Gelatine nimmt s^0 dagegen mit abnehmender Drehzahl zu, und zwar um so mehr, je größer das Molekulargewicht der Gelatine ist.

Zwischen der reziproken Sedimentationskonstante und der Konzentration der Gelatine besteht nur bei großen Drehzahlen ($N > 40\,000$ min^{-1}) eine lineare Beziehung. Aus der Steigung der Geraden ergibt sich, daß die Gelatinemoleküle unter diesen Versuchsbedingungen als relativ stark durchspülte Knäuel vorliegen. Bei Drehzahlen $N < 40\,000$ min^{-1} zeigt die Konzentrationsabhängigkeit der Sedimentationskonstante einen anomalen Verlauf.

Zur Untersuchung des hydrodynamischen Verhaltens der Gelatinemoleküle wurde zusätzlich die Molekulargewichtsabhängigkeit der Viskositätszahl, der Diffusionskonstante und der Sedimentationskonstante bei verschiedenen Drehzahlen ermittelt. Die erhaltenen Beziehungen lauten:

$$[\eta]_{40} = 7,25 \cdot 10^{-2}\ \text{ml/g} \cdot M_w^{0,51},$$
$$s_{40,w}^{0,w} = 0,65\ \text{Svedberg} \cdot M_w^{0,17}\ (\text{für } N = 60\,000\ \text{min}^{-1}),$$
$$s_{40,w}^{0} = 2,2 \cdot 10^{-2}\ \text{Svedberg} \cdot M_w^{0,50}\ (\text{für } N = 0),$$
$$D_{40,w}^{0} = 1,5 \cdot 10^{-4}\ \text{cm}^2/\text{sec} \cdot M_w^{-0,50}.$$

Die Größe der Exponenten dieser Beziehungen läßt sich wie die Drehzahlabhängigkeit der Sedimentationskonstante auf die unterschiedliche Durchspülbarkeit der Gelatinemoleküle zurückführen. Unseres Erachtens liegen die Gelatinemoleküle bei Viskositäts- und Diffusions-messungen und bei Sedimentationsversuchen mit sehr kleinen Drehzahlen als undurchspülte Knäuel vor. Bei Sedimentationsversuchen mit großen Drehzahlen werden dagegen die Gelatinemoleküle relativ stark durchspült.

Zur Molekulargewichtsberechnung nach der *Svedberg*schen Formel muß für die Sedimentationskonstante der hochmolekularen Gelatine deshalb der auf $N = 0$ extrapolierte Wert verwendet werden. Mit der bei hohen Drehzahlen erhaltenen Sedimentationskonstante wird das Molekulargewicht zu klein.

Summary

The dependence of the sedimentation constant on rotor speed in ultra-centrifuge experiments was studied for two gelatin samples in 0.15 m K-acetate buffer, pH 5.0, and for one polystyrene sample in methyl ethyl ketone, at various concentrations.

Whereas the sedimentation constant s^0 of polystyrene is independent of the speed, that of gelatin increases with decreasing speed, this increase being the more pronounced the higher the molecular weight of the gelatin.

A linear relationship between the reciprocal sedimentation constant and the concentration of the gelatin only exists at speeds $N > 40{,}000 \ \mathrm{min}^{-1}$. From the slope it can be deduced that the gelatin molecules under these experimental conditions are heavily drained coils. At speeds $N < 40{,}000 \ \mathrm{min}^{-1}$, the dependence of the sedimentation constant on concentration shows an anomalous profile.

To study the hydrodynamic behaviour of the gelatin molecules, the dependence of the limiting viscosity number, diffusion constant and sedimentation constant on molecular weight was determined at various speeds. The relationships obtained are:

$$[\eta]_{40} = 7.25 \cdot 10^{-2} \ \mathrm{ml/g} \cdot M_w^{0.51},$$
$$s_{40,w}^{0,w} = 0.65 \ \mathrm{Svedberg} \cdot M^{0.17} \ (\text{for } N = 60{,}000 \ \mathrm{min}^{-1}),$$
$$s_{40,w}^{0} = 2.2 \cdot 10^{-2} \ \mathrm{Svedberg} \cdot M_w^{0.50} \ (\text{for } N = 0),$$
$$D_{40,w}^{0} = 1.5 \cdot 10^{-4} \ \mathrm{cm^2/sec} \cdot M_w^{-0.50}.$$

Like the dependence of the sedimentation constant on speed, the exponents of these relationships can also be attributed to the different draining capacity of the gelatin molecules. It is our view that, in viscosity and diffusion measurements as well as in sedimentation experiments at very low speeds, the gelatin molecules are undrained coils. In sedimentation experiments at high speeds, however, the gelatin molecules are heavily drained.

For molecular weight determination by *Svedberg*'s formula, therefore, the value extrapolated to $N = 0$ must be used for the sedimentation constant of the high molecular weight gelatin. If the sedimentation constant obtained at high speeds is used, molecular weights obtained are too small.

Literatur

1) *Scholtan, W., H. Lange, H. Rosenkranz* und *F. Moll*, Colloid & Polymer Sci. **252**, 949 (1974).
2) *Volmert, B.*, Grundriß der Makromolekularen Chemie, S. 328, 329 u. 332 (Berlin-Göttingen-Heidelberg, 1962).
3) *Skaska, W. S.* und *W. M. Jamtschikow*, Plaste u. Kautschuk **20**, 24 (1973).
4) *Newman, S.* und *F. Eirich*, J. Colloid Sci. **5**, 541 (1950).
5) *Wales, M.* und *K. E. van Holde*, J. Polymer Sci. **14**, 81 (1954).
6) *Jamakawa, H.*, J. Chem. Physics **11**, 2995 (1962).
7) *Brandrup, J.* und *E. H. Immergut*, Polymer Handbook, S. IV-90 (New York-London-Sydney, 1966).

Anschrift der Autoren:

W. Scholtan und *H. Lange*
BAYER AG
IN-AP-CP 2-LEV
509 Leverkusen

Progr. Colloid & Polymer Sci. **58**, 159–163 (1975)

Deutsches Kunststoff-Institut, Darmstadt

Über die Copolymerisation von Acrylnitril, Methacrylsäure und Vinylidenchlorid

D. Braun, G. Mott und *F. Quella*

Mit 3 Abbildungen und 4 Tabellen

(Eingegangen am 16. Juli 1974)

Einleitung

Während binäre Copolymere aus den verschiedensten Monomeren seit langem bekannt und eingehend untersucht sind, finden in neuerer Zeit Multikomponentensysteme immer mehr Interesse. Für Systeme aus drei Monomeren wurde von *Alfrey* und *Goldfinger* (1) eine Beziehung zwischen der Reaktivität der Monomeren und der Zusammensetzung der Polymeren abgeleitet. Wegen ihrer interessanten Eigenschaften haben solche Terpolymere auch wirtschaftlich stark an Bedeutung gewonnen. Aber nicht nur technische, sondern auch theoretische Aspekte geben heute vielfachen Anlaß zu Untersuchungen der Copolymerisation ternärer Systeme.

In dieser Arbeit wurden Untersuchungen darüber angestellt, ob die bekannten Beziehungen zwischen Reaktivität der Monomeren und Zusammensetzung der Polymeren auch für Terpolymere aus Acrylnitril (AN), Methacrylsäure (MAA) und Vinylidenchlorid (VDC) gültig sind.

Aus der von *Alfrey* und *Goldfinger* (1) abgeleiteten Terpolymerisationsgleichung

AN/MAA mit r_{12} und r_{21},

AN/VDC mit r_{13} und r_{31},

MAA/VDC mit r_{23} und r_{32},

wobei r_{12} etc. die binären Copolymerisationsparameter sind. Da die Parameter von den experimentellen Bedingungen, z.B. Lösungsmittel (2) und Temperatur (3), abhängen, müssen bei der Parameterbestimmung im Hinblick auf die Terpolymerisation einheitliche Reaktionsbedingungen eingehalten werden. Da in der Literatur entsprechende Werte nicht vorliegen, wurden die binären Parameter unter identischen Bedingungen in Dimethylformamid (DMF) erneut ermittelt.

Experimenteller Teil

Monomere. Die verwendeten, stabilisatorfreien Monomeren wurden durch Destillation unter Stickstoff gereinigt:

Acrylnitril	Kp$_{760}$ 77 °C,	$n_D^{25} = 1,3880$,
Methacrylsäure	Kp$_{28}$ 84 °C,	$n_D^{25} = 1,4288$,
Vinylidenchlorid	Kp$_{760}$ 32 °C,	$n_D^{25} = 1,4248$.

$$d[M_1]:d[M_2]:d[M_3] = m_1:m_2:m_3 = M_1\left[\frac{M_1}{r_{31}r_{21}} + \frac{M_2}{r_{21}r_{32}} + \frac{M_3}{r_{31}r_{23}}\right]\left[M_1 + \frac{M_2}{r_{12}} + \frac{M_3}{r_{13}}\right]$$

$$:M_2\left[\frac{M_1}{r_{12}r_{31}} + \frac{M_2}{r_{12}r_{32}} + \frac{M_3}{r_{32}r_{13}}\right]\left[M_2 + \frac{M_1}{r_{21}} + \frac{M_3}{r_{23}}\right] \quad [1]$$

$$:M_3\left[\frac{M_1}{r_{13}r_{21}} + \frac{M_2}{r_{23}r_{12}} + \frac{M_3}{r_{13}r_{23}}\right]\left[M_3 + \frac{M_1}{r_{31}} + \frac{M_2}{r_{32}}\right]$$

geht hervor, daß man zur Berechnung der Zusammensetzung von Terpolymeren die Zusammensetzung des Monomergemischs sowie die Copolymerisationsparameter der drei möglichen binären Copolymerisationssysteme kennen muß. In diesem Fall sind dies:

Reagentien. Die Initiatoren Na$_2$S$_2$O$_4$ und (NH$_4$)$_2$S$_2$O$_8$ waren vom Reinheitsgrad p.A. und wurden direkt eingesetzt. Das Lösungsmittel DMF wurde auf die übliche Weise (4) gereinigt und fraktioniert unter Stickstoff destilliert.

Polymerisationsversuche. Jeder Probe wurden 50 ml DMF als Lösungsmittel und 0,05 Gew.-% (bezogen auf den gesamten Ansatz) des Initiatorsystems Na$_2$S$_2$O$_4$/

(NH$_4$)$_2$S$_2$O$_8$ (im Molverhältnis 1:1) zugefügt. Um nur bis zu möglichst geringen Umsätzen zu polymerisieren, wurden die Polymerisationen bei Auftreten einer erkennbaren Viskositätszunahme abgebrochen.

Die Proben des Systems AN/VDC wurden in Methanol ausgefällt, alle anderen, auch die Terpolymeren, in Methanol/Wasser (1:1) unter Zusatz von bis zu 10% 2 N HCl. Anschließend wurden die Proben im Vakuum bei 50 °C bis zur Gewichtskonstanz getrocknet und zweimal aus THF mit den gleichen Fällungsmitteln umgefällt.

Die Zusammensetzung der Polymeren wurde durch Chlor- und Stickstoffanalysen ermittelt.

Binäre Copolymerisationen

Die Copolymerisationsversuche wurden unter einheitlichen Bedingungen in DMF als Lösungsmittel bei 60 °C mit Na$_2$S$_2$O$_4$/(NH$_4$)$_2$S$_2$O$_8$ als Initiator ausgeführt. Für die Systeme AN/MAA (5) und AN/VDC (6, 7) liegen vergleichbare, in polaren Solventien ermittelte Parameter vor. Im System MAA/VDC wurden sie bisher nur in Substanz (8) ermittelt.

Die Parameter der binären Systeme wurden in dieser Arbeit nach einem auf der Kurveneinpaßmethode (9) beruhenden Verfahren (10) berechnet. Diese Methode nähert die für alle experimentellen Punkte günstigste Kurve mit Hilfe eines Fortran-IV-Rechenprogramms an. In den Tabelle 1, 2 und 3 sind die Daten für die drei binären Copolymerisationssysteme zusammengestellt, die zugehörigen Copolymerisationsdiagramme zeigen die Abb. 1a bis c.

Als Parameter berechneten sich für

	diese Arbeit	Literatur
AN/MAA	$r_{12} = 0,24$	$r_{12} = 0,35$ $r_{21} = 0,75$ (5)
	$r_{21} = 1,37$	(DMF, $T = 60$ °C)
AN/VDC	$r_{13} = 0,70$	$r_{13} = 0,91 \pm 0,1$
	$r_{31} = 0,43$	$r_{31} = 0,37 \pm 0,1$ (6)
		(tert. Butanol, $T = 50$ °C)
		$r_{13} = 1,2$ $r_{31} = 0,49$ (7)
		(tert. Butanol,
		$T = 43-47$ °C)
MAA/VDC	$r_{23} = 4,48$	$r_{23} = 3,0 \pm 0,45$
	$r_{32} = 0,36$	$r_{32} = 0,15 \pm 0,023$ (8)
		(Substanzpol., $T = 70$ °C)

Tabelle 1. Copolymerisation von AN mit MAA bei 60 °C in DMF

Nr.	Zusammensetzung des Monomergemischs		Umsatz [Gew.-%]	N [Gew.-%]	Copolymerzusammensetzung [Mol-%]		Pol.-Zeit [min]
	Mol AN	Mol MAA			AN	MAA	
1	0,153	0,706	5,3	2,2	0,129	0,871	45
2	0,382	0,590	3,9	4,8	0,267	0,733	50
3	0,535	0,476	3,7	6,8	0,356	0,644	65
4	0,765	0,354	3,8	9,4	0,476	0,524	55
5	0,918	0,236	3,3	12,6	0,595	0,405	75
6	1,144	0,118	2,7	16,8	0,733	0,267	85

Tabelle 2. Copolymerisation von AN mit VDC bei 60 °C in DMF

Nr.	Zusammensetzung des Monomergemischs		Umsatz [Gew.-%]	Cl [Gew.-%]	Copolymerzusammensetzung [Mol-%]		Pol.-Zeit [min]
	Mol AN	Mol VDC			AN	VDC	
1	0,153	0,627	0,8	59,6	0,294	0,706	55
2	0,610	0,627	3,5	44,9	0,536	0,464	120
3	0,765	0,438	1,5	37,2	0,638	0,362	70
4	0,918	0,314	1,7	30,0	0,724	0,276	170
5	1,144	0,125	2,6	15,2	0,875	0,125	75

Tabelle 3. Copolymerisation von MAA mit VDC bei 60 °C in DMF

Nr.	Zusammensetzung des Monomergemischs		Umsatz [Gew.-%]	Cl [Gew.-%]	Copolymerzusammensetzung [Mol-%]		Pol.-Zeit [min]
	Mol MAA	Mol VDC			MAA	VDC	
1	0,118	0,627	5,1	45,7	0,404	0,596	175
2	0,236	0,502	2,8	30,3	0,615	0,385	50
3	0,413	0,376	8,6	16,5	0,795	0,205	75
4	0,590	0,314	3,0	7,8	0,904	0,096	60
5	0,706	0,125	6,7	3,1	0,963	0,037	60

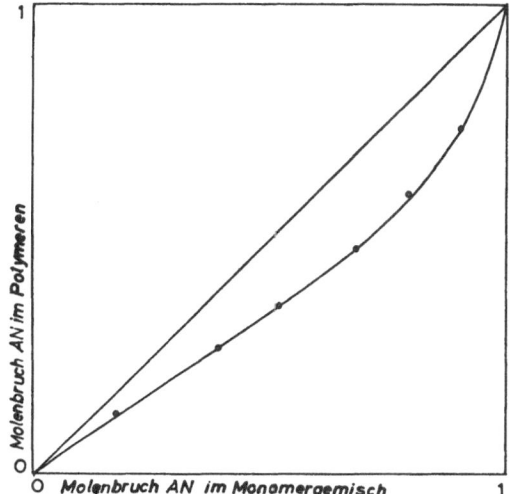

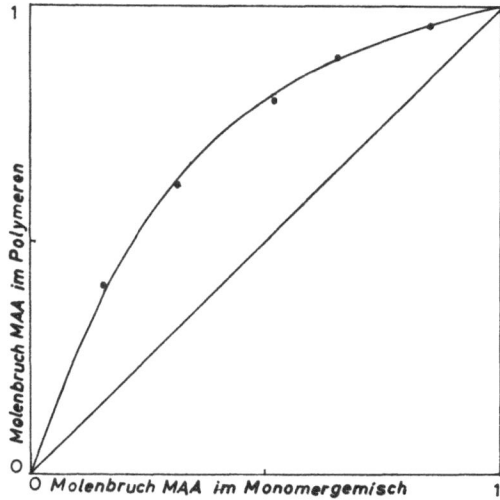

Abb. 1. Copolymerisationskurven.

a) System AN/MAA
— mit den Parametern
$r_{12} = 0,24$ $r_{21} = 1,37$
berechnete Kurve
○ Meßwerte

c) System MAA/VDC
— mit den Parametern
$r_{23} = 4,48$ $r_{32} = 0,35$
berechnete Kurve
○ Meßwerte

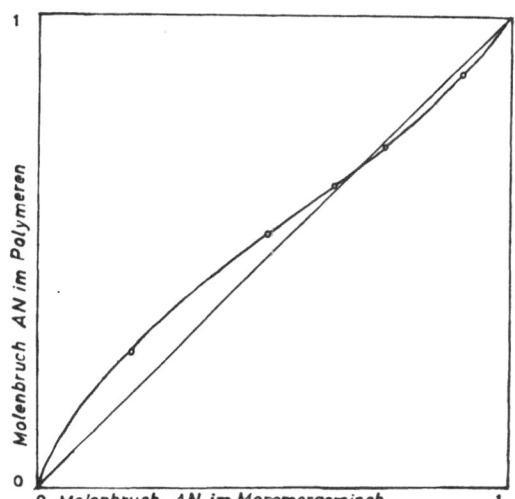

b) System AN/VDC
— mit den Parametern
$r_{13} = 0,70$ $r_{31} = 0,43$
berechnete Kurve
○ Meßwerte

Änderung der Zusammensetzung des Monomergemischs in Abhängigkeit vom Umsatz

Bei der Anwendung der binären Copolymerisationsgleichung und der Terpolymerisationsgleichung in ihrer differentiellen Form muß darauf geachtet werden, daß nur bis zu solchen Umsätzen polymerisiert wird, bei denen noch keine wesentliche Änderung in der Monomer-

zusammensetzung auftritt. Ein exakter Wert für diesen Umsatz kann aber nicht allgemein angegeben werden. Zur Klärung dieser Frage muß für jedes System eine Integration (11) der binären Co- bzw. der Terpolymerisationsgleichung vorgenommen werden.

Die Zusammensetzung der Polymeren in Abhängigkeit vom Gesamtumsatz α berechnet sich nach der Beziehung

$$M_{i(\alpha)} = M_{i(0)} - \theta_i M_{i(0)} \frac{\Delta\alpha}{100} \qquad [2]$$

mit $i = 1, 2$ für die binäre Copolymerisation bzw.
$i = 1, 2, 3$ für die ternäre Copolymerisation,
wobei $M_{i(0)}$ die molaren Konzentrationen der Monomeren M_i zu Beginn der Polymerisation sind und $M_{i(\alpha)}$ die bei einem bestimmten Umsatz α.

Die Größe θ_i ist die differentielle Änderung der Molenbrüche

$$\theta_i = \frac{\mathrm{d}M_i}{\mathrm{d}M_{\text{gesamt}}}$$

und errechnet sich auf der Basis der binären Co- und der Terpolymerisationsgleichung. Auf dieser Grundlage wurde ein Fortran-IV-Rechenprogramm (12) geschrieben, das die Änderung der Zusammensetzung im Monomergemisch und im Polymeren in Abhängigkeit vom Umsatz berechnet und mit Hilfe eines Plotters aufzeichnet. In den Abb. 2a bis c wird für jedes Polymerisationssystem ein charakteristisches Beispiel dazu gezeigt.

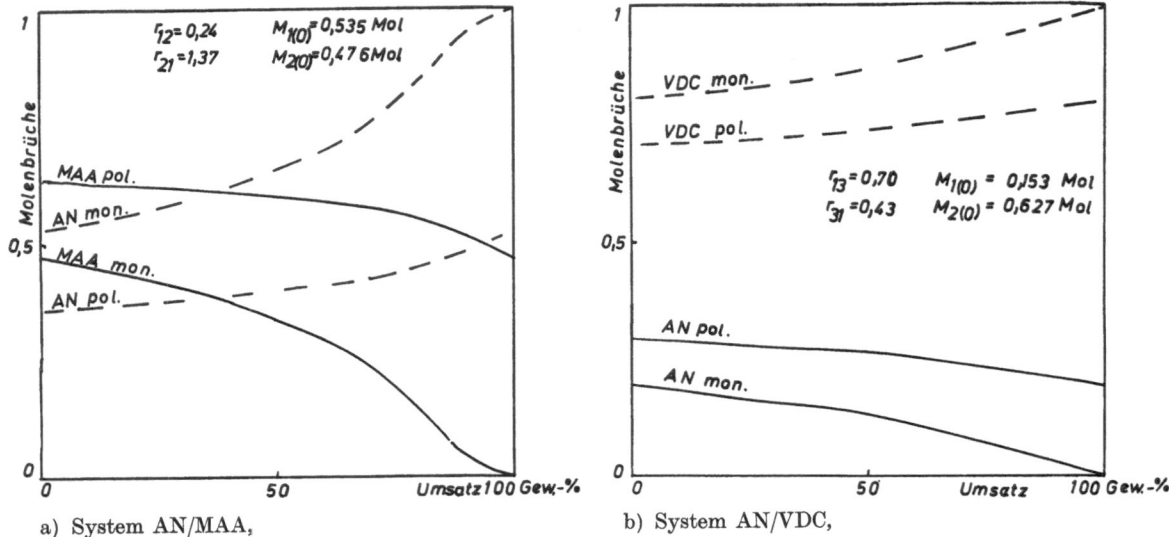

a) System AN/MAA,   b) System AN/VDC,

Abb. 2. Abhängigkeit der Monomer- und Copolymerzusammensetzung vom Umsatz (Beschreibungen siehe experimenteller Teil).

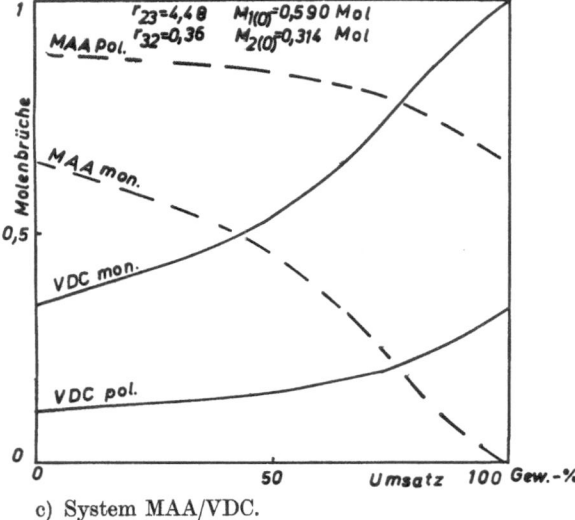

c) System MAA/VDC.

Wie man daraus sieht, ändert sich die Zusammensetzung der Monomergemische teilweise so stark, daß Umsätze von mehr als 10 Gew.-% schon beträchtliche Fehler bei der Parameterbestimmung ergeben können.

Terpolymerisation von Acrylnitril, Methacrylsäure und Vinylidenchlorid

Um die allgemeine Terpolymerisationsgleichung [1] in ihrer differentiellen Form anwenden zu können, muß wie bei der differentiellen Copolymerisationsgleichung bei nahezu konstanter Zusammensetzung des Monomergemischs poly-

merisiert werden. Daher wurde auch für die Terpolymeren die Änderung der Zusammensetzung im Monomergemisch und im Polymeren in Abhängigkeit vom Umsatz berechnet. In Abb. 3 ist ein Beispiel dafür dargestellt. Aus Gründen der Übersichtlichkeit wurde dort nur die Polymerzusammensetzung gegen den Umsatz aufgetragen. Wie aus Abb. 3 zu ersehen ist, ändert sich für diesen speziellen Ansatz die Polymerzusammensetzung bis zu 10 Gew.-% Umsatz nicht sehr stark. Auch in den anderen untersuchten Fällen findet man ähnliche Ergebnisse. Deswegen wurden Polymerisationsumsätze unterhalb dieses Wertes eingehalten.

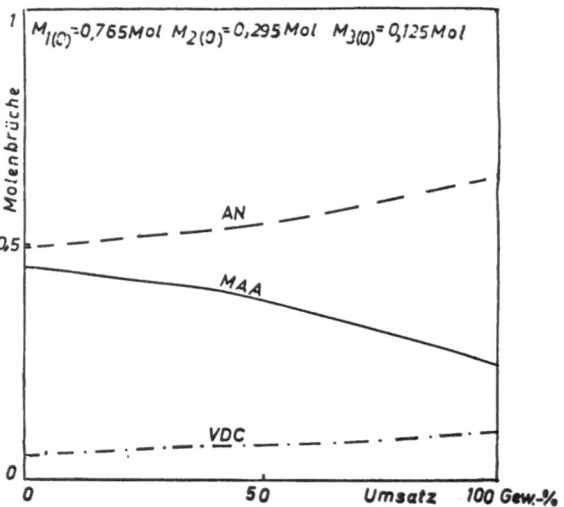

Abb. 3. Abhängigkeit der Terpolymerzusammensetzung vom Umsatz für das System AN/MAA/VDC.

Tabelle 4. Terpolymerisation von AN, MAA und VDC bei 60 °C in DMF

Nr.	Zusammensetzung des Monomergemischs			Um-satz [Gew.-%]	Pol.-Zeit [min]	N [Gew.-%]	Cl [Gew.-%]	Polymerzusammensetzung [Mol-%]					
	Mol AN	Mol MAA	Mol VDC					gemessen			berechnet		
								AN	MAA	VDC	AN	MAA	VDC
1	0,382	0,295	0,314	3,5	85	6,33	17,50	0,347	0,464	0,189	0,346	0,502	0,152
2	0,765	0,118	0,314	1,4	150	10,36	20,19	0,525	0,273	0,202	0,571	0,233	0,196
3	0,382	0,118	0,627	1,2	135	6,36	33,75	0,356	0,271	0,373	0,385	0,230	0,385
4	0,765	0,295	0,125	1,6	90	9,76	6,59	0,492	0,440	0,068	0,492	0,451	0,057

Mit den errechneten Parametern der binären Systeme wurde nun für verschiedene Ausgangs-zusammensetzungen des Monomergemischs die Terpolymerzusammensetzung berechnet und experimentell überprüft. Die Daten sind in Tabelle 4 zusammengefaßt. Man kann daraus erkennen, daß eine gute Übereinstimmung zwischen den experimentell ermittelten und den berechneten Werten besteht. Dies ist ein Beweis dafür, daß die allgemeine Terpolymerisations-gleichung auf dieses System angewendet werden kann, wenn die von *Alfrey* und *Goldfinger* gemachten Voraussetzungen gegeben sind.

Wir danken der Arbeitsgemeinschaft Industrieller Forschungsvereinigungen für die Förderung dieser Untersuchungen.

Zusammenfassung

Es wurden zunächst die binären Copolymerisations-parameter der Systeme Acrylnitril/Methacrylsäure ($r_{12} = 0,24$, $r_{21} = 1,37$), Acrylnitril/Vinylidenchlorid ($r_{13} = 0,70$, $r_{31} = 0,43$) und Methacrylsäure/Vinyliden-chlorid ($r_{23} = 4,48$, $r_{32} = 0,36$) in Dimethylformamid als Lösungsmittel mit 0,05 Gew.-% des Initiators $Na_2S_2O_4/(NH_4)_2S_2O_8$ (molares Verhältnis 1:1) bei 60 °C bestimmt. Sowohl für die binären Systeme als auch für die Terpolymeren wurde durch numerische Integration der binären und ternären Copolymerisationsgleichung die Änderung der Zusammensetzung der Monomergemische und der Polymeren in Abhängigkeit vom Umsatz berechnet. Damit konnte gezeigt werden, daß bei Umsätzen von mehr als 10 Gew.-% keine konstante Zusammensetzung der Monomergemische mehr gegeben ist. Unter Berücksichtigung dieser Befunde wurden verschiedene Gemische aus Acrylnitril, Methacrylsäure und Vinylidenchlorid polymerisiert. Die dabei experimentell ermittelten Terpolymerzusammensetzungen stimmen mit den aus der Terpolymerisationsgleichung berechneten Werten gut überein.

Summary

The radical copolymerization of the three binary systems acrylonitrile/methacrylic acid ($r_{12} = 0.24$, $r_{21} =$ 1.37), acrylonitrile/vinylidene chloride ($r_{13} = 0.70$, $r_{31} = 0.43$) and methacrylic acid/vinylidene chloride ($r_{23} = 4.48$, $r_{32} = 0.36$) in dimethylformamide as solvent at 60 °C is investigated. The initiating system is 0.05 weight-% of $Na_2S_2O_4/(NH_4)_2S_2O_8$ (molar ratio 1:1).

For the binary as well as for the ternary systems the change in monomer and polymer composition is calculated as a function of conversion by numerical integration of the binary and ternary copolymerization equation. The calculations showed that at conversions higher than 10% a constant composition of the monomer mixture cannot be expected.

With respect to these results different mixtures of acrylonitrile, methacrylic acid and vinylidene chloride have been polymerized. The experimentally obtained compositions of the terpolymers are in good agreement with the values calculated by the terpolymerization equation.

Literatur

1) *Alfrey, T.* und *G. Goldfinger*, J. Chem. Phys. **12**, 322 (1944).
2) *Ryabov, A. V., Y. D. Semchikov* und *N. N. Slavnit-skaya*, Vysokomol. Soedin., Ser. A **12**, 553 (1970).
3) *Bejnoravičjus, M. A., G. I. Bajoras* und *I. I. Vosljus*, Vysokomol. Soedin., Ser. B **15**, 535 (1973).
4) *Organikum*, S. 623 (Berlin, 1967).
5) *Ulbricht, J.* und *H. Herma*, Faserforschg. Textiltech. **16**, 387 (1965).
6) *Hill, E. H.* und *J. R. Caldwell*, J. Polymer Sci. **47**, 397 (1960).
7) *Parker, R. B.*, jr. und *B. V. Mokler*, J. Polymer Sci. B **2**, 19 (1964).
8) *Alfrey, T.*, jr., *J. Bohrer, H. Haas* und *C. Lewis*, J. Polymer Sci. **5**, 719 (1950).
9) *Alfrey, T., J. J. Bohrer* und *H. Mark*, Copolymerization, S. 16 (New York, 1952).
10) *Braun, D., W. Brendlein* und *G. Mott*, Eur. Polym. J. **9**, 1007 (1973).
11) *Skeist, I.*, J. Amer. Chem. Soc. **68**, 1781 (1946).
12) *Quella, F.*, Diplomarbeit, TH Darmstadt (1974).

Adresse des Autors:

D. Braun
Deutsches Kunststoffinstitut Darmstadt
61 Darmstadt
Schloßgartenstr. 6 R

Progr. Colloid & Polymer Sci. **58**, 164—168 (1975)

Department of Chemistry, Science University of Tokyo, Tokyo (Japan)

Simple determinations of the thickness and molecular orientations in some polymer monolayers by the Langmuir-Blodgett method

M. Ueno, Y. Dei and *K. Meguro*

With 3 figures and 2 tables

(Received April 5, 1974)

Introduction

Langmuir-Blodgett films (1—5) of a wide variety of long chain polar molecules have been used extensively in fundamental studies of surface properties such as monolayer thickness, molecular orientation and adhesion. In particular, molecular orientations of fatty acids in *Langmuir-Blodgett* multilayers have been estimated by means of electron diffraction (6), electron microscopy (7), and multiple beam interferometry (8).

Blodgett and *Langmuir* (4), and *Langmuir* and *Schaefer* (9) have shown that by depositing a protein monolayer on a barium stearate step-multilayer and by measuring the optical thickness of the combined layer, it is possible to check the molecular orientation of protein on the water surface. However, relatively few studies of multilayers and their thickness of synthetic polymers have so far been reported.

In this paper, the thicknesses of three synthetic polymer monolayers are determined by optical measurements on built-up films obtained by the *Langmuir-Blodgett* method and molecular orientations in polymer monolayers spread on the water surface or on metal plates are discussed taking into consideration possible orientations of molecular models based on the experimental results on the thicknesses and surface pressure-area curves of the polymers.

Experimental

Materials

Polyvinylalcohol (abbrev. PVA) required for synthesis of polyvinylbenzal and polyvinylbutyral was supplied from Nihon Gosei Kagaku Co. Ltd. and was pu-

rified by repeated precipitation from aqueous solution with addition of a large excess of methanol, then by extraction of a trace of polyvinylacetate with methanol in a *Soxhlet* extractor and finally dried under vacuum. Its purity was checked by infrared spectroscopy and elementary analysis, and it was found to be above 99.8%. The viscosity-average molecular weight was 7.96×10^4.

Polyvinylbenzal (abbrev. PVBenzal) and polyvinylbutyral (abbrev. PVButyral) were prepared by acetalization of the PVA in HCl solution with benzaldehyde and butyraldehyde, respectively. These polymers thus obtained were washed thoroughly with dilute ammonia to remove a trace of the acid, then with water. They were then purified by repeated precipitation from dimethylsulfoxide solution with addition of hot water for the former and from methanol solution with addition of water or petroleum ether for the latter, and finally dried under vacuum. The mole contents (mole %) of benzal and butyral groups in these polymers were 74.5% and 46.0%, respectively.

Methylacrylate monomer was supplied from Toa Gosei Kagaku Co. Ltd. Polymethylacrylate (abbrev. PMA) was prepared by solution polymerization with α, α'-azobisisobutyronitrile as an initiator. The polymer was purified by repeated precipitation from acetone solution with addition of a large excess of methanol or *n*-hexane. The average molecular weight was determined to be 65.9×10^4 by viscosity measurement in acetone solution at 30 °C.

Stearic and arachidic acids required for standard interference color gauge were purified by repeated crystallization from ethanol to give a purity above 99.8%. The water substrate was prepared by distillation of alkaline permanganate solution made up from distilled ion-exchange water.

The piston oils required for building up the fatty acids and the polymers on the metal plate were oleic acid (29.5 dyne/cm), ethylmyristate (20.7 dyne/cm), and mixtures of ethylmyristate and liquid paraffin; 16.35 dyne/cm for 32.8% ethylmyristate, and 8.6 dyne/cm for 10.0% ethylmyristate, respectively.

Clean mirror and chromium finished stainless steel metal plates required for bearing a multilayer of barium stearate or arachidate were degreased by extracting their contamination with hot toluene for about 4 hrs in a *Soxhlet* extractor and washed thoroughly with distilled water. In this way, hydrophilic plates were obtained.

Multilayers and Thickness Measurements

20×10^3 M benzene solutions of stearic and arachidic acids were used as the spreading solutions, respectively. The water substrate was 1.0×10^{-3} M $BaCl_2$ solution and was contained in a silica trough with a waxed thread as the moveable barrier.

The pH of the substrate was adjusted at about 9.2 by addition of a small amount of aqueous ammonia. At such pH, 90% of stearic acid in monolayers spread on $BaCl_2$ solution is known to be converted to barium stearate (10), (11).

A freshly prepared metal plate was immersed in the aqueous substrate, the surface of which had been carefully swept clean; the monolayer was spread and immediately pressed with appropriate piston oil; it was aged for 5—10 minutes to permit evaporation of the solvent, and then the plate was slowly withdrawn and subsequently dipped through the monolayer by means of synchronous motor. The rate of dipping and withdrawal was usually 1.2 cm/min.

Successive barium stearate layers were deposited up to 39 layers by repeated dipping and withdrawal processes. In all cases, Y-type film were produced, that is, films composed of alternating monolayers. On this plate bearing 39 layers of barium stearate, double layers of each polymer were deposited and the optical thickness of the combined layers was measured by comparing the interference color with corresponding color of the following standard color gauges of multilayers made of a combination of barium stearate and arachidate. The schematical diagram of the standard color gauges is shown in Fig. 1.

In combined systems of barium stearate and arachidate formulated by

$$[(C_{17}H_{35}COO)_2Ba_{39-2n}] - [(C_{19}H_{39}COO)_2Ba_{2n}],$$

where n takes a number from 1 to 9, even number ($2n$) of barium stearate layers were replaced by $2n$ layers of barium arachidate with $n = 1$ to 9, and the total number of layers thus obtained was maintained at 39.

Since barium arachidate (C_{20}) has additional two carbon atoms compared with barium stearate (C_{18}), the deposition of double layers of barium arachidate on the plate bearing 37 layers of barium stearate caused an increase in the thickness corresponding to the additional

Table 1. The multilayers of the combined systems of barium stearate and arachidate

Number of layers of barium stearate $(39 - 2n)$	Number of layers of barium arachidate $(2n)$[a]	Increase in the number of carbon atoms in 39 layers	Thickness corresponding to the increase of carbon atoms (Å)[b]
37	2	4	5
35	4	8	11
33	6	12	16
31	8	16	22
29	10	20	27
27	12	24	32
25	14	28	38
23	16	32	43
21	18	36	49

[a] $n = 1$ to 9.
[b] Decimals of these values are obtained from $5.4 \times n$ ($n = 1$ to 9), counting fractions of 0.5 and over as a whole number and disregarding the rest.

four carbon atoms over that of 39 layers of barium stearate, and the resulted multilayer showed an interference color change corresponding to the increase in the thickness of 5 Å. This value was equal to that (5.4 Å) obtained from the difference of double layer spacing thicknesses of barium stearate (50.2 Å) and arachidate (55.6 Å) (8). The combined systems of multilayers, the increase in the number of carbon atoms in 39 layers for each plate and the thickness corresponding to the increase were summarized in Table 1.

Surface Pressure Measurements

Surface pressure-area measurements were carried out by using a *Wilhelmy*-type surface balance at $20 \pm 0.5°$. The stainless steel trough was lightly waxed, and paraffin-wax-molded barriers were used to confine and compress the polymer monolayers. The surface balance was completely enclosed in a protective box. The sensitivity of the apparatus was 0.1 dyn/cm and surface areas were estimated within an accuracy of 0.1%.

The spreading solutions of polymers were prepared at 90 mg/100 ml for PVBenzal from a mixed solvent consisting of benzene and dimethylsulfoxide (60/40 v/v), at 70 mg/100 ml for PVButyral from a mixed solvent consisting of benzene and methanol (75/25 v/v) and at 40 mg/100 ml for PMA from a mixed solvent consisting of benzene and acetone (50/50 v/v), respectively. The spreading solutions were delivered precisely to 0.0002 ml from an Agla microsyringe. Usually pressure-area measurements were commenced 30 minutes after addition of the spreading solution. This allowed elimination of the solvent and complete spreading. In order that an equilibrium be assured, surface pressure measurements were made 7 minutes after each compression and the rate of compression was 1.0 cm/min. Under these conditions, the reproducibility was satisfactory.

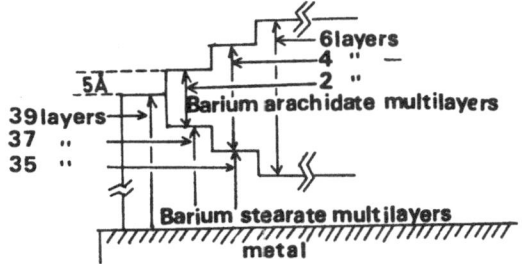

Fig. 1. The schematical diagram of the standard color guages

Results and Discussion

Measurements of the Thickness of the Polymer Monolayers

Double layers of PVBenzal were deposited by the dipping and withdrawal process on a prepared plate bearing 39 layers of barium stearate at different pressures exerted by the four piston oils, respectively. The double layer thickness of polymers on the plate thus obtained was measured by comparing the interference color changes produced by the double layer with the corresponding color of the standard gauges, the color change being observed by polaroid. The double and single layer thicknesses of three polymers are summarized in Table 2.

The interference colors of plates with double layers of PVBenzal were the same for those of the double layers which were built up at different pressures of the four piston oils. This indicates that all monolayers are equal in the thickness and the value coincides with the theoretical one of vertical orientation. From the fact that the orientation of PVBenzal monolayer is not affected by pressure change in the range from 8.6 to 29.5 dyne/cm, it is concluded that benzene rings are oriented vertically on the plate even

at low pressure due to strong intermolecular cohesion between them and this suggests that the polymers in the monolayers are oriented vertically at air-water interface as they are oriented vertically in the multilayer on the plate.

In general, when the molecular cohesion of a monolayer is very high as compared with its adhesion toward the substrate water, the monolayer molecules remain partly as three-dimensional clusters on the surface, and incomplete spreading results.

On the other hand, the high adhesion to the water substrate, coupled with a very low intermolecular monolayer cohesion, will result in an unstable, soluble film. Between these two extremes, a monolayer at the air-water interface may exist in a variety of phases with condensed, liquid expanded or gaseous states depending on the intermolecular cohesion forces in the monolayers.

The pressure-area curves of three polymers on the water substrate are shown in Fig. 2.

PVBenzal has a limiting area of 32 Å²/residue (at zero pressure). It is clear from Fig. 2 that PVBenzal gives a pressure-area curve of the condensed type, characterized by its linearity and low compressibility. By considering the pressure-area curve of PVBenzal, certain deductions may be made with respect to the orientation of the polymer at air-water interface. At low surface

Table 2. Thicknesses of double and single layer spacings of the polymers obtained from color gauges

Polymers	Pressure of piston oils (dyne/cm)	Double layer thicknesses (Å)	Single layer thicknesses (Å)
PVBenzal	29.50	22	11[a]
	20.70	22	11
	16.35	22	11
	8.60	22	11
PVButyral	29.50	above 22[b]	—
	20.70	16—22	8—11[c]
	16.35	16	8
	8.60	11	5 or 6
PMA	29.50	—	—
	20.70[d]	—	—
	16.35	16	8[e]
	8.60	11	5 or 6

[a]) This value corresponds to the thickness of the vertical type.

[b]) The thickness of the collapsed film.

[c]) This value corresponds to the thickness of the vertical type.

[d]) PMA monolayer is collapsed by the pressure above this.

[e]) This value corresponds to the thickness of the vertical type.

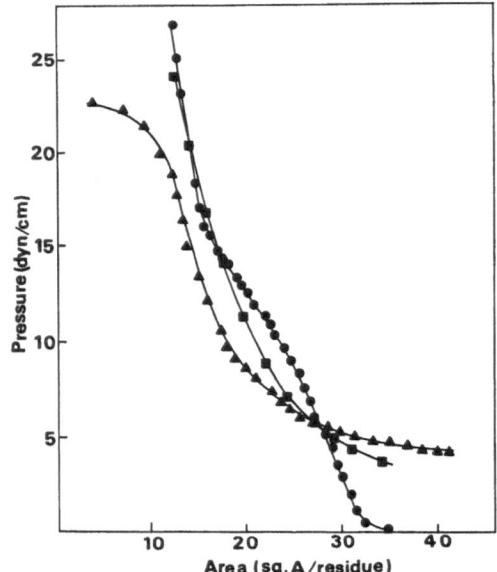

Fig. 2. Pressure-area curves of three polymers on the distilled water at 20°: PVBenzal; ●, PVButyral; ■, PMA; ▲

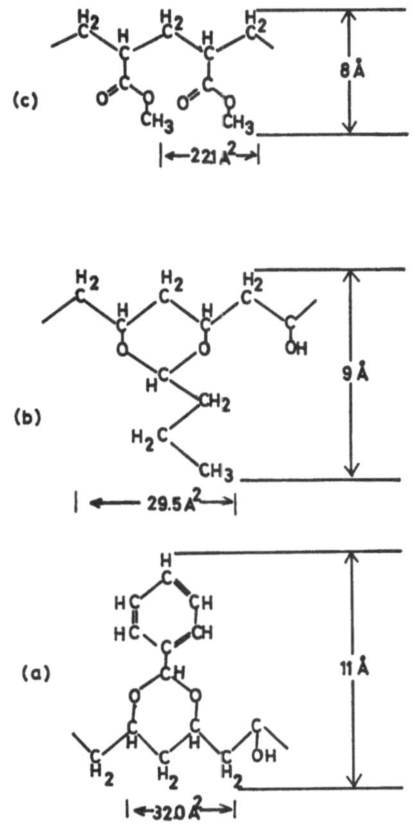

Fig. 3 (a), (b) and (c). Schematical diagrams of the vertical orientation of three polymers. (a): PVBenzal, (b): PVButyral, (c): PMA

pressures, the polymer is virtually completely extended at the surface. During compression, the benzene rings are oriented toward the vertical. Actually, its orientation on the water surface could be verified from the fact that the oriented benzene rings in the molecular model having the same theoretical area as a limiting area of 32 Å^2/residue corresponded to the vertical orientation as shown in Fig. 3 (a), schematically.

The experimental values of the thickness of PVButyral monolayers obtained by a method similar to that for PVBenzal, as shown in Table 2, decrease with the lowering pressure of piston oils, and its value at the pressure of 8.6 dyn/cm is much smaller than one of the molecular model corresponding to vertical type. This change in the thickness can be related with the change in the orientation for the polymer monolayer on the plate on which butyral groups in the polymer are oriented obliquely at low pressures and gradually tend to orient toward the vertical at high

pressures. Yet, the molecular orientation in the monolayers at the air-water interface can also be considered to be very similar to that in the multilayer at each pressure of the piston oils. As evident from a pressure-area curve of PVButyral in Fig. 2, this monolayer is only slightly more expanded than that of PVBenzal, and exhibits a behavior typical of a molecule of low intermolecular cohesion. PVButyral has a limiting area of 29.5 Å^2/residue (at zero pressure). The oriented butyral groups in the molecular model occupied by the same theoretical area as this limiting area obtained from the pressure-area curve of PVButyral corresponded to the vertical type as shown in Fig. 3 (b), and the thickness was equal to that of single layer of PVButyral on the plate. Therefore, this polymer on the water surface also seemed to be completely extended and butyral groups in the polymer are lying horizontally at low surface pressures. During compression, butyral groups were likely to tend to orient gradually toward the vertical.

A tendency of the thickness of PMA monolayer to decrease with lowering pressure of the piston oils is also very similar to the case of PVButyral as shown in Table 2. This suggests that methoxy groups in molecules of PMA are oriented horizontally at low pressures and vertically at high pressures on the plates. The orientation of PMA monolayers on the air-water interface can be presumed by considering the results on multilayers and a pressure-area curve of PMA, as shown in Fig. 2. PMA has a limiting area (at zero pressure) of 22.1 Å^2/residue as obtained from the pressure-area curve. This thickness corresponding to the closed-packed molecules and the vertical orientation at the air-water interface is shown in Fig. 3 (c), schematically.

Summary

The thickness of double layer spacings and possible molecular orientations of polyvinylbenzal, polyvinylbutyral and polymethylacrylate monolayers were estimated by comparing the interference color changes produced by the double layer spacings of polymers deposited on the top surface of 39 layers of barium stearate which were built up on a chromium finished metal plate, with a standard interference color of 39 layers consisting of a combination of barium stearate and arachidate multilayers. The single layer thickness of polyvinylbenzal was found to be 11 Å at all surface pressures of piston oils from low to high, while the thickness of the single layer spacing of the other two polymers varied with the increase in the surface pressure of piston oils. From these results, benzene rings in polyvinylbenzal monolayer were concluded to be vertically oriented under the applied surface pressures either on the water surface or on the metal plate. It was supposed that butyral groups in polyvinylbutyral monolayer tended gradually to orient vertically as the pressures of piston oils increased.

On the other hand, carboxymethyl groups in polymethylacrylate were considered to orient horizontally on the water surface or on the metal plate at low pressures, and to be submerged into water, while they were oriented vertically on the metal plate when the surface pressure increased.

Zusammenfassung

Die Dicke der Doppelschicht und die mögliche Orientierung der Polyvinylbenzal-, Polyvinylbutyral- und Polymethylacrylat-Oberflächenfilme wurden durch Vergleich der Interferenzfarbenänderungen, die an den Doppelschichten der Polymeren auf den über eine Metallplatte 39fach aufgebauten Bariumstearatschichten auftreten, mit den Standardinterferenzfarben der 39-fachen Oberflächenfilme von Bariumstearat und -arachidat bestimmt.

Es zeigt sich, daß die Dicke einer Schicht für den Polyvinylbenzaloberflächenfilm 11 Å ist, bei allen Oberflächendrucken des für das Aufbringen angewandten Pistonöls. Aus diesen Ergebnissen haben wir geschlossen, daß sich die Benzolringe in den Filmen auf der Wasseroberfläche oder der Metallplatte vertikal orientieren. Die Butyralgruppen in den Polyvinylbutyralfilmen orientieren sich nach und nach ebenfalls vertikal, wenn nämlich der Druck des Pistonöls gesteigert wird, während sich die Carboxymethylgruppen in den Polymethylacrylatfilmen auf der Wasseroberfläche oder zu den Metallplatten bei niedrigem Druck horizontal orientieren, indem die Polymethylacrylatmoleküle auf dem Wasser nach unten ins Wasser tauchen und sich auf der Metallplatte vertikal orientieren, wenn der Oberflächendruck zunimmt.

References

1) *Langmuir, I.*, Trans. Faraday Soc. **15**, 68 (1920).
2) *Blodgett, K. B.*, J. Amer. Chem. Soc. **56**, 495 (1934).
3) *Blodgett, K. B.*, ibid, **57**, 1007 (1934).
4) *Blodgett, K. B.* and *I. Langmuir*, Phys. Rev. **51**, 964 (1937).
5) *Langmuir, I., V. J. Schaefer* and *D. Wrinch*, Science **85**, 76 (1937).
6) *Germer, G. L.* and *K. H. Storks*, J. Phys. Chem. **6**, 280 (1938).
7) *Epstein, H. T.*, J. Phys. Colloid Chem. **54**, 1053 (1950).
8) *Shirahata, R.* and *G. D. Scott*, Applied Optics **10**, 2192 (1971).
9) *Langmuir, I.* and *V. J. Schaefer*, Chem. Rev. **24**, 181 (1939).
10) *Spink, J. A.* and *J. V. Sanders*, Trans. Faraday Soc. **51**, 1154 (1955).
11) *Ellis, J. W.* and *J. L. Paully*, J. Colloid Sci. **19**, 755 (1964).

Authors' Addresses:

M. Ueno and *K. Meguro*
Department of Chemistry
Science University of Tokyo
Kagurazaka Shinjuku-ku
Tokyo, Japan
Y. Dei
Dainippon Ink and Chemicals Inc.
Nihon bashi Chuo-ku
Tokyo, Japan

Progr. Colloid & Polymer Sci. **58**, 169—177 (1975)

Institute of Physical Chemistry, Florence (Italy)

Study of the bidimensional state conformation of poly-β-benzyl-L-aspartate

I. Spreading isotherms

Gabriella Gabrielli and *Armand Davidson*

With 5 figures and 2 tables

(Received March 5, 1974)

Introduction

It has been shown in previous papers (1) that it is possible to obtain parameters useful for the definition of macromolecular conformation at the interphase both with the virial coefficients of the equations of bidimensional state and with the thermodynamic spreading quantities.

Once this possibility had been established we subsequently proceeded with our study to establish the influence of both the type of interphase (2) and the constitution of the monomer unit (3) on such a configuration. We have not yet, however, investigated the effect of the spreading solvent, and therefore of the conformation in solution, on the polymeric conformation at a liquid surface. Several, often contradictory, papers have appeared in the literature (4) on the possibility of obtaining different macromolecular conformations of a given polymer by using different spreading solvents. This is found to be particularly important not only for the study of bidimensional phase transitions, but above all for the comparison of the energies and thermodynamic stabilities of macromolecules at an interphase with the corresponding quantities in solutions of various solvents and for the consequent establishment of whether tridimensional phase conformations can be modified or not in the bidimensional phase. The aim of this paper is to study bidimensional films of poly-β-benzyl-L-aspartate (PBLA) in various spreading solvents and to characterize them with the bidimensional state equations and the thermodynamic spreading quantities.

Our choice of polymer is justified by the fact that it can exist in various conformations in tridimensional phase, in particular as α and β helixes

which are stable in several solvents, and that it forms spreading films at the water/air interphase over a range of pressures that are easy to determine.

Experimental Part

The poly-β-benzyl-L-aspartate (PBLA) was prepared by Miles-Yede Laboratories, Israel, and had a mean molecular weight of 4,200. The solvents were: very pure chloroform (Riedel de Haen), pure pyridine (Riedel de Haen), and dichloroacetic acid (Fluka). Trials were run on all solvents to guarantee the absence of tensioactive substances; in particular, successive quantities of each were allowed to spread on the aqueous substrate. In each case a negligible surface pressure was measured. The solutions were prepared by dissolving a known quantity of polymer in the solvent mixtures and allowing them to stand 24 hours prior to use. The films were formed on supports of bidistilled water (whose surface purity was continuously checked) using a Scientific Glass Engineering Pty. LTD (Australia) microsyringe and allowing the microdrops to fall from a minimum height from the surface at time intervals that were sufficiently long to guarantee spreading. Particular care was taken in film formation with pyridine-containing solutions. In this case spreading was achieved very slowly with the microsyringe held almost parallel to the surface in order to avoid passage of the drops into the lower phase.

Solutions of various concentrations and in varying quantities were used for each solvent as a check for perfect spreading and the reproducibility of the results. Spreading time was set at 20 minutes for all solutions since this was considered sufficient for the attainment of perfectly homogeneous films. The surface pressure measurements were made with the *Wilhelmy* method using an apparatus already described in detail (1, 2, 3), consisting essentially of a bifilar tensiometer whose ring was replaced with a thin glass plate.

Temperature constancy was maintained with a previously described apparatus (1), (2).

The spreading isotherms were constructed by points; that is, for each surface area value the attainment of

equilibrium was assured by waiting until there were no
further variations of the surface pressure. The surface
pressure values were approximated to ± 0.1 dyne/cm
and the surface area values to ± 0.02 m²/mg.

Results and Discussion

We consider the results with the various
spreading solvents separately.

1. Spreading Solutions: Chloroform and Dichloro-acetic Acid

All spreading isotherms reported in graph 1
were obtained from solutions of the polymer in
chloroform containing 0.8% by volume of dichloroacetic acid. The latter was added to assure
complete solubilization of the polymer, the quantity mentioned being the minimum necessary to
attain this condition.

All the isotherms reported show an arrest in
surface pressure beyond 10 dynes/cm. It is thus
necessary to consider the curves as composed of
two parts, one superior and one inferior to the
arrest pressure.

From the experimental $\pi - A$ curves we calculated the following equations of bidimensional
state for the range of pressures lower than the
arrest pressure:

$$15\,°C \quad \pi A = 0.052 + 0.648\,\pi - 0.0048\,\pi^2,$$
$$20\,°C \quad \pi A = 0.035 + 0.681\,\pi - 0.0053\,\pi^2,$$
$$25\,°C \quad \pi A = 0.035 + 0.705\,\pi - 0.0056\,\pi^2.$$

With a procedure analogous to that used for
other macromolecules (1) we calculated the following parameters from the above equations:

1) B_1 coefficient of the first degree term in the
   bidimensional state equation theoretically deduced by *M. L. Huggins* (5);

2) Z' coordination number of the bidimensional
   pseudolattice of *Singer's* theory (6), not directly deduced from the latter theory but

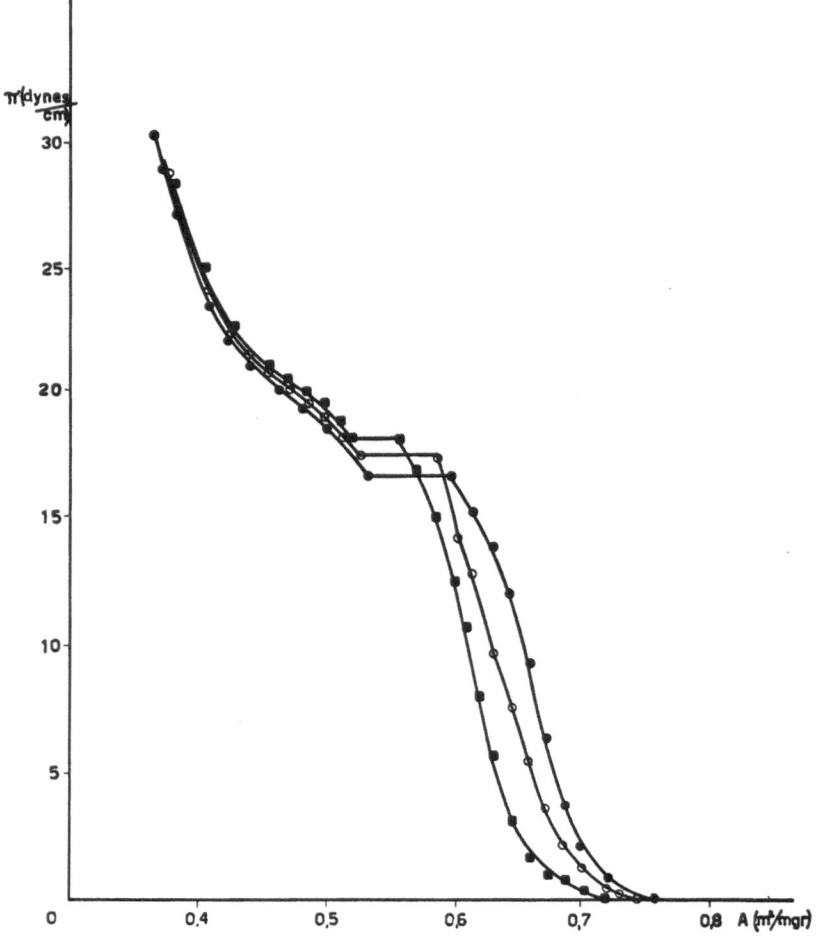

Fig. 1. Spreading isotherms
obtained from solutions of
PBLA in chloroform +
0.8% of dichloroacetic acid

rather from a comparison of the virial coefficients of the bidimensional state equations obtained from the experimental results with the approximate equation proposed by *Huggins* (5);

3) f_m partial submersion factor of the theory of *Frisch* and *Simha* (7), calculated in the same way as the preceding parameter.

The values of the above parameters for the three temperatures examined together with the values of A_0, limiting area, obtained by extrapolation of the curves in the range of pressures just inferior to the arrest pressures, are reported in Table 1.

Table 1. Parameters of PBLA in bidimensional state. Pyridine-free spreading solutions

Temperat.	A_0	B_1	Z'	ω	f_m
15 °C	0.650	0.997	2.22	11%	0.81
20 °C	0.680	1.000	2.22	11%	0.81
25 °C	0.698	1.010	2.22	11%	0.82

Spreading solutions containing 80% pyridine

Temperat.	A_0	B_1	Z'	ω	f_m
15—25 °C	0.530	1.018	2.23	11.5%	0.83

It is possible to make the following observations regarding this table:

a) the values of the limiting areas are in satisfactory agreement with those values previously reported by other authors, but do not constitute a valid criterion for deducing the most probable conformation assumed by the macromolecules at the surface. In fact, the same value is attributed to the α form by some authors (8) and to the β form by authors (9);

b) the values of Z' and ω (percentage of flexibility) even considering the approximations made in their deduction (1, 2), indicate that the macromolecule is moderately flexible at the water-air interphase. The values of the two parameters, which are not too high, are in agreement with a bidimensional helical configuration since, as is known, polypeptidic type macromolecules possess a greater flexibility at the interphase than even the more rigid forms of synthetic polymers. Naturally even these two parameters cannot definitively indicate the macromolecular conformation present at the water-air interface;

c) the f_m values show that at all temperatures studied about 20% of the macromolecule is submerged in the substrate. This fact is perfectly plausible and is verified in most macromolecular films of proteins and polypeptides, in which the presence and the distribution of numerous hydrophilic groups renders interaction between polymer and aqueous substrate plausible.

The above parameters would thus seem to indicate the presence of a bidimensional helical macromolecule, which is not very flexible and is partially submerged in the substrate, at the water-air interface.

A further contribution to the definition of the macromolecular conformation at the interphase is borne, as previously shown both by us (1) and by other authors (10), by the thermodynamic spreading functions.

In Table 2 are reported the spreading entropy ΔS_s and the spreading enthalpy ΔH_s for five surface area values. The thermodynamic functions are given only at 20 °C since it was deemed opportune to calculate the variation of π as the difference between the experimental value of the surface pressure at 25 °C and the corresponding value at 15 °C.

Table 2. The thermodynamic spreading quantities of PBLA from free-pyridine spreading solutions

Area m²/mg	$\Delta S_s \dfrac{\text{erg}}{\text{cm °K}}$	$\Delta H_s \dfrac{\text{erg}}{\text{cm}^2}$
0.72	+ 0.10	+ 28.7
0.70	+ 0.16	+ 45.6
0.68	+ 0.38	+ 108.8
0.66	+ 0.68	+ 193.7
0.64	+ 0.82	+ 231.9

As it is possible to ascertain in this table the values of both the entropy and the enthalpy of spreading are never negligible. This means that, as previously noted (1, 10), the polymer interface system cannot be considered to be athermic; that is, ideal. In particular the positive sign of ΔH_s would seem to indicate the existence of attractive energies between the macromolecular segments, naturally other contributions being taken as remaining constant. This deduction is confirmed by the fact that the solvation process is present (as shown by the f_m values) and that in such a process, if the hydrophilic and polar

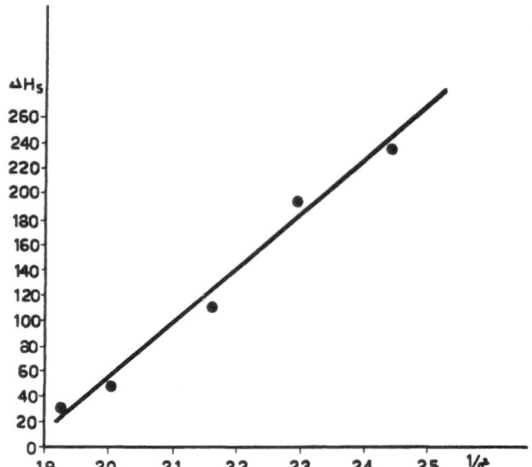

Fig. 2. Plot of spreading enthalpy versus $1/A^2$ for PBLA from free-pyridine spreading solutions

groups of the macromolecule and the length of the hydrophobic chain are considered, the prevalence of hydration of the polar groups, and as a consequence the negative sign of the corresponding enthalpic contribution, is seen to be evident. Further confirmation that the positive value of ΔH_s depends prevalently on the attractive energy between macromolecular segments is given by the fact that ΔH_s is nearly linearly dependent on $1/A^2$ (graph 2) (10).

From the above it results that in the range of surface pressures lower than the arrest pressure a partially submerged and rather rigid macromolecule, with non-negligible cohesive energies between macromolecular segments, is stable on the surface.

These deductions are in satisfactory agreement with the α helical form, thus concording with other authors for this and for other macromolecules with different methodes (4, 8, 11).

This means that the interphase does not modify the conformation, which is stable in tridimensional phase in the spreading solution (12), as shown by other authors (13). It must therefore be admitted that the interactions between macromolecular segments are greater than those between the polymer and the substrate. This hypothesis is confirmed by the fact that, the variation of the second virial coefficient with the temperature being small, the surface temperature ϑ at which in fact polymer-substrate interactions do not exist is found to be very low.

Let us now briefly examine the characteristics of the arrest of the surface pressure which is

present in all isotherms obtained from chloroform solutions containing small quantities of dichloroacetic acid.

In general an arrest in the spreading isotherm of only one component means a first order phase transition or a separation from the surface of part of the monolayer; that is, a collapse.

Previous authors have proposed for several polypeptides the mechanism of passage of α forms from mono-to bilayers to explain the presence of analogous arrests in the isotherms (4, 8, 14).

In our case it has been experimentally observed that from the arrest pressure on there is an evident decrease of the initial surface pressure after each compression, a phenomenon which is characteristic of collapse processes. Moreover, the very low and negative value of the transition entropy (-0.15 erg/cm^{2}K) is in agreement with the affirmations of other authors concerning the formation of the bilayer (4, 8, 14). Finally, if it is recalled that the arrest surface pressure corresponds to that of equilibrium between the monolayer and the collapsed form, the collapse process can be considered as an equilibrium transformation, as has already been noted by other authors (15).

General thermodynamic relations of the following type can consequently be applied:

$$\ln \pi_1/\pi_2 = -\frac{\Delta H}{R}\left[\frac{1}{T_1} - \frac{1}{T_2}\right]$$

where π_1 and π_2 are the surface arrest pressures at the temperatures T_1 and T_2, and ΔH is the transition enthalpy from monolayer to bilayer. In our case this enthalpy is about 60 erg/cm^2, taking it to be constant in the interval $15-25\,°C$ and taking as the surface area the mean area of the arrest at 20 °C. This value is in good agreement with the work values required to displace a unit area of monolayer from the surface of the water to form the bilayer, as calculated in another way by *Malcom* in the case of other polymers (4). This constitutes a further proof that the arrest must be considered as a collapse of the α form.

Thus the portions of the spreading isotherms existing at pressures higher than the collapse pressure must be considered as compression curves of collapsed monolayers, and therefore not interpretable with state equations which are valid only for monomolecular films.

2. *Spreading Solutions: Chloroform + Dichloroacetic Acid + 80% Pyridine*

In Fig. 3 are reported the isotherms at the temperatures 15—20 and 25 °C, obtained using a mixture of chloroform and pyridine (80%) with 0.8% of dichloroacetic acid as the spreading solvent. The line represents the mean of the curves at the three temperatures whereas the points represent the values at the single temperatures. It must first be noted that the particular case taken during spreading and the high reproducibility of the curves obtained with different surface concentrations and with spreading solutions of varying concentration allow us to exclude that the difference between the curves previously obtained from spreading solvents without pyridine and this curve can be attributed to loss of polymer from the interphase due to the solubility of pyridine in the aqueous substrate. This is con-

firmed by the fact that the polymer losses calculated to obtain the curve reported in graph 3 should be very high, around 20%, and that the isotherms obtained from pyridine-rich solutions never show a surface pressure arrest. It must thus be concluded that the form and the molecular orientation at the interphase are different from those previously defined and attributed to the α form.

In order to define the characteristic form of the isotherm reported in graph 3, since the surface pressure can be considered to be independent of the temperature, from the mean $\pi - A$ values we obtained, as previously done in the case of spreading solutions not containing pyridine, the corresponding bidimensional state equation $\pi A - \pi$:

$$\pi A = 0.044 + 0.539\,\pi - 0.0078\,\pi^2 + 4 \times 10^{-6}\,\pi^3\,.$$

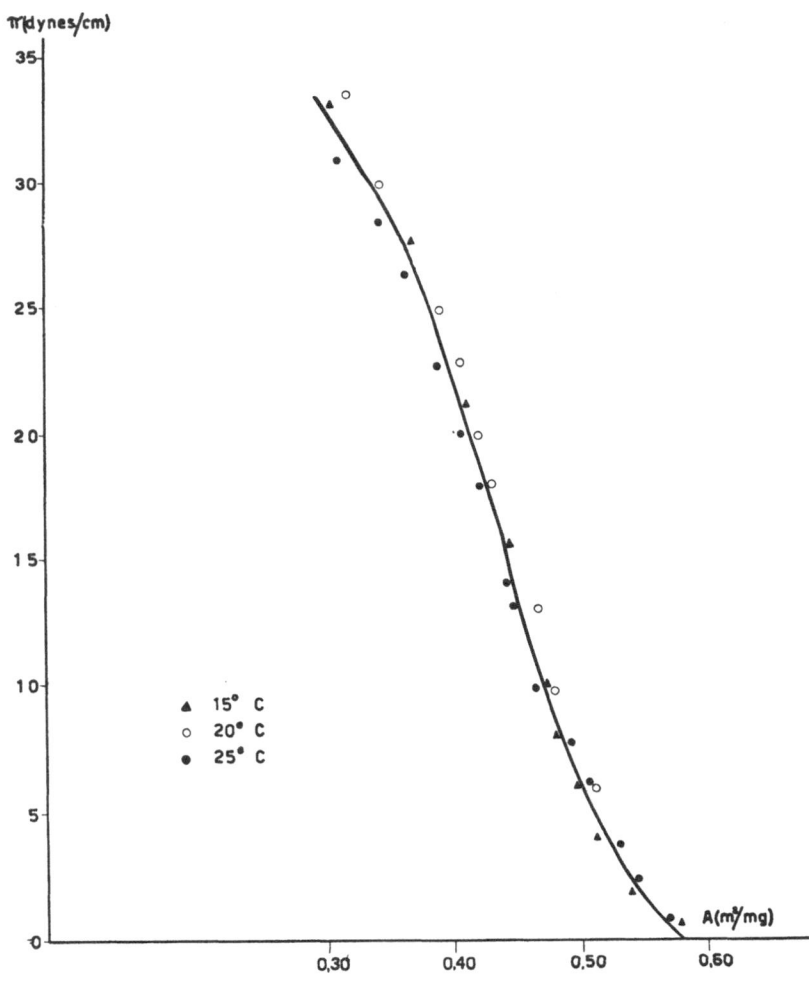

Fig. 3. Spreading isotherms obtained from solutions of PBLA in chloroform + dichloroacetic acid + 80% pyridine

From the virial coefficients of this equation we calculated the same parameters as before, which are reported in Table 1 together with the limiting area obtained by extrapolation of the $\pi - A$ curve in the highest pressure rectilinear portion. From this table the following observations can be made:

1) The value of the limiting area (18 A^2) is less than the corresponding value found for the form held to be α helical. A similar contraction of area effect has been observed by *Loeb* (4) for polymethylglutamate films obtained from spreading solutions composed of mixtures of chloroform and pyridine with a high percentage of the latter the author attributes lower value of the surface area to the β form.

Naturally, as has already been noted, the value of the limiting area alone cannot constitute a criterion for distinguishing the two forms.

2) The value of Z, and consequently of ω, is of the same order of size as that found in the case of the previous form, but due to the approximations made in its deduction this value cannot be considered to constitute a valid means for distinguishing two helical forms. It can only be affirmed that the low value constitutes a valid criterion for excluding the presence of a random-coil form.

3) The f_m value demonstrates, in this case too, the existence of the partial submersion process and a percentage of submerged macromolecule of the same order of size as the α form. This, too, hydrophilic groups being the same, could allow us to exclude the presence of the random-coil form.

4) Since the spreading isotherms are temperature independent, the coefficient of the first degree term is also temperature independent. This, besides constituting a further confirmation of the different forms present at the surface using pyridine-rich spreading solvents, means that the interaction energies between monomeric segments present at the surface can be considered to be negligible. This is in agreement with the existence of a form which is more extended than the α form. On the other hand, the fact that the flexibility is nearly the same for the two forms would induce one to consider a random-coil form as being improbable, as mentioned above. The most probable extended form would seem to be the β. Moreover, the first degree coefficient is of the same order of size, and even less than the

corresponding coefficient of the equations of state of the α form. This is in contrast with the existence of a random-coil form whose state equation is characterized, as noted by other authors (16), by virial coefficients of the first degree term higher than those of the equations corresponding to helical forms.

5) Since all spreading isotherms are temperature independent, the spreading entropy must be considered to be negligible and the corresponding enthalpy to be small and of negative sign. This means that the attractive interaction energies found in the case of the α form have decreased, even if it is not possible to conclude that repulsive energies exist, since, as we have already mentioned, a negative contribution to the spreading enthalpy derives without any doubt from the submersion process, as shown by the lower than unity f_m value.

6) A further, albeit approximate and indirect, confirmation of the existence of the β form can be obtained from a comparison of the spreading enthalpies of the two forms. In fact, the difference between the spreading enthalpy of the form obtainable from chloroform and the corresponding enthalpy of the form obtainable from chloroform with 80% pyridine can furnish a value of the transition enthalpy if it is considered that the enthalpic contribution due to the process of partial submersion can be taken to be approximately the same, as justified by the f_m value. The value of the transition enthalpy calculated in this way, taking the limiting areas of the two forms as the surface areas and considering the corresponding pressures, is about 3500 calories/m.u, which is in good agreement with the $\alpha - \beta$ bidimensional transition enthalpy for other compounds (17), if the difference in the hydrophobic chains is considered.

The above considerations lead to the conclusion that a form different from the α and the random-coil is present at the surface. The existence of the β form thus seems plausible when the spreading solutions contain high percentages of pyridine. This is in complete agreement with previous findings and is confirmed with other methods by *Loeb* (4) in the case of polymethylglutamate.

3. Spreading Solutions: Chloroform + Dichloroacetic Acid + Variable Quantities of Pyridine

In graph 4 are reported spreading isotherms at 25 °C obtained from solutions of chloroform

Fig. 4. Spreading isotherms obtained from solutions of PBLA in chloroform + 0.8% of dichloroacetic acid + variable quantities of pyridine

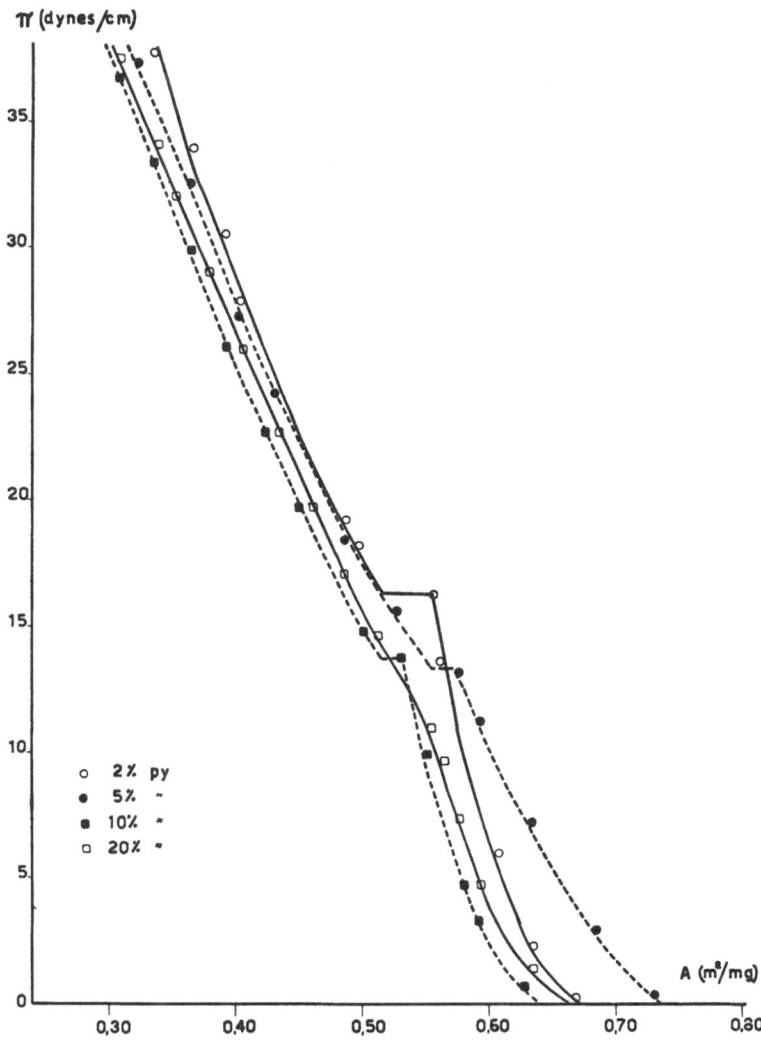

with 0.8% dichloroacetic acid and the following percentages of pyridine: 2, 5, 10, 20.

As can be ascertained therein, all isotherms show either an arrest in surface pressure or an inflection in the $\pi - A$ curve (20% pyridine) around the same surface pressure values at which arrest takes place in the isotherms obtained from spreading solutions not containing pyridine, from which we showed that the α form is obtained. This means that the interfacial macromolecules are at least partially present in the α form. To convalidate this deduction we considered the arrest variation as a function of the probable percentage of the α form: considering 100% to be in the α form in films obtained from solutions not containing pyridine and 0% to be in the α form in films obtained from solu-

tions containing 20% pyridine for which there is no notable surface pressure arrest.

The length of the arrests and the corresponding work $\pi \Delta A$ (where π is the arrest pressure, ΔA the difference between the areas at the beginning and the end of arrest) as a function of the percentage of the α form, calculated as above, are reported in graph 5. It can be seen that both graphs are rectilinear and thus constitute a confirmation that the pressure arrest is a property of the α form, which is therefore also present in films obtained from solutions containing quantities smaller than 20% pyridine.

In this case too the arrests can be attributed to a collapse of the α form. This is convalidated by the fact that the isotherms above the collapse pressure are completely different from the iso-

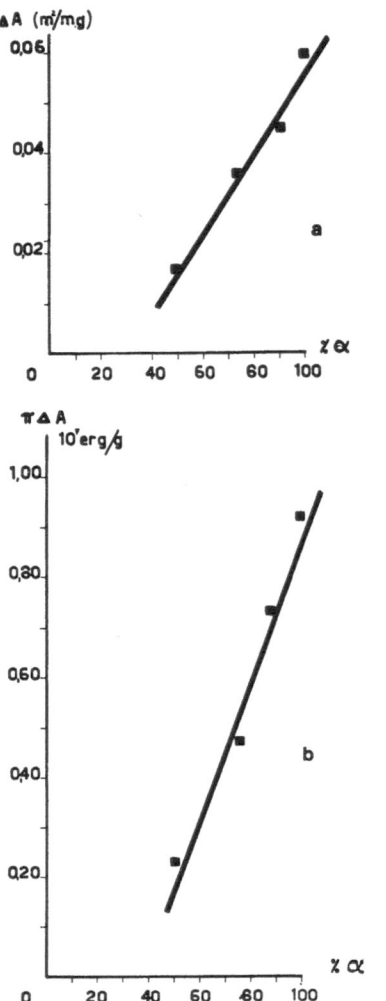

Fig. 5. Plot of: a) the length of the arrests ΔA; b) the corresponding work $\pi \Delta A$ versus the percentage of the α form

to whether solutions of chloroform or chloroform-pyridine mixtures (rich in the latter component) were used. The conformations, by means of the equations of bidimensional state and the thermodynamic spreading quantities, were attributed to the α- and β-helical forms, whose complete characterization is being studied by us with other methods.

The above means that on the one hand it is possible to induce different macromolecular conformations with the spreading solvent and on the other that, at least in the case of PBLA, the macromolecular conformation existing in solution may not be modified by the interphase.

This is particularly important since it permits us to compare polymer-solvent interaction in tridimensional phase with those in bidimensional state and thus to find substrate capable of conserving or modifying tridimensional conformations.

Summary

The pressure-area curves of poly-β-benzyl-L-aspartate (PBLA) are determined using various spreading solvents. The equations of bidimensional state and the thermodynamic spreading quantities agree in defining the probable presence of the α helix when chloroform containing small quantities of dichloroacetic acid is used as the spreading solvent, whereas spreading solutions containing about 80% of pyridine give rise to a different macromolecular form which is likely to be the β form. With spreading solutions containing intermediate quantities of pyridine bidimensional films containing variable quantities of the α form are obtained.

Zusammenfassung

Es werden die Kurven des Oberflächendrucks als Funktion des molekularen Flächeninhalts des Poly-β-benzyl-L-Aspartats (PBL) bestimmt, unter Verwendung von verschiedenen Spreitungslösemitteln. Die Gleichungen des zweidimensionalen Zustandes und die thermodynamischen Spreitungsgrößen stimmen darin überein, daß sie das Vorhandensein der α-Spirale als wahrscheinlich definieren, wenn man zur Spreitung Lösemittel benutzt, die aus Chloroform und kleinen Mengen von Dichloressigsäure bestehen, während man, wenn man von Lösungen ausgeht, die 80% Pyridin enthalten, eine andere Form der Makromoleküle im Film erhält, die plausiblerweise die β-Form ist. Auf Lösungen mit mittleren Mengen an Pyridin entstehen zweidimensionale Filme, die variable Mengen der α-Form enthalten.

therm of the β form. From the above it follows that in the monolayers obtained from spreading solvents containing rather low percentages of pyridine the α form is at least partially present.

Conclusions

The above experimental results demonstrate that, at least in the case of PBLA, the macromolecular conformation at the interphase depends, for a given substrate and macromolecule, on the spreading solvent, or better, on the conformation of the polymer in the spreading solution.

In particular it was possible to obtain two macromolecular forms on the surface according

References

1) *Gabrielli, G., M. Puggelli* and *E. Ferroni,* J. Colloid Interface Sci. **32**, 242 (1970).

Gabrielli, G., M. Puggelli and *E. Ferroni,* J. Colloid Interface Sci. **33,** 133 (1971).

Birdi, K. S., G. Gabrielli and *M. Puggelli,* Colloid & Polymer Sci. **250,** 591 (1972).

2) *Gabrielli, G., M. Puggelli,* J. Colloid Interface Sci. **35,** 460 (1971); **45,** 217 (1973).

3) *Gabrielli, G.* and *M. Puggelli,* J. Appl. Polymer Sci. **16,** 2427 (1972).

Gabrielli, G., M. Puggelli and *R. Faccioli,* J. Colloid Interface Sci. **37,** 213 (1971); **41,** 63 (1972).

4) *Ikeda, S.* and *T. Isemura,* Bull. Chem. Soc. Japan **34,** 416 (1961).

Malcom, B. R., Proc. Roy. Soc. **A 305,** 363 (1968).

Loeb, G. I. and *R. E. Baier,* J. Colloid Interface Sci. **27,** 38 (1968).

5) *Huggins, M. L.,* Makromol. chem. **87,** 119 (1965).

6) *Singer, S. J.,* J. Chem. Physics **16,** 872 (1948).

7) *Frisch, H. L.* and *R. Simha,* J. Chem. Physics **27,** 702 (1957).

8) *Malcom, B. R.,* Nature **219,** 929 (1968). — J. Polymer Sci. **34,** 87 (1971).

9) *Ikeda, S.* and *T. Isemura,* Bull. Chem. Soc. Japan **34,** 416 (1961).

10) *Llopis, J.* and *D. V. Rebollo,* J. Colloid Sci. **11,** 543 (1956).

Llopis, J. and *J. A. Subirana,* J. Colloid Sci. **16,** 618 (1961); Proc. 3rd Intern. Congr. Surface Activity (Köln, 1960).

11) *Loeb, G. I.,* J. Colloid Interface Sci. **26,** 236 (1968).

Kummer, J., J. M. Ruysschaert and *J. Jaffé,* VI. Internationaler Kongreß für grenzflächenaktive Stoffe, Band II (1), 284 (1973).

12) *Bradbury, E. M., A. R. Downie, A. Elliot* and *W. E. Hanby,* Proc. Roy. Soc. **A 259,** 110 (1960).

13) *Deboeck, Y.* and *J. Jaffé,* VI. Internationaler Kongreß für grenzflächenaktive Stoffe, Band II (1), 305 (1973).

14) *Malcom, B. R.,* Polymer **7,** 595 (1966).

15) *Joos, P.,* Bull. Soc. Chim. Belges **78,** 207 (1969).

16) *Jaffé, J., J. M. Ruysschaert* and *W. Heca,* Biochim. Biophysica Acta **207,** 11 (1970).

17) *Glazer, J.* and *A. E. Alexander,* Trans. Faraday Soc. **47,** 401 (1951).

Authors' address:

Prof. *Gabriella Gabrielli*
Instituto Chimica-fisica
Via Gino Capponi 9
I-50121 Firenze (Italy)

Progr. Colloid & Polymer Sci. **58**, 178—186 (1975)

Department of Physical Chemistry, Faculty of Pharmacy, University of Santigo de Compostela (Spain)

Action of polyvinyl pyrrolidone on the interaction of polysilicic acid with human serum albumin monolayers

J. Miñones, E. Iribarnegaray, S. García Fernández*) and P. Sanz Pedrero

With 6 figures and 4 tables

(Received August 24, 1974)

Introduction

Most foreign particles and bacteria taken into the human body by inhalation are ingested by phagocytic cells (macrophages) in the lungs and are enclosed in membrane-bounded vesicles called phagosomes. In the next step primary lysosomes fuse with the phagosomes and release their enzymes into them. Bacteria are killed and digested by this way. Non-toxic particles, such as the carbon ones, remain enclosed in residual bodies which may persist in the cell. However, certain inhaled particles, such as several crystalline forms of silica, induce the formation of pathological lesions in which the normal tissue is replaced by fibrotic tissue.

According to *Vigliani* and *Pernis* (1) it seems that the first step in the silicotic process is the damage of the phagosome membrane because of its contact with silica. The reactivity of silica particles appears to be due to the fact that silicic acid is formed on their surface. Silicic acid has hydroxyl groups (silanol) and can form powerful hydrogen bonds with suitable acceptor.

We have previously studied the action of polymerized silicic acid on protein monolayers (2), (3). It was observed the existence of a silicic-protein interaction due probably to the formation of hydrogen bonds between the silanol groups of the polysilicic acid and the secondary amide groups of the proteins. Keeping in mind that the proteins are characteristic of cellular membranes, such hydrogen bonding is sufficient to account for the disruption of lysosomal membranes, freeing the hydrolitic enzymes which kill

the cell, and lets silica particles to be taken up by other cells, causing their death also. The repeated death of phagocytic cells stimulates a reaction of fibroblasts, cells that synthesize and lay down nodules of collagen fibers (4).

In any case the agents responsible for the noxious effects of silica are the surface silanol groups. Therefore, it seems logic that the cure or prevention of the silicosis lies in the use of substances which can block the reactive (silanol) groups of silica, and thus prevent their interaction with proteins. Evidence in support of this interpretation comes from the observation made some years ago by *Schlipköter* and *Brockhaus* (5) that the polymer polyvinylpyridine-N-oxide (PNO) protects cells against the toxic effects of silica. This polymer is taken up into the lysosomes along with the silica. The oxygen atoms of PNO form hydrogen bonds with the silanol groups of silicic acid, and thereby prevent the latter from bonding with proteins, and thus attacking lysosomal membranes. Also the polymers of vinyl pyrrolidone can block the hydroxyl groups of silicic acid (6), (7).

As part of a general study of the silicic-protein interaction we have examined in this paper the action of PVP on the interaction between silicic acid and human serum albumin monolayers. The concentration of PVP necessary to produce the inhibition of the silicic-protein interaction, as well as the influence of polymerization time of silicic acid on it, has been investigated.

Experimental

Silicic acid. A solution of sodium silicate was prepared by heating at 200 °C 6 g of Aerosil (SiO_2) with 100 ml of NaOH 4 M. Once obtained

---

*) Present address: Department of Physical Chemistry. Faculty of Pharmacy. University of Barcelona. Spain.

a solution completely transparent, it was diluted to one liter with distilled water. This solution was treated with activated carbon for 12 h, and filtered before use to remove surface-active impurities. Its SiO_2 content was found by the gravimetric method of acid insolubilization, obtaining a concentration in SiO_2 of 0.1 M.

The silicate solution was brought to pH 6 (glass electrode) by adding 2 N acetic acid. Almost all silicic acid solutions were polymerized at pH 6 to a concentration of 0.08 M during 40 min at 20 °C.

Substrates. In order to obtain the different substrates on which the serum albumin monolayers were spread, the 0.08 M polysilicic acid solution was diluted in a buffer solution of acetic-sodium acetate of pH 5.4 and ionic strength 0.01. The concentration of polysilicic acid in the substrate was always 5×10^{-3} M (concentrations have been expressed as molarities respecting to SiO_2). At the same time to the substrate prepared by that way, it was added PVP in different proportions in such a way that the concentration of it should be comprehended between the limits of 50 to 90 mg/l. PVP of molecular weight 24.000 was supplied by Fluka A.G.

Spreading solution. The human serum albumin (HSA) was a crystalline sample supplied by Fluka A.G., and it had puriss. grade. Its isoelectric point, measured electrophoretically, was of 5.1. The solution used to spread the monolayers was prepared dissolving the protein in a buffer solution of acetic-sodium acetate of pH 5.1 and $\mu = 0.01$. Amyl alcohol (0.5% v:v) was added to this solution to increase the spreading rate of the monolayer. All protein solutions had a concentration of 0.1 mg/ml.

The $\pi - A$ isotherms were measured with an automatic *Langmuir* apparatus. The speed of compression in all experiments was 0.16 cm/sec. Pressure measurements were made after enough time, usually 5 min, that allows to reach the equilibrium.

Colorimetric method for determining soluble silica. The experimental determination of ortho, oligo and polysilicic acids that there are in the solution of polymerized silicic acid has been carried out spectrophotometrically being based on the formation of a yellow complex between the orthosilicic acid and the molybdic acid (8) which obeys the *Lambert-Beer* law (9). At a wave length of 400 mm the absorbance of the formed silicomolybdic complex measured in function of time

gives the neccesary data for the knowledge of the silica amount in its different forms (10).

Molybdic acid was prepared just before being used adding 40 ml of a 10% ammonium molybdate solution to 500 ml of water, then adding 100 ml of 1 N sulfuric acid and diluting the entire solution to 1 liter.

The silicomolybdate yellow complex is formed by mixing 2 ml of the polymerized silicic acid solution diluted up to a concentration in SiO_2 of 5×10^{-3} M, with 48 ml of molybdic reactive acid. Under these conditions the reading of the optical density, realized after 6.5 min, gives the present silica amount as orthosilicic acid form, while the difference between the absorbance measured when a constant value is reached and the one obtained at 6.5 min allow us to calculate the amount of SiO_2 as oligosilicic acids form. The silica that is found in form of polysilicic acids is obtained by the difference between the total silica and the one measured spectrophotometrically.

The value of 6.5 min taken as the time to determinate the amount of orthosilicic acid is due to the fact that the development of the yellow colour of the silicomolybdic complex depends on time, being obtained as a constant value of absorbance at such time in the case of founding the silicic acid in its ortho form (Fig. 1). The ortho-

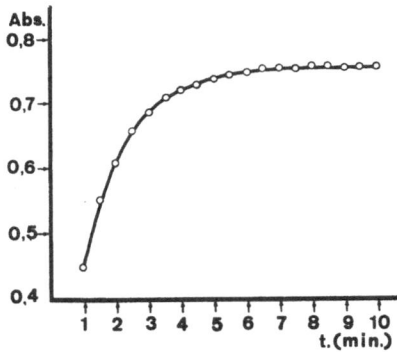

Fig. 1. Plot of absorbance vs. time for silicomolybdic yellow complex

silicic acid used in this comprobation was formed by adding the sodium silicate solution in a very thin stream from a pipet to violently agitated cold sulfuric acid in an amount to give a final pH around 2.5 and a silica concentration about 0.1%. Under these conditions both polymerization and depolymerization of silicic acid are quite slow (11).

Results

When the HSA was spread on a silica-free substrate the compression and decompression isotherms of protein monolayer show the existence of a noticeable hysteresis, typical of protein films (12). To the pH from which the monolayers are obtained (pH = 5.4) the second curve of compression coincides practically with the first one and the same thing happens with the decompression ones (Fig. 2 A, dotted curves). The limiting area is 0.88 m²/mg, which shows an acceptable spreading (13), and the collapse pressure is very close to the value of 18 dyn/cm.

When polysilicic acid is added to the substrate the isotherms obtained (Fig. 2 A. Unbroken curves) show the existence of a notable interaction between the silicic acid and the protein. The most remarkable modifications produced in the

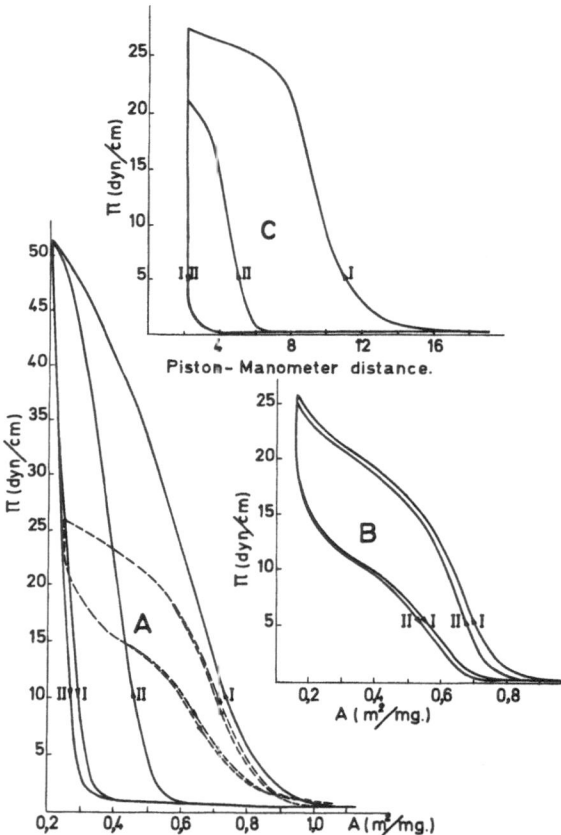

Fig. 2. π-A isotherms of HSA on substrate of pH = 5.4.
A) ----- substrate μ = 0.01 without silicic acid.
——— μ = 0.03 and polysilicic acid.
B) Substrate μ = 0.03 and 80 mg/l of PVP-24.
C) PVP-silicic acid adsorption film. Substrate: 40 mg/l of PVP-24 and polysilicic acid

monolayer as a consequence of the interaction are: a decrease in its compressibility, an increase in its stability and hysteresis area and a notable deviation between the first and second compression curve.

This silicic-protein interaction can be inhibited by incorporating to the substrate an adequate quantity of polyvinyl pyrrolidone.

Due to the fact that PVP is surface-active and is adsorbed in the air water interface forming a film by adsorption, the HSA has been spread on a substrate containing 80 mg/l of PVP with the aim of observing if the protein monolayer thus formed shows notable differences with the normal one. The results obtained are shown in Fig. 2 B, in which can be observed that the shape of the isotherms is very similar to that corresponding to the dotted curves of Fig. 2 A. The presence of PVP in the substrate, in the concentration mentioned before, or next to it, produces on human serum albumin monolayers a slight reduction in its specific area and at the same time a slight increase in its hysteresis area. So that the results show the existence of a very weak interaction between the HSA monolayer and the PVP adsorbed film.

As results of the interaction between polysilicic acid and PVP one adsorbed monolayer appears of which the isotherm lines of compression and decompression are shown on the Fig. 2 C. In this figure, the surface pressure was used as a function of the piston-manometer distance, since measurement of the amount of material in the surface layer is difficult and makes impossible to calculate the specific film area. The film obtained is very rigid, with a collapse pressure of 32 dyne/cm, a great hysteresis and a great deviation is observed between the curves that correspond to the first and second compression. As a result, the silicic acid causes the same type of modifications on the PVP monolayers as are produced on HSA. Therefore it is logical to suppose that when the three substances are present there is a competition between the polymer and the protein to bind the silicic acid.

Fig. 3 shows the results obtained when a human serum albumin monolayer is spread on a substrate containing PVP, in smaller quantities than the inhibition ones, and polysilicic acid. Curves were recorded after 30 min before the spreading of protein, without verifying the surface cleaning of the substrate. This period of time permits to diffuse the PVP molecules to the surface. Under these conditions, the mixed mono-layer obtained will be formed by PVP-silicic acid and by HSA and the silicic acid that has not interacted with PVP. In the abscissa axis the specific area was only referred to the amount of serum albumin spread, without considering that in the surface there is also a certain quantity of adsorbed PVP which is difficult to determine. In the low-pressure region the results obtained are similar to those recorded when a PVP-silicic acid monolayer is compressed. When the pressure is very close to the value of 16 dyne/cm a sharp break occurs and, probably, the silicic-PVP complex is squeezed out, leaving the monolayer formed by silicic acid-human serum albumin. Under these conditions, in the high-pressure region, the film compressibility and

stability depends on the larger or smaller inter-action between both components (silicic acid-serum albumin), which is a function of the ini-tial concentration of PVP in the substrate.

When this concentration is low, 50 mg/l (Fig. 3 A), the film shows a great stability (high col-lapse pressure) and rigidity. So, the silicic-serum albumin interaction is very remarkable. The con-trary happens when the concentration of PVP is high, 70 mg/l (Fig. 3 B). It is convenient to remark the fact that the recompression curves (curves II) are similar to those obtained when the HSA is spread on substrates containing silicic acid (A) or on silica-free substrates (B), which suggests that under these conditions the remaining monolayer in the interface is the HSA one interacted (A) or not (B) with the silicic acid.

When the spreading of HSA was verified im-mediately after cleaning the surface of substrate and the isotherms were recorded 5 min after spreading the protein, curves of Fig. 4 were ob-tained. Monolayers show different features ac-cording to the PVP concentration in the sub-strate. The collapse pressure, the extent of hys-teresis and the deviation between the first and the second compression curves decrease as the PVP concentration is increased, and when this reaches the value of 80 mg/l a monolayer is ob-tained whose characteristics are similar to those of films spread on silica-free substrates (Fig. 4 D).

The differences observed in the specific or limit-ing area of different monolayers are very small, and so the above mentioned criteria are applied in order to value the degree of HSA-silicic inter-action.

The course followed by the successive com-pression and expansion curves depends on the pressure reached in the first isotherm. By this way, in order to compare the hysteresis areas of the different monolayers, these have been com-

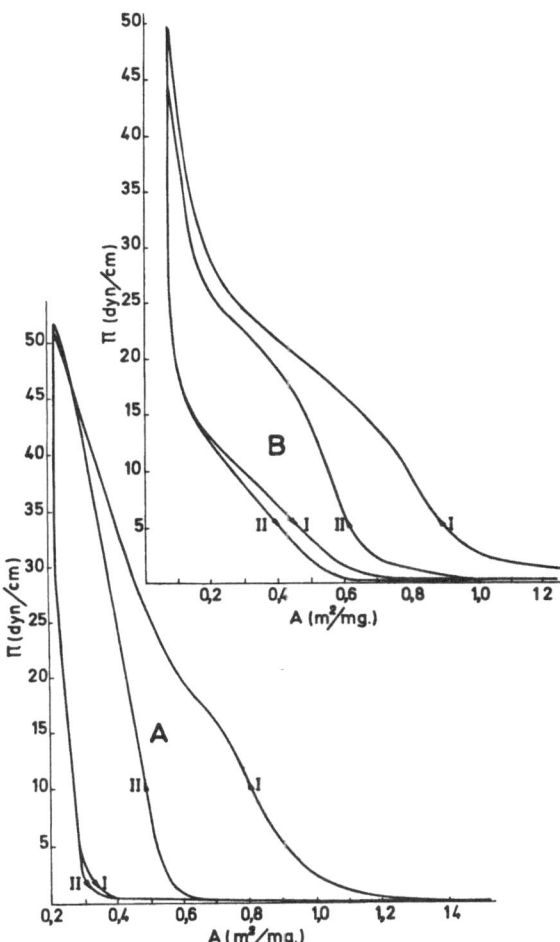

Fig. 3. π-A isotherms of HSA on substrate pH = 5.4, μ = 0.03 with polysilicic acid and PVP-24.
A) concentration of PVP, 50 mg/l,
B) concentration of PVP, 70 mg/l

Table 1. Effect of the substrate PVP concentration on the area of hysteresis cycle for HSA-silicic acid films

Concentration	Area of hysteresis	Area comprehended between 1st and 2nd compression curve
(mg/l)	(cm²)	(cm²)
50	61.0	34.0
55	48.0	24.0
65	44.0	8.5
80	18.0	4.0

pressed up to a pressure of 52 dyne/cm. Nevertheless, in the case of the curves D, it has not been possible to carry out the compression up to the mentioned value. Areas of hysteresis cycle, as well as the areas comprehended between the first and the second compression curve are shown in Table 1. Both areas decrease as the PVP concentration of substrate is increased, i.e., as the inhibition caused by PVP on the interaction between the silicic acid and HSA increases.

Curves of Fig. 4 correspond to the average of 3 measurements performed successively on the same substrate. The time elapsed between the recording of the first and the third isotherm is about 1 h. In this interval of time, and even if it is more, the reproducibility of the measure-

ments is quite acceptable, such as it is shown comparing the curves A of Fig. 5, obtained after 1 h 35 min of adding the silicic acid to the substrate and the curves C of Fig. 4, that were recorded under the same conditions as the former, but after 20 min of adding the silicic acid. Nevertheless, when the measurements are performed long after adding the silicic acid to the substrate, the characteristics of isotherms obtained depend on the same one. As the time increases, the isotherms obtained show a diminution of the interaction between the HSA and the silicic acid (Fig. 5). Even though in this figure the three compression curves do not show great differences between them, the expansion curves are the ones that clearly manifest the changes and differences

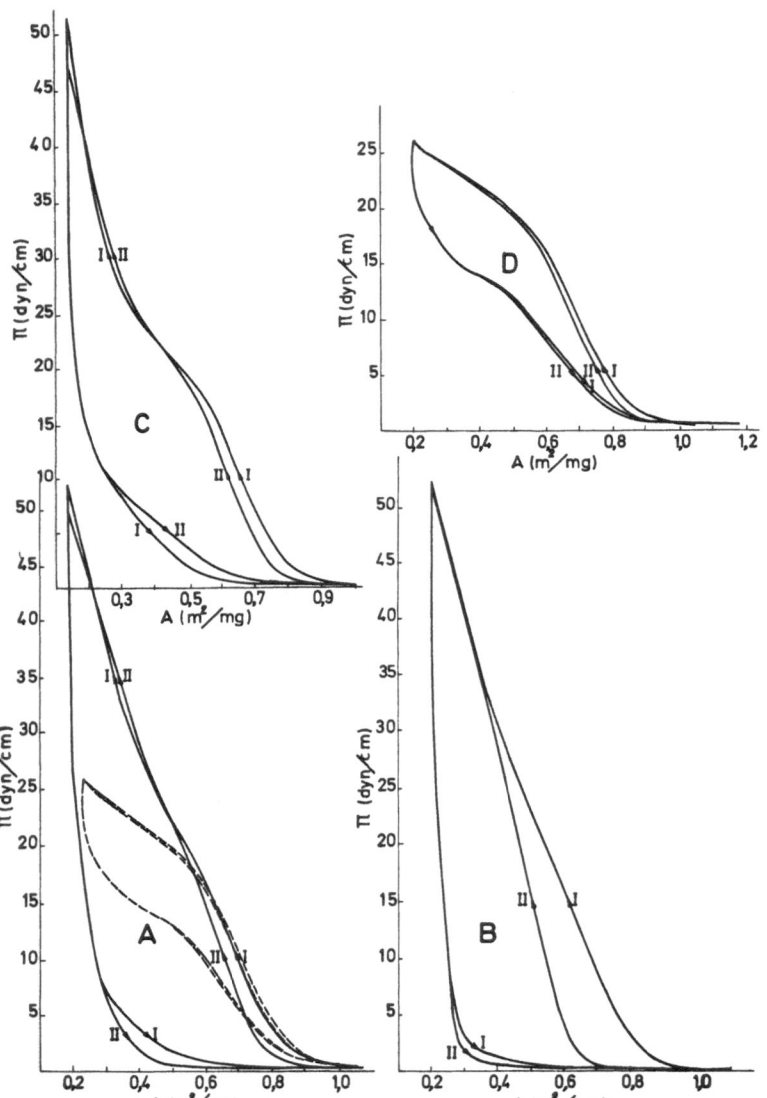

Fig. 4. Inhibition of silicic-HSA interaction by PVP-24. π-A isotherms of HSA on substrate pH $= 5.4$, $\mu = 0.03$ with polysilicic acid and PVP-24 at concentrations:

A) 50 mg/l,   B) 55 mg/l,
C 65 mg/l,   D) 80 mg/l

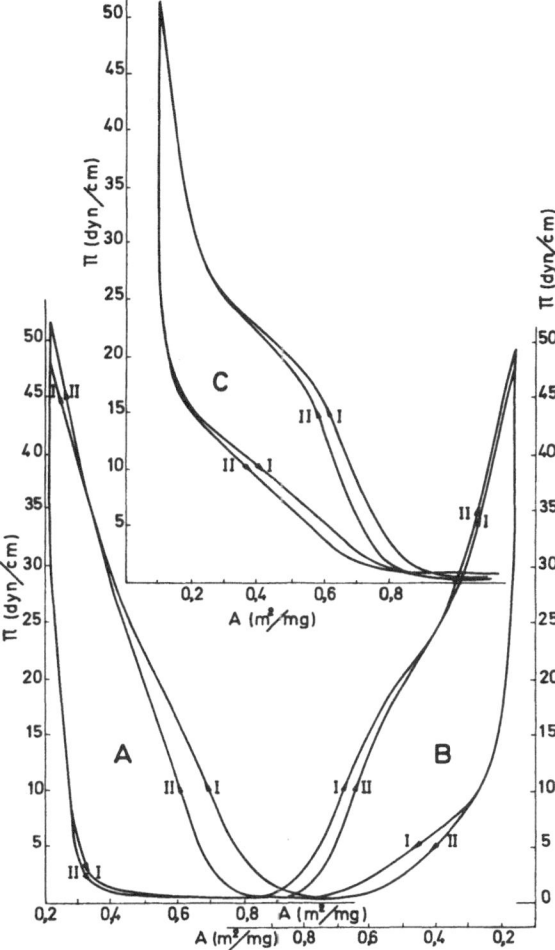

Fig. 5. π-A curves of HSA on substrate pH $= 5.4$, $\mu = 0.03$ with polysilicic acid and 65 mg/l of PVP-24. Curves A were recorded after 1 h 35 min of adding the silicic acid to the substrate. B) 7 h 43 min. C) 22 h 53 min

tained (Fig. 6 A, dotted curves) were about the same as those obtained on silica-free substrates. As the polymerization time of silicic acid is increased, its interaction with HSA monolayers increases, being obtained a maximum at $t =$

Table 2. Effect of silicic acid polymerization time on the area of hysteresis cycle for HSA-silicic acid films. PVP concentration of substrate: 70 mg/l

Time of polymerization (min)	Area of hysteresis (cm²)	Area comprehended between 1st and 2nd compression curve (cm²)
5	13.0	2.0
30	44.0	6.0
40	48.0	13.0
60	37.5	4.5
90	15.0	2.5

Table 3. Proportion of ortho, oligo and polysilicic acid as function of polymerization time

Time of polimerization (min)	Orthosilicic acid %	Oligosilicic acid %	Polysilicic acid %
0	95.8	4.2	0
2.5	76.6	5.9	17.5
5.0	62.0	8.0	30.0
7.5	52.0	6.0	42.0
10.0	39.0	5.6	55.4
15.0	29.1	3.5	67.4
20.0	24.3	3.3	72.4
25.0	28.0	3.0	76.2
30.0	18.7	2.1	79.2
35.0	14.2	2.5	83.3
40.0	9.7	1.3	89.0

in the interaction. On the other hand the area of hysteresis cycle passed from a value of 42.8 cm² in the isotherm A to 41.6 cm² in the B and 32.2 cm² in the C, i.e. they decrease as the time elapsed is increased, what shows a diminution in the silicic acid-HSA interaction.

Monosilicic acid does not interact with HSA monolayers. In order that interaction takes place, the silicic acid must be polymerized (3). So, the protein-silicic acid interaction and its inhibition caused by PVP depends on the degree of polymerization of silicic acid, which in turn is a function of time.

When the substrate contains 70 mg/l of PVP and silicic acid (polymerization time, 5 min) the compression isotherms of HSA monolayers ob-

Table 4. Depolymerization of polysilicic acid throughout the time. Concentration of solution: 5×10^{-3} M pH $= 6$

Time (min)	Orthosilicic acid %	Polysilicic acid %
5	10.6	88.3
30	16.4	83.2
60	19.2	78.6
105	23.3	76.0
120	27.5	71.6
180	29.0	70.0
210	34.2	65.0
240	36.2	63.0
300	39.0	60.0
360	39.5	60.0
420	41.5	58.0

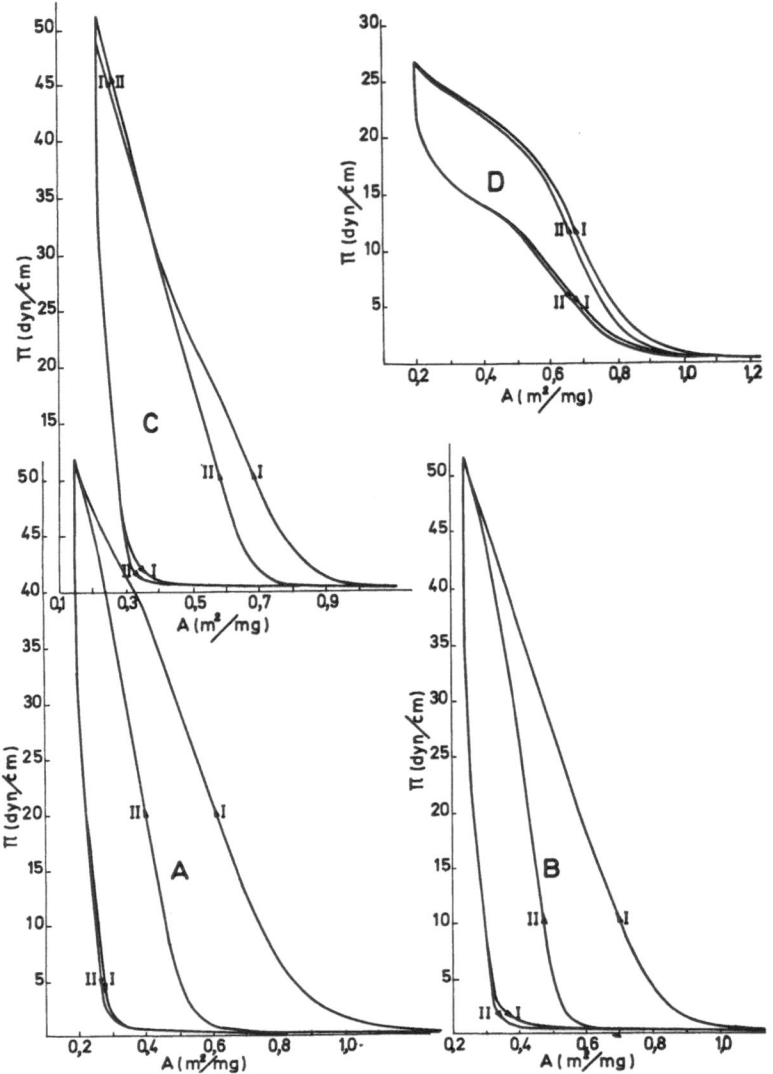

Fig. 6. π-A curves of HSA on substrate pH = 5.4, μ = 0.03 with 70 mg/l of PVP-24 and silicic acid.
Polymerization time of silicic acid:

A) ---- 5 min, —— 30 min.
B) 40 min. C) 60 min.
D) 90 min

40 min (Fig. 6B). When the time of polymerization is larger than 40 min, the silicic acid-HSA interaction diminishes, and if $t = 90$ min, there is no interaction (Fig. 6D). According to this, the area of hysteresis cycle, as well as the one comprehended between the first and second compression curves increases with the polymerization time until getting a maximum value to decrease afterwards. The values of the above mentioned areas are shown in Table 2.

When the sodium silicate solution was brought to pH 6 by adding 2 N acetic acid, and silicic acid solutions were polymerized during different intervals of time, the spectrophotometric results on the proportion of ortho, oligo and polysilicic acids, corresponding to the different times of polymerization, are shown in Table 3.

The depolymerization by dilution of polymerized silicic acid for 40 min was also followed spectrophotometrically, the results being the ones that are shown in Table 4.

Discussion

In order that the silicic-protein interaction takes place, the silicic acid must be polymerized (3), i.e., the presence of silanol groups in the silicic acid molecule is not enough, but these must be distributed in a proper way, as it happens in the polysilicic acid molecule, to combine with the protein through the formation of hydrogen bonds. According to these results when the monosilicic acid interacts with HSA monolayers the structure of protein is unchanged. So the curves obtained are similar to those recorded when the HSA is spread on silica-free substrates. In this way it could be thought that there exists an exchange of the water molecules, bound to the protein, by the monosilicic acid, without producing an apparent modification in the protein structure. Nevertheless, when the silicic acid is polymerized, a polysilicic acid molecule would react with several protein molecules binding them together into aggregates with a close-packed structure, what explains the rigidity and stability of the monolayers thus obtained.

Silicic acid also interacts with polymers of vinyl pyrrolidone, even though in this case it is not necessary that it will be polymerized (*E. Iribarnegaray and J. Miñones*, unpublished). The interaction was thought to be due with the formation of hydrogen bonds between the oxygen or nitrogen atoms of PVP which contain lone-pair electrons and the silanol groups of silicic acid.

Logically it could be thought that an interaction between PVP and HSA monolayers should also exist by the formation of the same type of bonds between the oxygen or nitrogen atoms of PVP and the secondary amide groups of the protein. Nevertheless the results of Fig. 2 B show the existence of a very weak, or nearly zero, interaction. This is probably due to the fact that the concentration of PVP used in this experiment was very small (80 mg/l), and lower than the one that is normally used in the interaction of HSA or PVP with silicic acid, since the concentration in SiO_2 in these cases was 300 mg/l.

Even though that concentration of PVP (80 mg/l) is not enough to produce notable modifications in the HSA monolayers, it is enough to block the silanol groups of polymerized silicic acid, and thereby prevent its interaction with HSA monolayers (Fig. 4 D). So the inhibition produced by PVP on the silicic-protein interaction is probably due to a competition between the polymer and the protein to bind the silanol groups of silicic acid. When the concentration of the former is enough to block such silanol groups, there is no possibility of interaction between the protein and the silicic acid. According to this hypothesis the inhalation of PVP could lead to the prevention of silica toxic effects.

Silicic-protein interaction decreases as the time elapsed since the moment that polysilicic acid is added to the substrate is increased (Fig. 5). The *possible interpretation* could be founded in the

spectrophotometric results on the depolymerization of polysilicic acid by dilution (Table 4). These results show that, though slowly, the diluted solutions of polysilicic acid ($C = 5 \times 10^{-3}$ M) become depolymerized as the time is increased. For this reason the interaction with HSA becomes smaller, since, as we have mentioned, monosilicic acid does not interact with this protein. Besides, the presence of PVP in the substrate causes an additional diminution of the silicic-protein interaction. After 7 h of diluting the polysilicic acid in the substrate, its proportion measured spectrophotometrically is 58% (Table 4), which should provoke a strong interaction with HSA (*E. Iribarnegaray and J. Miñones*, unpublished). Nevertheless the mentioned interaction is weak (Fig. 5 A), because the presence of PVP in the substrate blocks a part of the silanol groups of silicic acid, leaving only a small fraction of them which are able to combine with the protein.

In a similar way can be explained the existence of a maximum in the silicic-protein interaction when the polymerization time of silicic acid is 40 min, and this is found when in the substrate there are 70 mg/l of PVP (Fig. 6). The existence of the maximum is due to the fact that there are still silanol groups non-blocked by the polymer and so able to interact with the protein. When the polymerization time is increased a progressive substitution of non-blocked silanol groups is produced by siloxane groups. So the interaction with the protein becomes less intense till disappearing. When the polymerization time is lower than 40 min, the interaction diminishes as time does, due to the fact that even though there are progressively more non-blocked silanol groups, in turn the proportion of silicic acid as polymer form is smaller (Table 3); so, the interaction also is.

Summary

The interaction between polysilicic acid and human serum albumin monolayers can be inhibited by the action of polyvinyl pyrrolidone (PVP). This effect is produced by a competition between the polymer and the protein for the silanol groups of polysilicic acid. The results obtained show that in a substrate containing 0.30 g/l of SiO_2 in form of polysilicic acid a concentration of 80 mg/l of PVP is enough to produce the inhibition of the silicic-protein interaction.

The inhibition caused by a certain quantity of PVP depends on the polymerization time of silicic acid and on the time elapsed from the moment ˙in which the acid is added to the substrate until the record of the isotherms starts.

Zusammenfassung

Die Wechselwirkung zwischen Polykieselsäure und Humanserumalbumin in monomolekularer Schicht kann durch die Wirkung von Polyvinyl-Pyrrolidon (PVP) verhindert werden, weil eine Wechselwirkung des Polymeren mit den Silanolgruppen mit deren Wechselwirkung mit dem Protein konkurriert. Die erhaltenen Ergebnisse zeigen, daß in einem 0,30 g/l SiO$_2$ in Form von Polykieselsäure enthaltenden Substrat eine Konzentration von 80 mg/l PVP ausreicht, um die Wechselwirkung Kieselsäure-Protein zu unterbinden.

Die von einer gewissen Menge PVP erzeugte Unterbindung dieser Wechselwirkung hängt einerseits von der Polymerisierungszeit der Kieselsäure ab sowie andererseits von dem Zeitraum zwischen der Säurezugabe zum Substrat und dem Beginn der Isothermen-Aufzeichnung.

References

1) *Vigliani, E. C.* and *B. Pernis,* J. Occup. Med. **1,** 319—1329 (1959).
2) *Miñones, J., S. García Fernández* and *P. Sanz Pedrero,* in Proc. First. Eur. Biophys. Congr., Vol. I, pp. 203—207 (Wien, 1971).
3) *Miñones, J., S. García Fernández, E. Iribarnegaray* and *P. Sanz Pedrero,* J. Colloid Interface Sci. **42,** 503—515 (1973).
4) *Heppleston, A. G.,* in Inhaled Particles III, pp. 357 to 369 (1971).
5) *Schlipköter, H. W.* and *A. Brockhaus,* Klin. Wschr. **39,** 1182—1189 (1961).
6) *González Carrero, J.,* Pharm. Acta. Helvet. **38,** 529 to 538 (1963).
7) *González Carrero, J.,* Med. Segur. Trab. **20,** n° 80, 5—24 (1972).
8) *Dienert, F.* and *F. Wandenbulcke,* Compt. Rend. **176,** 1478—1480 (1923).
9) *Strickland, J. D. H.,* J. Amer. Chem. Soc. **74,** 868 to 871 (1952).
10) *Weitz, E., H. Franck* and *M. Schuchard,* Chem. Z. **74,** 256—257 (1950).
11) *Alexander, G. B.,* J. Amer. Chem. Soc. **75,** 5655 to 5657 (1953).
12) *Dervichian, D. G.,* Kolloid-Z. **126,** 15—20 (1952).
13) *Miñones, J., E. Iribarnegaray, S. García Fernández* and *P. Sanz Pedrero,* Colloid & Polymer Sci. **250,** 318—324 (1972).

Authors' address:

S. García Fernández
Department of Physical Chemistry
Faculty of Pharmacy
University of Barcelona, Spain

Progr. Colloid & Polymer Sci. **58**, 187—194 (1975)

Instituut voor Chemie-ingenieurstechniek, Katholieke Universiteit te Leuven, Heverlee (Belgium)

A force field description for entanglements in polymer systems

J. Mewis and *G. Schoukens*

With 3 figures and 1 table

(Received August 15, 1974)

1. Introduction

The basic molecular models for describing the mechanical behaviour of polymer fluids assume the physical picture of isolated molecules (1, 2). Except for dilute solutions this assumption must be regarded as unrealistic. Several investigators have tried to incorporate the effect of entanglement in their models. These attempts are partly handicapped by the lack of knowledge about the exact nature of intermolecular interaction (3, 4). Some success has been obtained with a procedure suggested by *Bueche* (5), which expresses the increased resistance to flow due to the presence of neighbouring molecules by a larger value for the molecular friction factor. The procedure, later modified by *Ferry et al.* (6), gives an improvement but still fails at conditions where a high degree of entanglement prevails (7). Thus no accurate theory is available for almost all situations of practical interest.

Recently other possibilities for expressing the interaction have been applied; these are entanglement density (8), a friction factor varying with distance (9, 10), and additional local forces along the chain (11, 12). None has led to a solution of the problem, though progress has been made. In the present work attention has been concentrated on simple methods of super-imposing entanglement effects on existing molecular models. The interaction effects to be incorporated can be made function of stress, strain or strain rate, resulting in non-linear behavior (13). Attempts to develop non-linear models on this basis have proved partly successful (8, 14) and a similar extension of the proposed model will be explored.

2. A Force Field Model for Entanglement Effects

Even if the real nature of the interaction forces is sufficiently known, their incorporation in a molecular model would entail considerable mathematical problems (15). However such a physical picture is not available at present and it is not even evident that the accuracy of the basic molecular models warrants such a detailed modification at all (7).

That a detailed picture is not even required is supported by *Bueche*'s approach, which often gives reasonable results; also, it supports the possibility of expressing entanglement as a force field. In addition there is ample evidence that systems with quite divergent molecular structure show the same general feature in their mechanical behaviour (16).

Therefore, a simplified and general description of interaction effects seems justified.

Usually the analyses of *Rouse* and *Zimm* (1, 2) are taken as the starting point for subsequent modifications. In their approach a molecular chain is divided into a discrete number of sub-elements, beads and springs. Similarly the forces are made discrete and their action is localized in the sub-elements. A normal mode analysis of the resulting system of dynamic equations describes the rheological behavior by a discrete relaxation spectrum. Entanglements obviously cause coordinate movements of different molecules. Thus the system of dynamic equations to be solved becomes correspondingly larger (12).

The following argument leads to an expression of the forces exerted by the surrounding molecules in terms of a force field applied on the molecular chain under consideration. This enables one to reduce the system of dynamic equations to the size of that for a single non-interacting molecule. Local fluctuations of interaction with time are replaced by their time average. Thus the entanglement probability is incorporated in the intensity function; this could lead eventually to a loss in dispersion of relaxation times, a shortcoming which is not considered important at this stage.

Any other procedure would require knowledge of entanglement kinetics, which is not available. Models based on kinetic arguments have been suggested (8, 7), but then other sources of dispersion are neglected.

In the present approach the variation of entanglement intensity throughout the chain must be given. Contradictory opinions have been put forward in this respect (9, 10, 11). A simple physical picture will be applied here. As we are dealing with fluids, no permanent links between molecules should exist. Consequently the time averaged force field should be a smooth function of place. It is possible to emphasize the momentarily discrete nature of entanglement forces, even in a fluid. Indeed, the two approaches do not exclude each other. Therefore the basic, smooth, force field will be investigated first.

It is normally accepted that interaction forces are short range forces and consequently presuppose close contact between chain elements of different molecules. Therefore the outer beads have, statistically speaking, a larger chance of being subjected to forces from other molecules. The probability will decrease in a yet unknown manner, towards the center of the molecule. The change with MW could be used to estimate the distribution (10), but then an additional assumption has to be made about the local change with MW as only the total effect is known. Before trying more elaborate fields a linear change of intensity of entanglement with distance from the center of gravity will be assumed.

The effect of surrounding molecules will be divided in two parts. First, the interpenetration of chains causes the beads of different chains to interact with each other even in the absence of any motion.

Second, when the material is subjected to shear the same force field is assumed to persist but besides, the interaction will cause an additional increase in viscous drag.

Hence the total entanglement effect will be described by two superimposed force fields. The first has spherical symmetry with radially oriented forces, increasing linearly with distance from the center of gravity; the second, a viscous drag effect, changes linearly with the distance from the flow line of the center of gravity, amounting to the description given by *Bueche* (5).

The suggested modification could be used in connection with several molecular models. As an illustration the *Rouse* model will be adopted in the subsequent analysis as it offers frequently a reasonable approximation to the mechanical behaviour of real polymer fluids (7).

3. Derivation of Model Equations

Since we want to follow the *Rouse* analysis, the individual molecules are represented by $N + 1$ identical beads interconnected by N *Gaussian* springs to form linear chains. The beads are located by the position vectors r_i $(i = 0, N)$, whereas the condition of the springs is determined by spring vectors s_i $(i = 1, N)$ (8):

$$s_i = r_i - r_{i-1}. \qquad [1]$$

Under equilibrium and without any interaction forces the system is characterized by a *Gaussian* distribution,

$$\Psi_0 = c \exp\left(-b^{-2} \sum_{i=1}^{N} |s_i|^2\right), \qquad [2]$$

where b is a measure for the mean end-to-end distance of the molecule.

Departure from the equilibrium condition changes the distribution from Ψ_0 to Ψ. The corresponding increase in free energy creates forces in the springs:

$$F_{s,i} = kT \frac{\partial \ln \dfrac{\Psi}{\Psi_0}}{\partial s_i} = -kT \frac{\partial \ln \varphi}{\partial s_i}$$
$$= \text{const.} (S_i - S_{i,0}), \qquad [3]$$

where $F_{s,i}$ represents the spring force in the spring i and the subscript 0 refers to equilibrium.

In addition to interval spring forces the molecule is subjected to hydrodynamic interaction with the solvent. This viscous drag $F_{v,i}$ is concentrated in the beads, and is characterized by a friction coefficient f_0;

$$F_{v,i} = -f_0[\dot{r}_i - (v_0 + \alpha \cdot r_i)], \qquad [4]$$

where v_0 is the solvent velocity at the origin and α signifies the rate of shear tensor.

The additional viscous drag due to entanglements will be expressed as an increased friction factor, $Q_e f_0$.

Now the interaction force field must be introduced. In the presence of entanglement the chains are intertwined and the position of the beads will be changed. The centers of gravity are assumed to be unaffected. Transforming the new position vectors to a frame with the center of gravity as origin, they will satisfy the condition.

$$\sum_{i=0}^{N} \boldsymbol{r}_i' = 0. \qquad [5]$$

The interaction force on the bead i can be written as

$$\boldsymbol{F}_{i,i} = - n' \cdot \boldsymbol{r}_i', \qquad [6]$$

where n' represents the intensity of the linear force field.

Through the particular shape of this force field the change in equilibrium position of the beads will be a proportional shift. Hence it can be absorbed in the distribution constants of eq. [2]. Therefore we will go on using eq. [2] for the equilibrium knowing the changed meaning of the constants. An eventual non-*Gaussian* distribution will not be considered here (19). The remaining force originating in interactions under flow will be

$$\boldsymbol{F}_{i,i} = - n'(\boldsymbol{r}_i' - \boldsymbol{r}_{i,0}'). \qquad [7]$$

We are now in a position to write the force balance on the beads, complying with the conditions of the *Rouse* model as for instance negligible inertia. Under shear the total force balance for bead i, with entanglements present, is given by

$$Q_e f_0[(\boldsymbol{v}_0 + \boldsymbol{\alpha} \cdot \boldsymbol{r}_i') - \boldsymbol{r}_i'] + kT \left(\frac{\partial \ln \varphi}{\partial \boldsymbol{s}_{i+1}} - \frac{\partial \ln \varphi}{\partial \boldsymbol{s}_i} \right) \\ - n'(\boldsymbol{r}_i' - \boldsymbol{r}_{i,0}') = 0. \qquad [8]$$

Taking the difference between the balance equations for beads i and $i-1$ eliminates the position vectors and gives

$$Q_e f_0(\boldsymbol{\alpha} \cdot \boldsymbol{s}_i - \dot{\boldsymbol{s}}_i) + kT \left[\frac{\partial \ln \varphi}{\partial \boldsymbol{s}_{i+1}} \right. \\ \left. - (2 - n'') \frac{\partial \ln \varphi}{\partial \boldsymbol{s}_i} + \frac{\partial \ln \varphi}{\partial \boldsymbol{s}_{i-1}} \right] = 0, \qquad [9]$$

where position vectors have been transformed to spring vectors and n'' designates (eq. [3])

$$n'' = \frac{n'}{\text{const.}}.$$

The behaviour of the entire chain is determined by a system of N similar equations. The resulting matrix equation can be written as [10]

$$Q_e f_0(\boldsymbol{\alpha} \cdot \boldsymbol{s} - \dot{\boldsymbol{s}}) + kT(\boldsymbol{A} - n''\boldsymbol{I}) \frac{\partial \ln \varphi}{\partial \boldsymbol{s}_i} = 0.$$

In this notation $\boldsymbol{A}$ designates the $N \times N$ *Rouse* component matrix (1):

$$A = \begin{bmatrix} 2 & -1 & 0 & 0 & . & . & 0 \\ -1 & 2 & -1 & 0 & . & . & 0 \\ 0 & -1 & 2 & -1 & 0 & . & 0 \\ 0 & . & . & . & . & . & 0 \\ 0 & . & . & . & 0 & -1 & 2 \end{bmatrix}. \qquad [11]$$

As far as the form of the equation is concerned the introduction of a spherical force field amounts to replacing $\boldsymbol{A}$ by $\boldsymbol{A} - n''\boldsymbol{I}$. The presence of the additional isotropic tensor $n''\boldsymbol{I}$ does not interfere with the normal mode analysis but will alter the eigen values of the solution. It is compatible with our assumption of high MW to assume a large value for N. Consequently the eigen values for the *Rouse* model in which we are interested become (7)

$$\mu_{p,R} = \frac{p^2 \pi^2}{N^2} \quad (p = 1, N). \qquad [12]$$

For our modified model the eigen values then are

$$\mu_p = \frac{p^2 \pi^2}{N^2} - n'' \quad (p = 1, N). \qquad [13]$$

The further analysis leading to a discrete relaxation time spectrum is similar to that for the basic model in terms of the eigen values. The normal *Rouse* relaxation times,

$$\tau_p = \langle l \rangle^2 Q_e f_0 / 6 kT \mu_p, \qquad [14]$$

in which $\langle l \rangle^2$ is the mean square length of a statistical chain, are now extended to a spectrum of the form

$$\tau_p = \frac{K}{p^2 - n} \quad (p = 1, N), \qquad [15]$$

where $n = (n'' N^2)/2$.

As in the other bead-spring theories the relaxation times are used together with a spring constant C. No attempt is made to derive the model parameters from first principles, a derivation that, to-date, remains unsolved. The following discussion will investigate whether the suggested modification provides a suitable means to describe the mechanical behaviour of high MW polymers, a case where the simple bead-spring models do not apply.

4. Discussion of the Model

4.1. Linear Behaviour

The type of behaviour predicted by the model can be visualized by calculating the dynamic storage (G') and loss (G'') moduli as a function of frequency (ω):

$$G'(\omega) = C \sum_{p=1}^{N} \frac{\omega^2 \tau_p^2}{1 + \omega^2 \tau_p^2} \quad \text{and}$$

$$G''(\omega) = C \sum_{p=1}^{N} \frac{\omega \tau_p}{1 + \omega^2 \tau_p^2} , \qquad [16]$$

where τ_p is given by eq. [15].

Under pronounced entanglement, i.e. τ_1 predominant (eq. [15]), it follows from eq. [16] that

$$G'(1/\tau_1) = G''(1/\tau_1) = C/2 = G''_{max} . \qquad [17]$$

At frequencies between $1/\tau_1$ and $1/\tau_2$, G'' goes through a minimum and G' shows a plateau zone, where

$$G'(\text{plateau}) = C . \qquad [18]$$

The extent of the plateau in a logarithmic plot will be determined by the relative values of the relaxation times.

As can be seen from eq. [14] the introduction of an increase in viscous drag brings about a proportional increase in relaxation times.

The spherical force field on the contrary alters the ratio between subsequent relaxation times. As in a fluid the largest relaxation time τ_1 can neither be infinity nor negative; n is bound to be smaller than unity (eq. [15]). This suffices, at least mathematically, for the particular force field under consideration to yield arbitrarily large values for τ_1. At the same time the ratios between subsequent relaxation times become larger although the largest possible increase decays quickly with increasing values of p. The global effect of entanglements is thus introduced as a telescoping of the relaxation times. This means that modes of deformation, involving coordinated motion of smaller parts of the chain are less affected by entanglement. Therefore a correction, as suggested by *Ferry et al.* (6) for the *Bueche* approach, is not necessary here. Roughly speaking pronounced entanglement causes the material to be described approximately by a

single relaxation time over a larger time or frequency domain. This is particularly so for the steady state flow where, even for the *Rouse* model, τ_1 is dominant (7).

For the zero shear viscosity and normal stress coefficient we find respectively

$$\eta = C \sum \tau_p \cong C \tau^2 (n \to 1) \quad \text{and}$$

$$\theta = 2 C \sum \tau_p^2 \cong C \tau_1^2 (n \to 1) . \qquad [19]$$

To evaluate the usefulness of the model its predictions will be compared with experimental data.

Equations [16] to [19] will be applied to data collected by *Vinogradov* and co-workers (20) to check their model for entangled molecules (11). As curves are presented for different molecular weights, the effect of this parameter can be estimated at once. Importantly, the present analysis is valid for monodisperse polymers only.

The polybutadienes, investigated by *Vinogradov*, give results that show a plateau zone of varying width. The level of the plateau however is independent of MW. From eq. [18] it is concluded that C must be independent of MW.

For the whole series

$$C = 6.3 \times 10^6 \text{ dyne/cm}^2 .$$

Consequently values for τ_1 can be calculated from the steady state viscosity (eq. [19]). In Table 1 the resulting relaxation times are reproduced. Similarly the position of G''_{max} can be employed to find τ_1 (eq. [17]). It was impossible to define the location of G'' accurately from the published figures. Nevertheless some estimated values are included in Table 1 for comparison. The correspondence is satisfactory.

The data do not extent far enough in the high frequency range to allow for a discussion of changes in τ_2.

Therefore the parameter K in eq. [15] has been considered constant and the dynamic moduli

Table 1. Terminal relaxation times for polybutadienes (20) calculated from eq. [17] and [19]

MW ($\times 10^{-5}$)	η (P)	$\tau_1(\eta)$ (sec)	$\tau_1(G''_{max})$ (sec)
3.2	3.14×10^7	4.98	
2.4	1.12×10^7	1.78	
2.0 (4)	7.1×10^6	1.13	0.9
1.51 (3)	1.6×10^6	0.25	0.2
1.03 (2)	6.3×10^5	0.1	0.08

have been calculated from τ_1 and a constant small value for τ_2. In effect n only is employed to describe the effect of MW. The predicted curves are represented in Fig. 1 and 2.

The simple version of the model, which has been used above, renders a good prediction of the $G'(\omega)$ curves. The pronounced minimum in $G''(\omega)$ is typical for discrete relaxation spectrum approximations of monodisperse systems. The discrepancy with experiment is mainly due to a persisting degree of dispersion in MW. Indeed a rough calculation, using a linear mixing rule, shows that a most probable MW distribution, with $M_w/M_n = 1.01$, results already in a considerable raising of the minimum as well as a broadening of the shoulder of the $G'(\omega)$ curves. Interestingly the prediction in the terminal zone

seems to be better than that obtained by *Vinogradov* using a more complex force field for the entanglement effects (Fig. 3).

The concentration of interaction phenomena in the longest relaxation must also be perceptible in relaxation tests.

Relaxation, in the linear region, after cessation of flow at shear rate $\dot{\gamma}$ is given by

$$\sigma(t) = C\dot{\gamma}\sum_{p=0}^{N}\exp(-t/\tau_p).\qquad[20]$$

With the large differences between τ_1 and τ_2 applied above, a considerable part of the relaxation curve would be represented by the first term only. Suitable experiments have been performed by *West* on polystyrenes (21). His results confirm the previous conclusion.

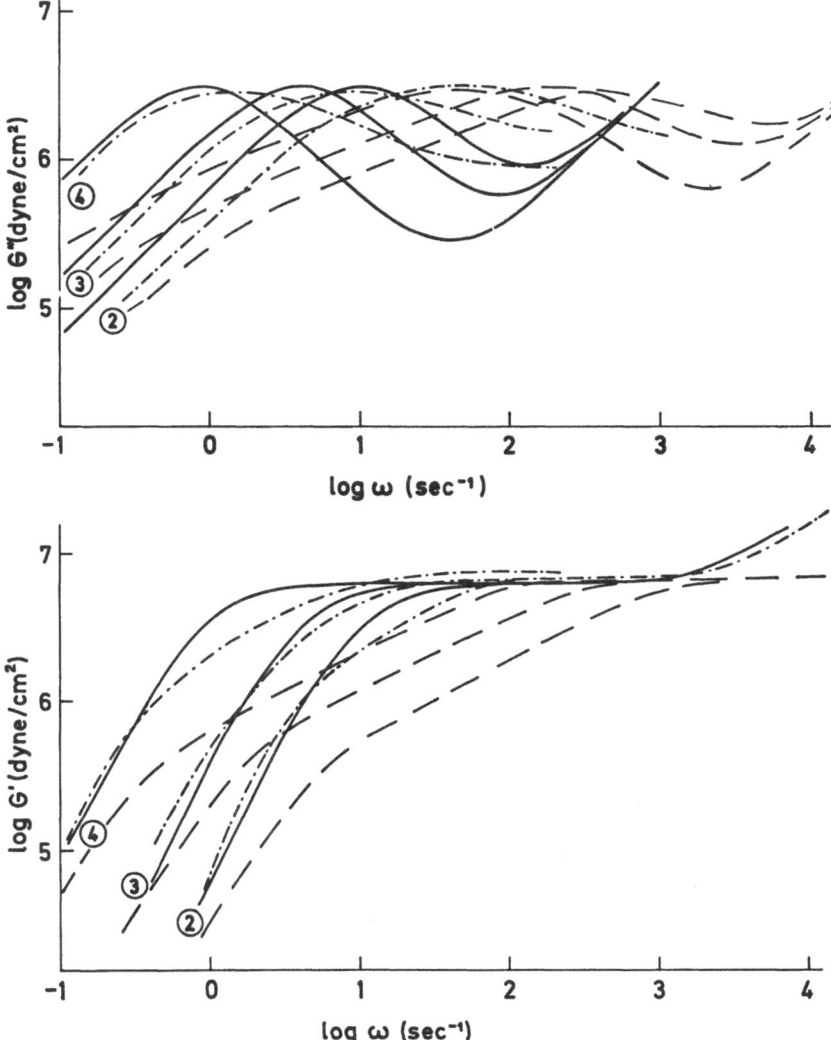

Fig. 1. Shear storage moduli, comparison with experiment;

$-\cdot-\cdot-$ experimental data taken from ref. (10),

——— calculated as a one parameter family from the model,

$-----$ calculated by means of *Vinogradov's* theory (numbers refer to Table 1)

Fig. 2. Shear loss moduli, comparison with experiment (legend vide Fig. 1)

A comparison with experimental data, as given above, does not demonstrate conclusively the advantage of the applied model or any other one (7). We conclude only that the three-parameter model shows the general format of a suitable spectrum for highly entangled chains, whereas the *Rouse* model for such conditions fails to describe the effect of MW on the rheological behavior of real systems (7).

The predominance of τ_1 as employed in the description of the polybutadienes causes the shear viscosity, the location of $G''_{\max}$ and the width of the plateau to change similarly with MW, viz. as τ_1. *Ferry* (7) and *Vinogradov* (10) give several references which support this conclusion.

In addition eq. [19] leads to an expression for the steady state shear compliance, J_e;

$$J_e = \frac{\theta}{2\,\eta^2} = \frac{1}{C}\,. \qquad [21]$$

As C is independent of MW, the same must hold for J_e. Experimentally an upper limit for J_e has been found at high MW for several polymers (7, 22—24). Evidently this is only true for monodisperse systems. We note that, irrespective of the physical picture employed for the model, the typical feature result from flexibility in adapting parameters to the data and from a large ratio τ_1/τ_2, which is brought about here by introducing the additional parameter n.

Clearly the procedure is not unique and different basic assumptions might lead to similar results. In this respect model can be compared with the analysis of *Forsman* and *Grand* (12), who find a shifting of the odd relaxation times as compared with the even ones.

Their model is based on a head-tail symmetry argument for interacting molecules. This argument is debatable as can be seen from the configuration of two entangled molecules. The telescoping of relaxation times necessary to describe their experiments (25) is also provided by our model. In many cases their results will be indistinguishable from ours.

The experimental evidence for their theory (25) also supports the present analysis.

4.2. Non-Linear Behaviour

In systems with considerable entanglement the material response becomes readily nonlinear as shown by the viscosity function. Spectral descriptions of the non-linear region have received attention recently (8, 10, 14, 26—28). Such an approach is useful as it provides a general picture of the rheological behaviour in a way that may be related to the underlying molecular mechanisms.

The suggestion has been made that deviations from linearity could be described by the shear rate dependence of the interaction force field (8, 13). The expression for the rheological properties derived above remain valid if the model parameters are made variable. If we consider steady state conditions, shear rate or shear stress could be employed as the independent variable. Generally K and n would be the shear dependent parameters as they express the effect of interaction, with, as concluded above, n predominant in the case of pronounced entanglement. Hence from eq. [21] one expects the apparent shear compliance to be independent of shear rate. Constant values of $J_e(\dot{\gamma})$ have been reported for polymers with narrow MW distribution (23, 24), although deviating behaviour has been encountered also (29).

One can also consider situations in which the dynamic conditions change with time. Then the parameter n, which has been used to describe non-linearity in the previous case, would also vary with time. For a full description the kinetics of the interactions needs to be taken into account.

No direct knowledge of the kinetics is available. Yet one could imagine that under certain circumstances of periodic testing an average value of n is reached after some time.

This could be explained as a situation where the recovery of entanglement proceeds much slower than breakdown.

Consequently the extent of entanglement does remain virtually constant during a cycle.

In such a case moduli can be determined, but they will differ from the values in the linear region. *Vinogradov* reports this kind of determination (10). In Fig. 3 his data for G' are compared with a model based on a change in n calculated from the given value for the viscosity. His description is included for comparison although it is obtained in a completely different manner.

Contradictory to *Vinogradov*'s analysis (10), his data do not show a change of the plateau level under finite shear. In our model this corresponds to a constant value of C as indicated by the linear data. A shift of the terminal relaxa-

Fig. 3. Effect of finite shear on storage modulus;
1: amplitude of deformation rate = 0 s^{-1},
2: amplitude of deformation rate = 2 s^{-1} (legend vide Fig. 1)

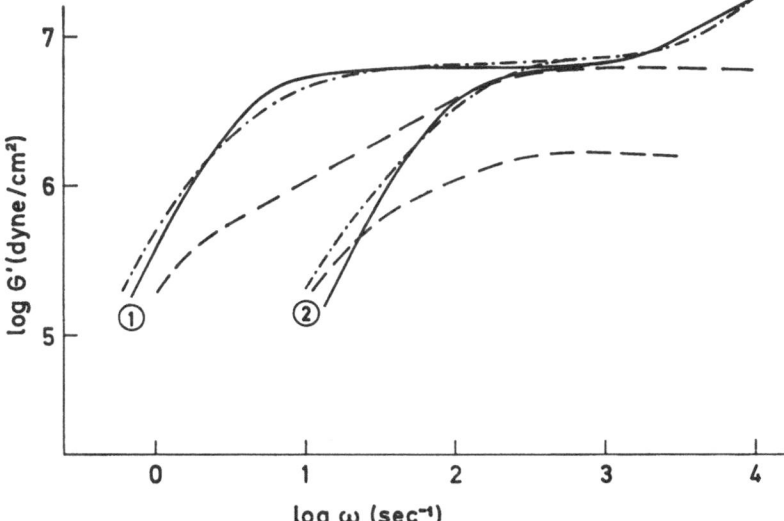

tion time results in a satisfactory description of the experimental data. We conclude that linear and non-linear data obtained using polybutadienes are consistent with a possible set of model parameters.

5. Conclusions

On the basis of a bead-spring model (*Rouse-Bueche*) and an additional spherical interaction force field, a rheological model for strongly interacting, monodisperse, macromolecules is developed. The three model parameters are, for the purpose of this work, considered as adjustable parameters to be determined from experimental data.

The model is shown to describe reasonably well some typical features of the materials to which model assumptions pertain to. In particular some shortcomings of the simple *Rouse* model can be overcome. A suitable description of MW effect is possible varying a single model parameter. Under these conditions the model predicts the same MW dependence for zero shear viscosity, the terminal relaxation time and the width of the plateau zone, whereas the equilibrium shear compliance is found to be independent of MW.

In the non-linear region the apparent shear compliance is found to be constant. Some effects of amplitude under dynamic testing could also be reproduced well.

The model renders the given behavior at one end and includes a *Rouse*-type spectrum at the other end. Therefore we conclude that it might be useful in describing polymer systems in general with a small number of parameters.

Acknowledgement

One of us (G. S.) wishes to acknowledge financial support provided by the Nationaal Fonds voor Wetenschappelijk Onderzoek (Belgium).

Summary

An extension of rheological bead-spring models that would encompass concentrated solutions and melts of high molecular weight polymers is investigated. As in some previous attempts the forces due to the interaction with surrounding molecules are represented by a force field. In a simple manner this force field takes into account the partial interpenetration of neighbouring molecules. A normal mode analysis renders a discrete relaxation spectrum for the model.

In further treatment the model parameters are considered as a set of adjustable variables. It is verified that the resulting form of the model describes the relevant features of the polymer systems under consideration.

The discussion is limited to monodisperse materials. The results are compared with experimental data taken from the literature as well as with other theories. The comparison covers the linear dynamic moduli, steady state shear flow and non-linear behaviour.

Of the molecular characteristics only molecular weight is considered. Its effect on the model parameters is investigated, particularly to find out whether the known deviations from the *Rouse* model could be described.

Zusammenfassung

Eine Ausweitung des rheologischen Kügelchen-Feder-Modells, die konzentrierte Lösungen und Schmelzen von hochmolekularen Polymeren mit umfaßt, wird untersucht. In einigen vorausgegangenen Versuchen dieser Art wurden die Wechselwirkungskräfte mit den umgebenden Molekülen durch ein Kraftfeld repräsentiert. In einer vereinfachten Weise berücksichtigt dieses Kraftfeld die teilweise Durchdringung benachbarter Moleküle. Eine Analyse auf normaler Basis ergibt ein diskretes Relaxationsspektrum für das Modell.

In weiteren Behandlungen werden die Modellparameter als ein Satz von justierbaren Variablen behandelt. Es wird verifiziert, daß die resultierende Form des Modells die wesentlichen Züge des betrachteten Polymersystems beschreibt.

Die Diskussion wird auf monodisperses Material beschränkt. Die Resultate werden mit experimentellen Daten aus der Literatur verglichen und auch mit anderen Theorien. Der Vergleich deckt die linearen dynamischen Moduln, die stationäre Scherung und nichtlineares Verhalten.

Von den molekularen charakteristischen Parametern wird nur das Molekulargewicht betrachtet. Sein Effekt auf die Modellparameter wird untersucht, besonders im Hinblick darauf, inwieweit die bekannten Abweichungen vom *Rouse*-Modell beschrieben werden können.

References

1) *Rouse, P. E.*, J. Chem. Physics **21**, 1272 (1953).
2) *Zimm, B. H.*, J. Chem. Physics **24**, 269 (1956).
3) *Ziabicki, A.* and *R. Takserman-Krozer*, J. Polymer Sci. **A 2**, 7, 2005 (1969).
4) *Thirion, P.*, VIth Intern. Congr. Rheol. (Lyon, 1972).
5) *Bueche, F.*, Physical Properties of Polymers (1962).
6) *Ferry, J. D., R. F. Landel* and *M. L. Williams*, J. Appl. Phys. **26**, 359 (1955).
7) *Ferry, J. D.*, Viscoelastic Properties of Polymers (1970).
8) *Graessley, W. W.*, J. Chem. Physics **43**, 2696 (1965).
9) *Chompff, A. J.* and *W. J. Prins*, J. Chem. Physics **48**, 235 (1968).
10) *Vinogradov, G. V., V. H. Pokrovsky* and *Yu. G. Yanovsky*, Rheol. Acta **11**, 258 (1972).
11) *Graessley, W. W.*, J. Chem. Physics **54**, 5143 (1971).
12) *Grand, H. S.*, Ph. D. Thesis, Univ. of Pennsylvania (1969).
13) *Chen, I. J.* and *Bogue, D. C.*, Trans. Soc. Rheol. **16**, 59 (1972).
14) *Tanner, R. I.*, A. I. Ch. E. Journ. **15**, 177 (1969).
15) *Chompff, A. J.* and *J. A. Duiser*, J. Chem. Physics **45**, 1505 (1966).
16) *Tager, A. A.* and *V. E. Dreval*, Russ. Chem. Rev. **36**, 361 (1967).
17) *Marrucci, G., G. Titimanlio* and *G. S. Sarti*, Rheol. Acta **12**, 269 (1973).
18) *Lodge, A. S.* and *Y. J. Wu*, Rheol. Acta **10**, 539 (1971).
19) *Tschoegl, N. W.*, J. Phys. Chem. **40**, 473 (1964).
20) *Vinogradov, G. V., Yu. G. Yanovsky, A. I. Isayev, V. P. Shatalov* and *V. G. Shalganova*, Intern. J. Pol. Mat. **1**, 17 (1971).
21) *West, G. H.*, Polymer **10**, 751 (1969).
22) *Tobolsky, A. V., R. Schaffhauser* and *R. Böhme*, Polymer Letters **2**, 103 (1964).
23) *Mieras, H. J. M. A.* and *C. F. H. van Rijn*, Nature **218**, 865 (1968).
24) *Graessley, W. W.* and *J. S. Prentice*, J. Polymer Sci. **A 2**, 6, 1887 (1968).
25) *Wolkowicz, R. I.* and *W. C. Forsman*, Macromol. **4**, 184 (1971).
26) *Bogue, D. C.*, Ind. Eng. Chem. Fundam. **5**, 253 (1966).
27) *Yamamoto, M.*, Trans. Soc. Rheol. **15**, 783 (1971).
28) *Leonov, A. I.*, in: Progress in Heat & Mass Transfer, Vol. 5 (Oxford, 1972).
29) *Sakai, M., T. Fujimoto* and *M. Nagasawa*, Macromol. **6**, 786 (1972).

Authors' address:

Prof. Dr. *Ir. J. Mewis* and *Ir. G. Schoukens*
Instituut voor Chemie-ingenieurstechniek
Katholieke Universiteit Te Leuven
de Croylaan, 2
B-3030 Heverlee, Belgium

Progr. Colloid & Polymer Sci. **58**, 195—200 (1975)

Division of Textile Physics CSIRO, 338 Blaxland Road, Ryde. N.S.W. 2112 (Australia)

X-Ray scattering by wool and polyacrylonitrile graft copolymers

R. B. Beevers

With 4 figures

(Received July 30, 1974)

1. Introduction

Vinyl monomers can be incorporated into wool by a variety of initiatory systems and the properties of the graft copolymers produced have been the subject of a number of investigations particularly in respect of the search for modified and improved wool textiles. Some studies of the X-ray scattering by these wool copolymers have been made (1—3).

Ingram et al. (1) have made electronmicroscopic and X-ray scattering studies of wool + polystyrene graft copolymers produced by gamma radiation initiation. This method was found to produce homopolymer in addition to grafted material. Electronmicroscopic examination of fibre sections showed some increase in contrast about the cell membranes and in the nuclear remnant sections. This was found to be mainly due to the presence of ungrafted polystyrene which could be extracted with benzene. At grafted polymer levels of less than 100% it was concluded that the deposition of polymer was uniform throughout the fibre section. X-ray scattering by the graft copolymer, which had been freed of homopolymer and isolated from ungrafted wool by enzymatic treatment, was identical with scattering from atactic polystyrene.

Simpson (3) has made an examination of wool fibres, graft copolymerized with polyacrylonitrile (PAN). X-ray studies, which were not reported in detail, indicated that incorporation of polymer produced a blurring of both the meridional and equatorial low angle spacings of the wool fibre and also an increase in the diffuse X-ray scattering at low angles.

In this paper the result of an examination of the high angle X-ray scattering by wool + PAN graft copolymers covering the composition range up to 87% PAN and of a horsehair + PAN graft copolymer containing 125% PAN is reported. This set of copolymers has been the subject of other studies (4, 5) and has been shown to contain little free homopolymer (6). Electronmicroscopic examination has not revealed any significant preferential deposition of polymer within any of the histological components of the wool fibre (7) and the graft copolymer system is similar, in this respect, to the wool + polystyrene graft copolymers.

2. Experimental

2.1. Samples

Fibres of Corriedale wool (batch SW1) which had been modified by graft copolymerization with acrylonitrile using the ferrous ion-hydrogen peroxide initiation system were available from the work of *D'Arcy, Hall* and *Watt* (8). Weight uptakes of polymer were determined on the basis of the dry weight of the wool. In addition a graft copolymer using horsehair and containing 125.2% PAN was prepared by Dr. *I. C. Watt*. A polyacrylonitrile sample, prepared by redox polymerization methods, was used to provide a powder sample for the X-ray work.

2.2. X-ray Method

X-ray measurements were made with CuK_α radiation collimated by a 0.2 mm diameter lead glass capillary and recorded on a flat film camera using Agfa-Gevaert Structurix FW film type D7. The specimen-to-film distance was about 4.5 cm and calibration was obtained by recording the 0.42 nm diffraction ring from a sample of crystalline polyethylene. Corrections for polarization and absorption of the X-rays were neglected since it was intended to use the data for comparative purposes. Specimen thickness of ca. 1 mm was close to the optimum value calculated by *Skertchly* and *Woods* (9) if a value of 12.5 is taken for the absorption coefficient of wool fibres *(Lincoln)*.

2.3. *Analysis of the X-ray Film*

Films were photometered along the equatorial scattering direction with a *Joyce-Loebl* double beam microdensitometer (Type Mk III B). The derived values of the optical density were adjusted to be relative to that of the film taking a region which was not exposed to X-rays. Correction was also made for air scattering. The relative optical density was plotted as a function of the reciprocal spacing (s^{-1}) given by,

$s = 0.5 \, \lambda/\sin \theta$

where λ is the wavelength and 2θ is the scattering angle. Derived equatorial scattering curves are shown in Fig. 1

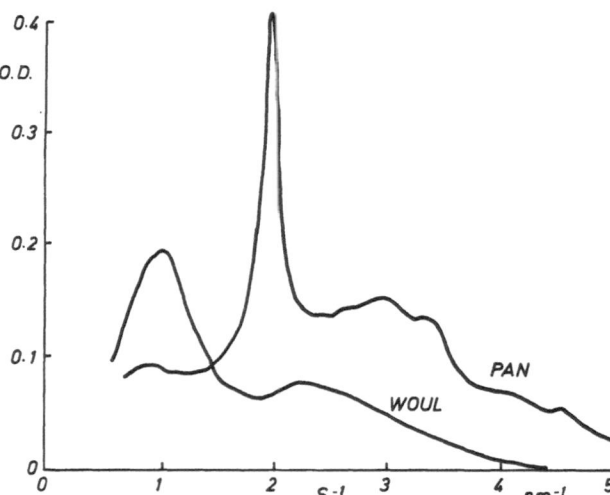

Fig. 1. Computed equatorial X-ray scattering shown as a function of the reciprocal spacing for a PAN powder sample and a Corriedale wool fibre

for a PAN powder sample, which has a prominent ring at 1.9 nm^{-1}, and for the untreated Corriedale wool with prominent scattering at 1.02 nm^{-1} characteristic of α-keratin. A diffuse region of scattering from the wool at 2.2 nm^{-1} partly overlaps the PAN ring. The X-ray diagrams are shown in Fig. 2.

2.4. *Determination of the Effect of Graft Copolymerization*

An examination of the effect of graft copolymerization has been made making the assumption that both components of the copolymer, i.e. the wool backbone and the PAN grafts, scatter X-rays independently. Also it has been assumed that the scattering functions for the two components are not modified by the copolymerization. The contribution to the X-ray scattering by the polymer, which is the main interest, may then be obtained from the difference between the scattering by the copolymer and the untreated wool fibre. This procedure serves to give a semi-quantitative analysis of the data but, as will be shown, the X-ray scattering by the wool fibre undergoes some modification by the incorporation of polymer so that for this copolymer system the devolution procedure can only be considered to be approximate.

In the practical application of the method the optical density data are adjusted to equal that for the untreated wool at $s^{-1} = 4.8$ nm^{-1}. This is close to the side of the X-ray film and as far removed as practicable from the main X-ray scattering by either component. The differential optical density curves obtained by this procedure are given in Fig. 3.

2.5. *Enzymatic Digestion*

A digest was prepared using 1% w/w papain and 0.1 M sodium bisulphite in 4 M urea. A trace of ethylenediamine tetraacetic acid was added to remove metal ions. The solution was filtered and adjusted to pH 7.0 with sodium hydroxide and then saturated with N_2. Treatment of the fibres with the mixture was carried out for 12 h at 50 °C.

3. Results and Discussion

X-ray diffraction diagrams are shown in Fig. 2 for, (a) horsehair (α-keratin); (b) a steam-drawn PAN fibre (see ref. 10); (c) horsehair + (125%) PAN graft copolymer and, (d) a diagram typical of a PAN powder sample. In Fig. 2(a) the prominent 0.98 nm equatorial spots and the sharp meridional scattering at 0.51 nm characteristic of α-keratin are shown together with an indication (on the original) of low angle meridional arcs and equatorial scattering. The prominent halo in the copolymer X-ray diagram shown in Fig. 2(c) in large part arises from the scattering

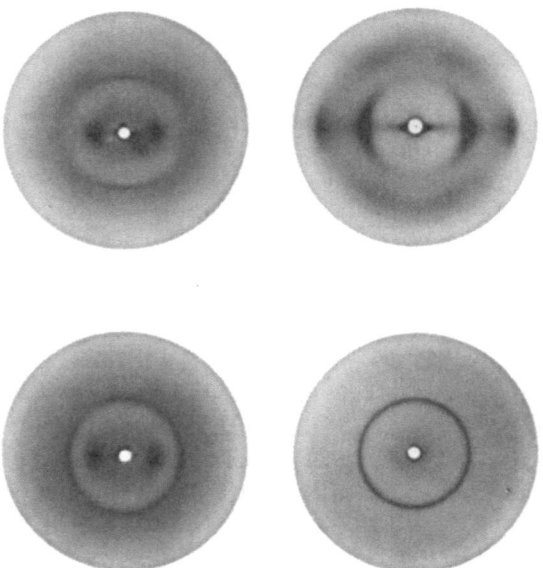

Fig. 2. X-ray diagrams for, (a) horsehair (α-keratin); (b) steam-drawn PAN fibre; (c) horsehair + (125.2%) PAN graft copolymer; (d) PAN powder sample. All diagrams were obtained with the same specimen-to-film distance

by the PAN grafts and is in the position which would be occupied by the ring from the PAN powder sample (cf. Fig. 2 (d)). In the meridional direction the PAN and wool scattering overlap, but along the equator the main reflections are reasonably well separated, so that equatorial scans are the most satisfactory to determine the scattering from the two components. There is however no region of the X-ray diagram where the scattering from the wool and polymer grafts can be unequivocally determined.

The PAN fibre diagram given in Fig. 2 (b) was obtained for an unannealed fibre spun from dimethylformamide into toluene and steam drawn × 6.5 (10). Strong but extended arcs are shown at 0.52 nm and 0.30 nm and significant low angle scattering can be detected (11, 12). There is also an indication of weak and diffuse meridional scattering.

The X-ray diagram for the horsehair + (125%) PAN graft copolymer given in Fig. 2 (c) shows the 0.98 nm scattering to be still a prominent feature and it is clear that graft copolymerization has not destroyed the α-keratin structure. There is some increase in the amount of diffuse low angle scattering as also reported by *Simpson* (3) and the weak meridional arcs shown in Fig. 2 (a) are not observed. Examination of Fig. 2 (c) shows that the polymer ring is not uniform in intensity and the X-ray scattering forms extended equatorial and meridional arcs. The X-ray diagram appears as a summation of the diagrams given in Fig. 2 (a) and (b) except that the medium strength equatorial arcs at 0.30 nm in the PAN fibre diagram are not discernible (see section 3.2). The polymer grafts are producing a fibre diagram with the fibre axis parallel to that of the wool fibre.

3.1. The Effect of Variation of Copolymer Composition

Differential optical density results for a selection of wool + PAN graft copolymers are given in Fig. 3, the top curve being for the horsehair + PAN graft copolymer. The weight per cent polymer forming the grafts is shown against each curve and for clarity these have been shifted by 0.1. All the results show a negative difference in the region of 1 nm^{-1} and this may be taken to indicate some decrease in the intensity of the α-keratin scattering. Even at a low level of graft copolymerization (10.5% PAN) the most significant effect is the change in the intensity at

0.98 nm. No reliable conclusions can be drawn from the relative magnitude of the differential effects since the amount of scattering material in the X-ray beam could not be maintained constant. The results do however suggest that although the major part of the polymer is formed in the matrix some may also be formed within the microfibrillar regions. Some dislocation of the microfibrillar regions is possible since this region has some non-helical content (13) and there will be regions more accessible to monomer and the growing polymer graft.

Palmer (2) has reported that the crystalline regions of wool + propiolactone graft copolymers become disrupted when the polymer content became greater than 60%. The polymer level at

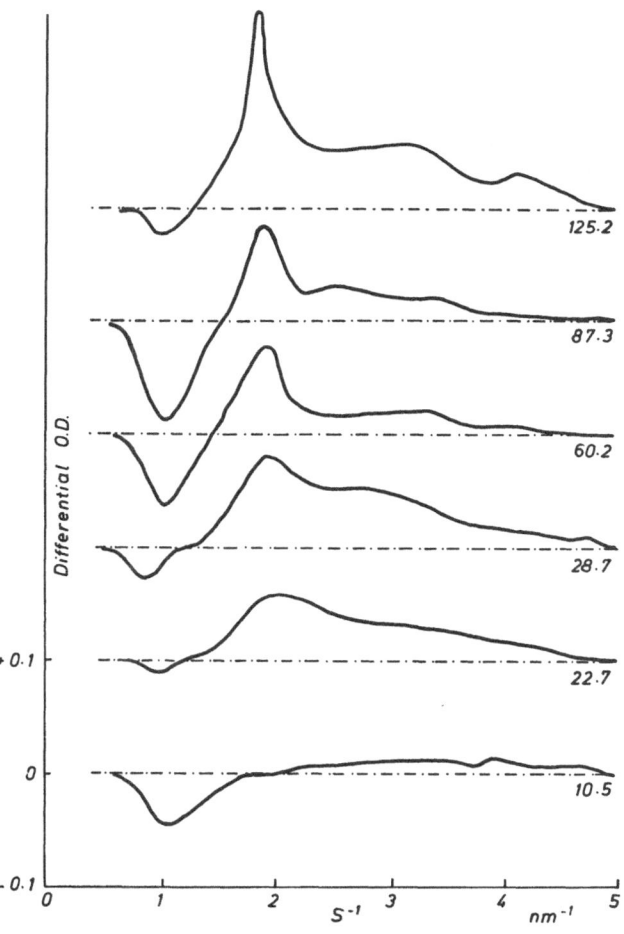

Fig. 3. The differential optical density for X-ray scattering by wool + PAN graft copolymers containing different amounts of grafted polymer shown as a function of reciprocal spacing. The polymer content is shown against each curve. Top curve is for the horsehair + PAN copolymer and the lower curves are for Corriedale wool + PAN copolymers

which such disruption becomes detectable will depend to some extent on the bulkiness of the side groups in the monomer and the lower figure for propiolactone would not be unreasonable. *Ingram et al.* (1) have also shown, from low angle X-ray scattering studies, that the microfibrillar structure of wool is disrupted by graft copolymerization with styrene. The 8.3 nm spacing in wool, which arises from the packing of the microfibrils in the cortical cells (14) was found to increase with the weight of polymer grafted thus indicating that the microfibrils are pushed apart by the polymer chains.

For reciprocal spacings greater than 1.5 nm^{-1} the differential optical density curves given in Fig. 3 show the contribution to the X-ray scattering which arises predominantly from the PAN grafts. Minor changes in the X-ray scattering in this region may however be expected on the part of the wool component particularly in the region of 3.2 nm^{-1}. This will be in part a consequence of the change in microfibrillar structure causing modification to the 0.98 nm scattering, and also as a result of change in the equilibrium regain of the fibre as a result of the incorporation of polymer (15). These effects are difficult to assess and have been neglected in the construction of Fig. 3.

At low polymer levels (10.5% PAN) the X-ray scattering is extremely diffuse, as shown by the lower curve in Fig. 3, and it is not until the polymer content reaches about 22.7% that the X-ray scattering from the polymer can be considered to be at all comparable to that from a PAN powder sample, see Fig. 1. In order to produce such diffuse scattering the initiation of the graft copolymerization must have occurred at a large number of sites both within the matrix and the microfibrillar regions of the wool and so result in the production of many very short polymer branches. These short grafts will cause disruption in the microfibrillar region, affecting the packing and hence the intensity of the α-keratin peak at 0.98 nm, but have insufficient coherence to produce significant X-ray scattering. In a series of poly(methyl methacrylate) + PAN graft copolymers which have been examined (16) the X-ray scattering from the PAN grafts was, at all levels of PAN, clearly identifiable with that from a PAN powder sample. For the wool + PAN graft copolymers shown in Fig. 3 it is not until the polymer level reaches 125.2% PAN that the width of the peak at 1.9 nm^{-1} becomes comparable to that of the PAN powder sample given in Fig. 1. Thus it must be adduced that the polymer is widely distributed throughout the wool fibre.

3.2. The Effect of the Keratin Structure on the Formation of the Graft Copolymer

In the horsehair + (125.2%) PAN graft copolymer which has been examined the X-ray scattering diagram given in Fig. 2 (c) shows orientation effects in the polymer halo. Although the equatorial arcs at 0.3 nm, which have a medium strength in the PAN fibre diagram given in Fig. 2 (b), are not discernible in Fig. 2 (c) the differential optical density plot given by this copolymer (top curve in Fig. 3) does show some evidence of this peak at 3.3 nm^{-1}. This would indicate that the polymer grafts are arranged to produce a fibre diagram with the fibre axis parallel to that of the wool fibre. Thus at a high level of polymer grafting the polymer chains are parallel to the microfibrils in the keratin structure. Since the microfibrils are relatively close packed in the matrix (13) they will form channels down which the polymer grafts may grow during polymerization (17) so inducing an orientation in the polymer chains. However polymer produced within the matrix will have a random orientation so that the combined result is a much poorer PAN fibre diagram than might have been anticipated.

3.3. Enzymatic Digestion of Wool Copolymers

During the gamma radiation grafting of styrene in wool, *Campbell, Williams* and *Stannett* (18) have shown, using electron spin resonance, that a large modification of the radical spectrum from wool occurs during the growth of the grafts. At least one of the radicals has been identified as a cystine radical arising from the rupture of the disulphide bonds. *Arai, Negishi, Komine* and *Takeda* (19) have also demonstrated in the graft copolymerization of styrene in wool that the site of initiation is most probably a cystine residue. This was established by digestion of the wool which left the vinyl polymer terminated at each end with an amino acid residue, one of which was usually cystine. Thus the graft should more correctly be considered to be an extended intermolecular crosslink. It is probable that as a general rule the graft copolymerization of vinyl monomers in wool involves the disulphide bond.

In the digestion of wool by papain it is necessary for the disulphide bonds to be broken before proteolysis occurs, and this is most suitably carried out by action of sodium bisulphite. If, however, disulphide bonds have previously been involved in the formation of polymer grafts, then the thiol group in the active papain (20) will no longer be able to react with a cystine radical along the wool backbone of the copolymer. Thus it may be expected that the observed degradation of the wool copolymer under the action of the enzyme will depend on the extent to which disulphide bonds have become the site of grafting.

The microphotographs shown in Fig. 4 are of single fibres each sealed-off in a small tube containing a 4 M urea solution of papain and sodium

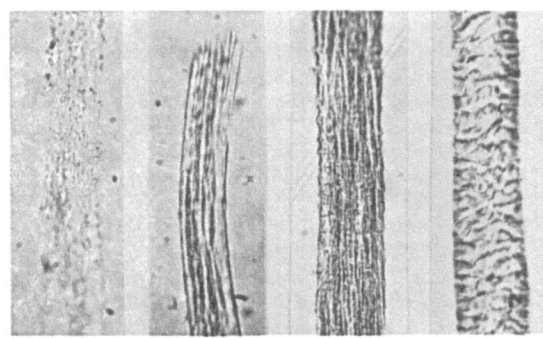

Fig. 4. Microphotographs of single Corriedale wool fibres in a solution of 1% w/w papain and 0.1 M sodium bisulphite in 4 M urea after treatment for 12 h at 50 °C. Polymer contents (% PAN) are, (a) 0; (b) 19.6; (c) 22.7; (d) 68.0. The bar indicates 50 micron

bisulphite. The almost complete digestion of the wool fibre as shown in Fig. 4(a) confirms that the preparation is active. *Milligan, Holt* and *Caldwell* (21) have shown that this treatment does not hydrolyse peptide crosslinks between side chains containing lysyl and glutamyl or aspartyl residues, with the result that some 5% of the wool remains as a residue. These residues and other miscellaneous components, viz. melanin particles, are sufficient to indicate the position of the fibre in Fig. 4(a).

The other micrographs are for wool copolymers containing increasing amounts of polymer as indicated. Incorporation of 19.6% PAN stabilizes part of the structure. The major part of the fibre has been digested and there is no indication of a scale structure. It is to be noted

that the residues are partly aligned along the fibre direction. Increase in the polymer content to 22.7% PAN produces a marked change. There has been little reduction of diameter and there is evidence along the edges of the remnants of the scale structure. In both Fig. 4(b) and (c) no difficulty was encountered in focussing both the edge and centre of the subject at the same time which would indicate that the microphotographs are of internal deposits. At much higher levels of polymer as shown in Fig. 4(d), the fibre is indistinguishable from an enzymatically untreated fibre under the same optical conditions.

The differential optical density results in Fig. 3 correlate in an interesting manner with the enzymatic examination. The polymer level required to stabilize the fibre as judged from Fig. 4 is in the region of 20 to 24% PAN and it is also in this range when the X-ray scattering begins to show the characteristics of the homopolymer. Thus it seems probable that the cystine residue is involved in the graft copolymerization with acrylonitrile.

4. Conclusion

It is probable that the cystine residues are the site for the graft copolymerization of vinyl monomers in wool. Since there are a large number of these, the number of polymer grafts will be large. At low polymer levels there can only be a large number of very short chains. Although these may be disruptive, especially where they occur within the microfibrillar regions, they will have insufficient coherence and only produce a diffuse X-ray scattering. It is not until the amount of polymer exceeds about 20% that the X-ray scattering becomes identifiable with that of PAN.

At a level of 125% PAN the microfibrillar organization of the wool fibre imposes a restraint on the direction of growth of the polymer chains with the result that the observed X-ray scattering is a superposition of the wool and PAN fibre diagrams, the fibre axes being parallel.

Acknowledgements

The wool + polyacrylonitrile graft copolymers and the horsehair + polyacrylonitrile graft copolymer were kindly provided by Dr. *I. C. Watt.* Drs. *L. J. Lynch* and *E. G. Bendit* have drawn attention to a number of important points and their advice is gratefully acknowledged.

Summary

X-ray scattering by wool + polyacrylonitrile graft copolymers prepared by the ferrous ion-hydrogen peroxide initiatory system is reported. Scattering which may be attributed to the polyacrylonitrile (PAN) grafts is not observed until the level of grafting has reached about 20% PAN. At high levels (125% PAN) the oriented molecular structure of the wool fibre has an effect on the formation of the graft copolymer and the component of the X-ray scattering arising from the polymer grafts shows a measure of orientation. It is also observed that at graft polymer levels above 23% PAN the grafted wool fibre becomes resistant to enzymatic hydrolysis. This provides evidence that the site of initiation of graft copolymerization using acrylonitrile as the vinyl monomer is on a cystine residue, as has been reported in respect of graft copolymerization of styrene and methyl methacrylate in wool.

Zusammenfassung

Es wird über die Röntgenstreuung von Wolle-Polyakrylnitril Pfropfkopolymeren, die mit dem Ferroion-Wasserstoffperoxid-Initiator-System hergestellt wurden, berichtet. Bis zu einem Pfropfgrad von 20% Polyakrylnitril (PAN) wird keine Streuung beobachtet, die dem PAN zugeschrieben werden konnte. Bei hohen Graden (125% PAN) beeinflußt die orientierte Molekülstruktur der Wollfaser die Bildung des Pfropfkopolymers, und die durch dieses bedingte Röntgenstreuungskomponente zeigt eine gewisse Orientierung. Es wird auch beobachtet, daß bei Pfropfgraden von über 23% PAN die Wollfaser gegen enzymatische Hydrolyse beständig wird. Das beweist, daß die Initiierungsstelle für die Pfropfkopolymerisation, mit Akrylnitril als das Vinylmonomer, an einem Cystinrest liegt, wie es schon im Zusammenhang mit der Pfropfkopolymerisation von Styrol und Methyl-methakrylat in Wolle berichtet wurde.

References

1) *Ingram, P., J. L. Williams, V. Stannett* and *M. W. Andrews*, J. Polymer Sci. A-1, **6**, 1895 (1968).

2) *Palmer, K.*, Proc. Intern. Wool Textile Res. Conf. Australia **3**, 374 (1955).

3) *Simpson, W. S.*, Congr. Intern. Recherche Textile (Paris) **3**, 359 (1965).

4) *Watt, I, C.*, J. Macromol. Sci. Chem. A **4**, 1079 (1970).

5) *Beevers, R. B.*, Kolloid-Z. u. Z. Polymer **252**, 367 (1974).

6) *Andrews, M. W., R. L. D'Arcy* and *I. C. Watt*, J. Polymer Sci. B, **3**, 441 (1965).

7) *Andrews, M. W.*, J. Roy. Microscop. Soc. 84, 439 (1965).

8) *D'Arcy, R. L., W. B. Hall* and *I. C. Watt*, J. Textile Inst. **57**, T137 (1966).

9) *Skertchly, A.* and *H. J. Woods*, J. Textile Inst. **51**, T518 (1960).

10) *Beevers, R. B.*, Macromol. Rev. **3**, 113 (1968).

11) *Hinrichsen, G.* and *H. Orth*, J. Polymer Sci. B, **9**, 529 (1971).

12) *Hinrichsen, G.* and *H. Orth*, Kolloid-Z. u. Z. Polymer **247**, 844 (1971).

13) *Fraser, R. D. B., T. P. MacRae, G. R. Millward, D. A. D. Parry, E. Suzuki* and *P. A. Tulloch*, Appl. Polymer Symp. 18, 113 (1971).

14) *Bendit, E. G.* and *M. Feughelman*, Encyl. Polymer Sci. Technol. 8, 1 (1968).

15) *Bendit, E. G.*, Textile Res. J. **30**, 547 (1960).

16) *Beevers, R. B., E. F. T. White* and *L. Brown*, Trans. Faraday Soc. **56**, 1535 (1960).

17) *Arai, K.* and *M. Negishi*, J. Polymer Sci. A1, **9**, 1865 (1971).

18) *Campbell, D., J. L. Williams* and *V. Stannett*, Advanc. Chem. Ser. **66**, 221 (1967).

19) *Arai, K., M. Negishi, S. Komine* and *H. Takeda*, Appl. Polymer Symp. 18, 545 (1971).

20) *Brocklehurst, K.* and *M. P. J. Kierstan*, Nature **242**, 167 (1973).

21) *Milligan, B., L. A. Holt* and *J. B. Caldwell*, Appl. Polymer Symp. 18, 113 (1971).

Author's address:

Dr. *R. B. Beevers*
Division of Textile Physics CSIRO
338 Blaxland Road
Ryde N.S.W. 2112
(Australia)

Progr. Colloid & Polymer Sci. **58**, 201—210 (1975)

Istituto di Chimica Fisica, Università di Firenze (Italia)
and the Arcadia Institute for Scientific Research, Woodside, California (U.S.A.)

Applications of polymer solution theory to monolayers

G. Gabrielli, E. Ferroni and *M. L. Huggins*

With 9 figures and 1 table

(Received October 7, 1974)

1. Introduction

In previous papers (1—3) two-dimensional equations of state obtained from experimental measurements on monolayers of linear macromolecules on (or between) liquids have been compared with corresponding theoretical equations published by various authors (4—9). These comparisons, on the one hand, tested the applicability of the theoretical equations. On the other hand, they led to parameters useful in defining the distributions, shapes and orientations of the polymer molecules in the interfaces.

One of us has recently modified his statistical thermodynamic treatment of solutions of linear polymers (10—12) to make it applicable to polymers in monolayers (13). This constitutes a refinement of his earlier treatment (14) of the thermodynamics of monolayers. The other authors have previously applied the earlier theory to their data on monolayers of various macromolecular substances and their mixtures (1—3, 15—18).

The purpose of the present paper is to compare experimentally deduced spreading isotherms with those given by the new theoretical treatment. Comparisons are made for the same polymers at interfaces between water and air and between water and other liquids (here termed "oils"), in order to demonstrate the applicability of the equations for both types of interface and to study the effect of the type of interface on the magnitudes of the structural and energetic parameters. The results show the general applicability of the theory and serve as a starting point for relating the behavior of polymers in two dimensions to that in three dimensions.

The comparisons have been made for three polymers for which the experimental data have

already been compared with previous theoretical treatments. These polymers are the following:

A 74 polypropylene grafted with 10% methyl acrylate,

A 79 polypropylene grafted with 13% methyl acrylate,

PMMA poly(methyl methacrylate).

2. Experimental

2.1. Polymers

Polymers A 74 and A 79 were provided by the Institute of Industrial Chemistry of the Polytecnico of Milano. Their preparation, purification and properties have been reported in detail elsewhere (2—19). The PMMA sample was prepared by Monomer-Polymer Laboratories, Bordon, Inc., Chemical Division, Philadelphia, Pennsylvania. Its characteristics have previously been reported (15).

2.2. Interfaces

Spreading isotherms have been determined at the following interfaces:

A 74 and A 79: water-air, water-*n*-hexane, water-toluene, water-(petroleum ether).

PMMA: water-air, water-(petroleum ether).

The liquid subphase used was twice-distilled water, purified from colloidal impurities by activated charcoal. Its surface purity was continually controlled over the whole field of surface area under examination.

The oils used were chosen because of the high stability of the interfaces they form with water. These liquids were extremely pure: toluene *(Riedel de Haën)*, *n*-hexane *(C. Erba)*, petroleum ether *(Rudi Pont*, PE 40—70 °C).

The spreading solvent used at all the interfaces was extremely pure benzene *(Riedel de Haën)*.

2.3. Measurement Methods

The surface pressure at the water-air interface was measured by *Wilhelmy* method, as reported in several previous papers (1—3, 15—18). The surface pressure at a water-oil interface was obtained from the difference between the water-oil interfacial tensions before and after addition of successive amounts of the polymer solution in the spreading solvent. The measurement procedure has previously been reported (3). Measurements were made using different surface concentrations of polymer in order to guarantee perfect spreading and reproducibility.

The surface pressure isotherms were obtained, point by point, after waiting for a time sufficient to achieve surface equilibrium, as indicated by invariance of the surface pressure.

Some of the experimental results with polymers A 74 and A 79 have been previously reported (2); others are here reported for the first time. The PMMA results have previously been presented (15, 16, 18, 20).

The probable errors of the surface area and surface pressure measurements are estimated to be 0.01 m²/mg and 0.05 dyne/cm, respectively.

3. Calculation of the Theoretical Parameters

3.1. Theoretical Equation

The theoretical eq. [13] for the dependence of surface pressure (Π) on surface area (A) is

$$\Pi = \frac{RT z_\beta}{a_\alpha^0} \left\| \psi \left[g_K z_\alpha + \frac{z_\beta - z_\alpha}{(1 + K' z_\alpha z_\beta)^{1/2}} \right] \right.$$
$$+ \left[\frac{1}{z_\beta} \ln \left(1 + \frac{r_{a/\sigma} z_\beta}{z_\alpha} \right) - \frac{(1 - 1/n\, r_a)\, r_{a/\sigma}}{1 + (r_{a/\sigma} - 1) z_\beta} \right]$$
$$+ \frac{1}{4} \left\{ \left[\frac{z_\alpha - z_\beta}{(1 + K' z_\alpha z_\beta)^{1/2}} - z_a\, g_K \right] \ln K \right. \tag{1}$$
$$+ \left. \left. \frac{1}{z_\beta} \ln \left(\frac{z_\alpha}{1 - z_\beta\, g_K} \right) \right\} \right\| ,$$

where

$$z_\beta = \frac{m/A}{\varrho_m\, r_{a/\sigma} + (1 + r_{a/\sigma})\, m/A} = 1 - z_\alpha, \tag{2}$$

$$\psi = \left(\frac{K_s}{r_\sigma} - \frac{\varepsilon_\Delta}{RT} \right), \tag{3}$$

$$g_K = \frac{-2}{K' z_\alpha z_\beta} \left[1 - (1 + K' z_\alpha z_\beta)^{1/2} \right], \tag{4}$$

$$r_a = a_\beta^0/a_\alpha^0 \quad r_\sigma = \sigma_\beta^0/\sigma_\alpha^0, \quad r_{a/\sigma} = r_a/r_\sigma, \tag{5}$$

$$K' = 4 \left(\frac{1}{K} - 1 \right). \tag{6}$$

K is an equilibrium constant, determining the relative contact lengths of the three types:

$$K = \frac{\sigma_{\alpha\beta}^2}{4\, \sigma_{\alpha\alpha}\, \sigma_{\beta\beta}} . \tag{7}$$

and n is the average number of segments per polymer molecule.

The parameters to be determined for each isotherm are K' (or K), a_α^0, r_a, $r_{a/\sigma}$ and ψ.

The procedure can be outlined as follows:

3.2. Apparent Molar Masses

Using the experimental π and A values and plotting πA vs. π, good agreement with a rectilinear relation was found at low π values. Shown in Fig. 1 are the rectilinear courses of the $\pi A - \pi$ curve used in the extrapolation of $\pi \to 0$. The points are experimental values and the straight lines are those calculated by means of the least square method.

According to the theory (see eq. [39] of Reference 13), extrapolation to $\pi = 0$ should give

$$\left(\frac{\pi}{A} \right)_{\pi=0} = \frac{RT m}{M_2} \tag{8}$$

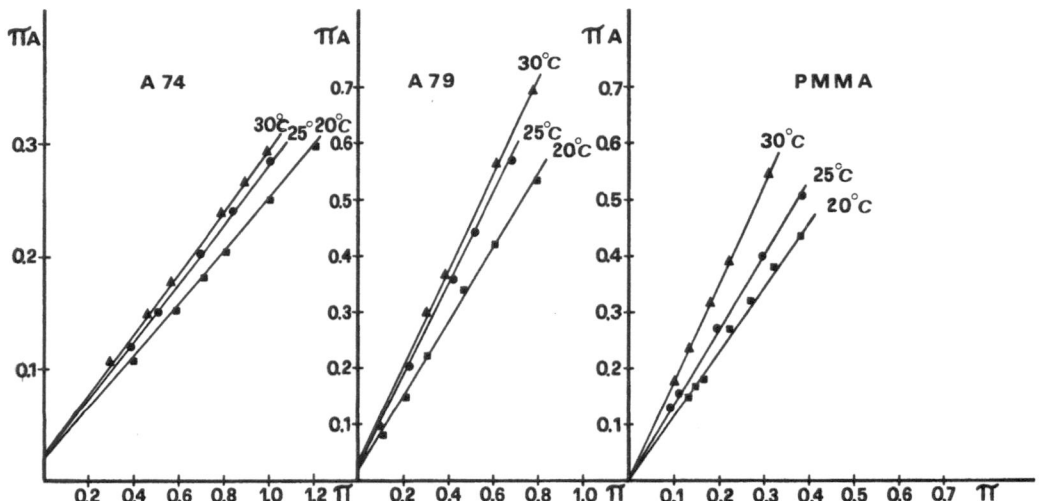

Fig. 1. Graph of πA vs. π for A 74, A 79, PMMA at water/air interface in the low π range

where m is the mass of polymer in the monolayer and M_2 is the apparent average molar mass (molecular weight) of the polymer in the surface.

Because of partial submersion of portions of the polymer molecules in the subphase, this is not necessarily equal to the true average molar mass. The average is a number-average, differing from the weight-average obtained from light scattering measurements on solutions or the viscosity-average obtained from solution viscosities.

The M_2 values were obtained from the water-air interface data. For each polymer and temperature the equations of the straight lines giving the best correlation coefficients over the straight line portions of the curves were determined by least squares. From each equation the intercept was calculated, and then M_2.

The M_2 values obtained for a given polymer at the three temperatures agreed within 10%, which is the estimated probable error of the value at each temperature, so the average of the values for the three temperatures was taken as the true value for all.

3.3. Molar Areas

The molar area of the subphase liquid (water) was approximated on the assumption of close-packing of spheres, of such size as to give the actual density (ϱ_1) of the subphase, by the equation

$$a_\alpha^0 \approx \frac{3^{1/2} N_A^{1/3} M_1^{2/3}}{2^{2/3} \varrho_1^{2/3}} = 0.9214 \times 10^8 \frac{M_1^{2/3}}{\varrho_1^{2/3}} . \quad [9]$$

Making the corresponding assumption with regard to the polymer segments, the "segment area ratio" was calculated by the equation

$$r_a = \frac{a_\beta^0}{a_\alpha^0} \approx \left(\frac{M_2}{M_1 n}\right)^{2/3} \frac{\varrho_1}{\varrho_2} , \quad [10]$$

using the apparent average M_2 obtained from the intercept of the ΠA vs. Π curve as described above.

The "contacting segment outline ratio" was similarly approximated as the ratio of the circumferences of spheres in hypothetical three-dimensional close-packed structures having the densities of the pure components. Then

$$r_\sigma = \frac{\sigma_\beta^0}{\sigma_\alpha^0} \approx \left(\frac{M_2}{M_1 n}\right)^{1/3} \left(\frac{\varrho_1}{\varrho_2}\right)^{1/3} . \quad [11]$$

3.4. Outline Fraction and Gibbs Energy Parameter

For each experimental value of surface area, the corresponding value of the contacting out-

line fraction z_β was calculated by [2]. Then, for each pair of z_β and Π values of each isotherm and for a series of values of K' (from -4 to 30 inclusive, at intervals of 0.2), the corresponding value of the *Gibbs* energy parameter ψ (see [3]) was computed. For these and later calculations a C. I. I. 10070 computer was used.

For each isotherm and each value of K', the mean of the individual values of ψ was calculated. The most probable value of K' was chosen as that leading to the smallest average deviation of the individual ψ values from their mean. The mean ψ corresponding to this most probable K' value was then adopted as the most probable for this isotherm.

3.5. Theoretical Spreading Curves

Using the K' and ψ parameters so deduced for each isotherm, Π was calculated by [1], at z_β intervals of 0.2, for the range of concentrations corresponding to the experimental measurements. The theoretical Π vs. A isotherms were then compared with the experimental points.

3.6. Entropy and Enthalpy Parameters

Values of k_s' and ε_Δ entering into the two terms, k_s'/r_σ and ε_Δ/RT, in the ψ parameter were calculated from the temperature dependence of Π. The spreading entropy and enthalpy were first computed for each temperature and for various values of the surface areas, in the range for which the experimental and theoretical data had been found to be in good agreement. For each system ε_Δ was then calculated with the aid of the equation

$$\varepsilon_\Delta = \frac{a_\alpha^0 \left[T \left(\frac{\partial \Pi}{\partial T}\right)_{N_2} - \Pi \right]}{z_\beta \left[z_\alpha g_K + \frac{z_\beta - z_\alpha}{(1 + K' z_\alpha z_\beta)^{1/2}} \right]} . \quad [12]$$

From [3], k_s'/r_σ and then k_s' were computed. The ε_Δ and k_s' values tabulated are means of the individual values calculated for the experimental concentrations.

4. Results and Discussion

4.1. Parameters

The molar masses, M_2, for the polymers, averaged for the three temperatures, are

A 74: 141,000, A 79: 150,000,
PMMA: 336,000.

Table 1

Polymer Interface	Temperat. °C	r_a	r_σ	K'	K	ψ	ε_Δ 10^8 ergs	$\varepsilon_\Delta/a_\alpha^o$ erg/cm²	k'_s
A 74 water/air	20	2.059	1.3734	3.8	0.51	− 31.64	5015	791	− 15.15
water/air	25	2.056	1.3730	3.6	0.53	− 17.96	2803	442	− 9.12
water/air	30	2.045	1.3725	3.7	0.52	− 21.78	1129	178	− 23.74
water/n-hexane	20−25−30	2.056	1.3730	0.40	0.91	0.98	− 44	− 70	1.09
water/toluene	20−25−30	2.056	1.3730	0.40	0.91	1.09	− 18	− 2.8	1.33
water/(petroleum ether)	20	2.059	1.3734	0.40	0.91	1.05	760	120	5.69
	25	2.056	1.3730	0.40	0.91	1.01	765	120	5.62
	30	2.045	1.3725	0.40	0.91	0.94	760	120	5.46
A 79 water/air	20	2.118	1.4304	3.2	0.56	− 6.29	6824	1076	29.83
water/air	25	2.115	1.4297	3.2	0.55	− 5.46	4858	766	19.39
water/air	30	2.113	1.4390	3.8	0.51	− 8.50	3122	492	5.34
water/n-hexane	20−25−30	2.115	1.4397	0.20	0.95	0.63	− 44	− 69	0.27
water/toluene	20−25−30	2.115	1.4297	0.40	0.91	0.95	− 50	− 7.9	0.90
water/(petroleum ether)	20	2.118	1.4304	0.80	0.83	1.05	2338	371	13.96
	25	2.115	1.4297	0.80	0.83	0.90	2337	371	13.55
	30	2.113	1.4290	0.40	0.91	0.64	2381	371	13.25
PMMA water/air	20	2.656	1.6759	5.6	0.417	− 6.57	11250	1775	66.4
water/air	25	2.653	1.6752	6.9	0.367	− 8.16	12325	1944	63.0
water/air	30	2.649	1.6745	4.8	0.451	− 3.07	11184	1763	67.2
water/(petroleum ether)	20−25−30	2.653	1.6752	3.8	0.510	− 1.07	−104	− 16.4	− 2.5

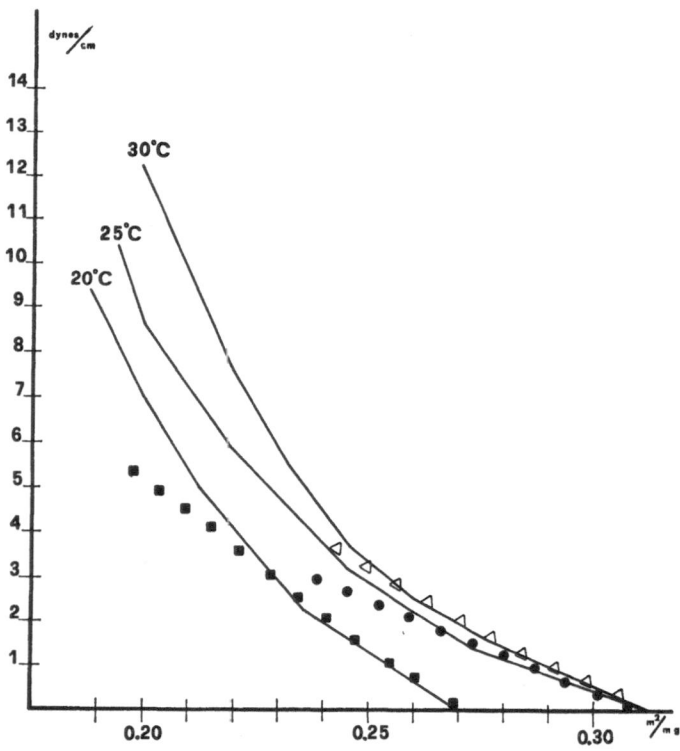

Fig. 2. Spreading isotherms of polymer A 74 at a water/air interface

The value for PMMA is close to that (340,000) deduced from viscosity measurements of (3-dimensional) benzene solutions (15).

From [9],

$$a_\alpha^0 = 6.341 \times 10^8 \text{ cm}^2.$$

The other parameters, computed as described in the preceding section, are listed in Table 1.

For A 74 and A 79 at water/hexane and water/toluene interfaces and for PMMA at a water/(petroleum ether) interface, the isotherms are identical, within the estimated probable error, for the three temperatures, hence only one set of parameters is given for each of these systems.

4.2. Theoretical and Experimental Curves

Figs. 2—9 show comparison of points on the theoretical pressure vs. area curves, calculated by [1] using the parameters a_α^0, r_a, r_σ, K' and ψ, with the corresponding experimental curves. Considering the theoretical approximations involved, the agreement is quite satisfactory, especially at the lower surface pressures. The surface area at which the theoretical and experimental curves begin to differ significantly is, in each case, approximately the limiting area, below which the monolayer begins to collapse. For lower areas (higher pressures) the assumptions

of the theory are of course no longer valid. As the limiting area is approached, moreover, any possible tendencies of adjacent chains to become parallel or to assume any preferential relative orientations would lead to departures from the assumed theoretical equations.

4.3. Equilibrium Constant K

The values of K calculated for all these systems at all three temperatures are less than one, indicating a preference for like contacts (polymer-polymer and water-water or oil-oil) over unlike contacts (polymer-water or polymer-oil). See [7]. This preference is in no case large, however. The K values are closer to unity for monolayers at water-oil interfaces than for those at water-air interfaces, as would be expected. Compare (3) and (21—23). The difference is less for the relatively polar polymer PMMA than for the polypropylene polymers. This is also reasonable. As previously demonstrated (15, 16, 18), PMMA (but not A 74 or A 79) at a water-air interface is partially submerged in the water phase. Corresponding K values for the two polypropylene samples are nearly equal, doubtless a result of their similar composition. In all these systems the value of K is nearly independent of temperature.

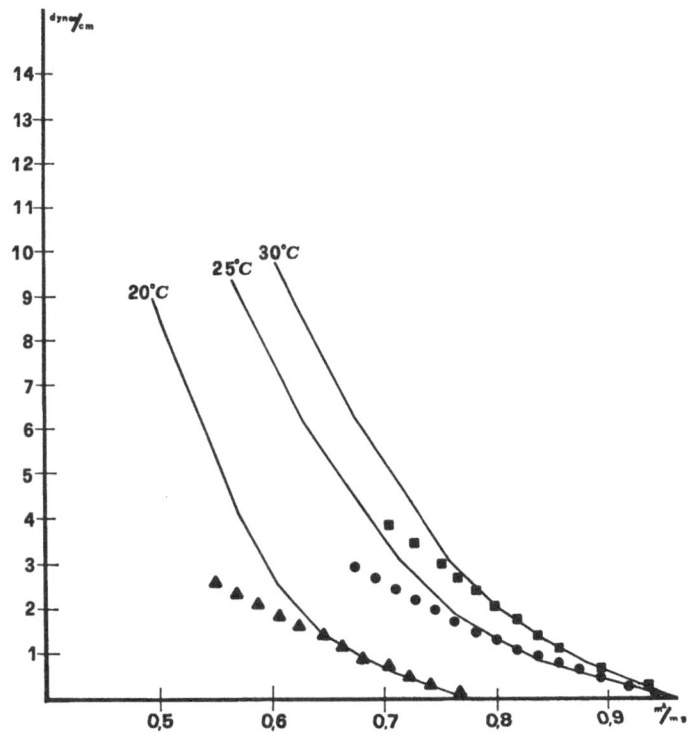

Fig. 3. Spreading isotherms of polymer A 79 at a water/air interface

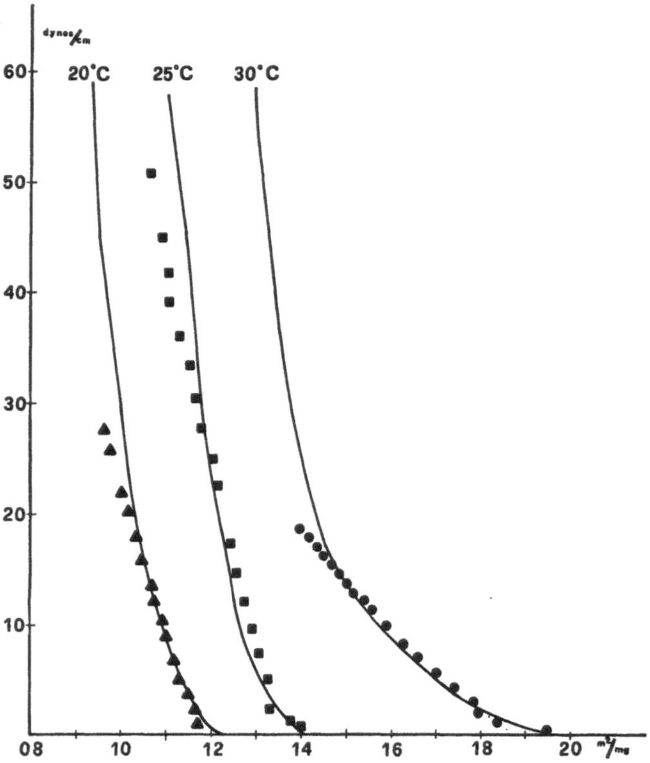

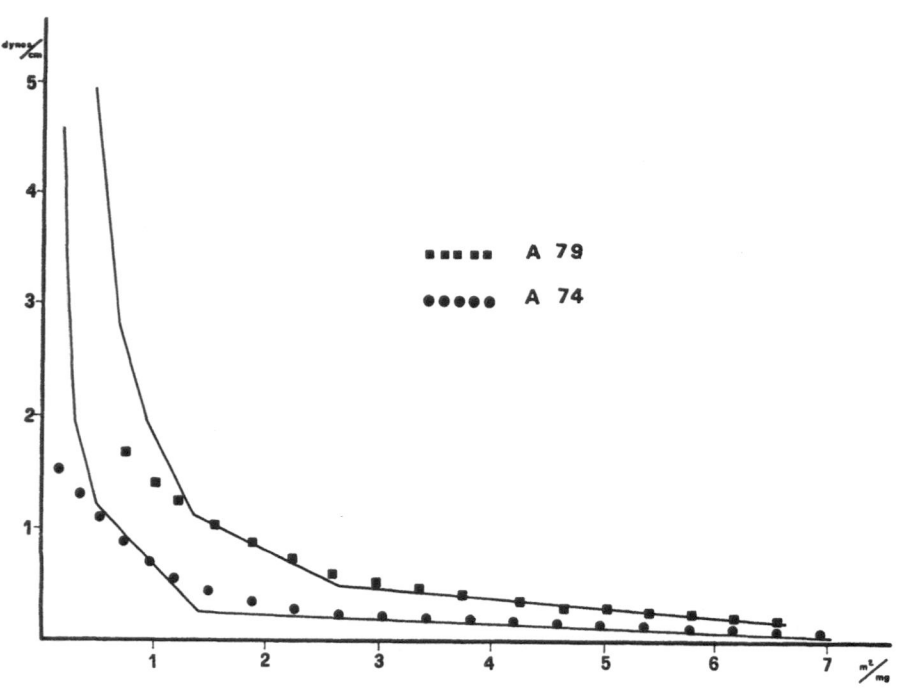

Fig. 4. Spreading isotherms of PMMA at a water/air interface

Fig. 5. Spreading isotherms of polymers A 74 and A 79 at a water/toluene interface

Fig. 6. Spreading iso-
therms of polymers A
74 and A 79 at a wa-
ter/*n*-hexane inter-
face

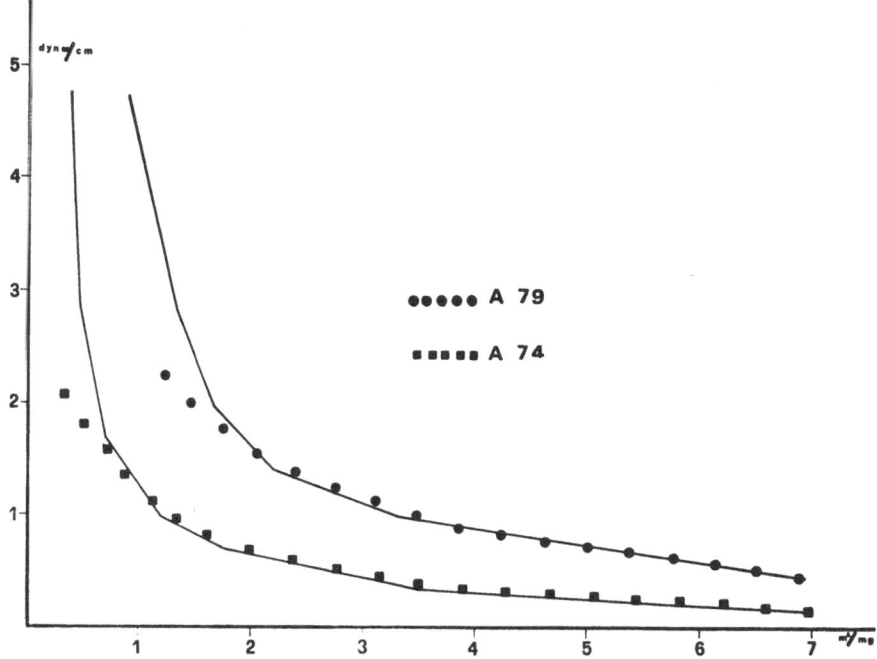

4.4. Energy Parameter ε_Δ

ε_Δ is related to the contact energies per unit contact length $\varepsilon_{\alpha\alpha}$, $\varepsilon_{\beta\beta}$, and $\varepsilon_{\alpha\beta}$ (negative for attraction energies) by the equation

$$\varepsilon_\Delta = \frac{\sigma_\alpha^0}{2}\left(2\,\varepsilon_{\alpha\beta} - \varepsilon_{\alpha\alpha} - \varepsilon_{\beta\beta}\right). \qquad [13]$$

The large ε_Δ values found for the monolayers at water-air interfaces therefore mean that the average of the water-water and polymer-polymer attraction energies (per unit contact length) is greater than the water-polymer attraction energy. This is reasonable. For the graft copolymers A 74 and A 79 the magnitude of the difference decreases as the temperature rises, presumably reaching zero at somewhat higher temperatures than those of the experiments.

Since A 79 contains a larger fraction of methyl acrylate mers than A 74, the average polymer-polymer attractions should be larger for the former than for the latter. This seems to be a reasonable explanation for the fact that the calculated ε_Δ value at a water-air interface is larger for A 79 than for A 74, at each temperature.

The magnitudes of ε_Δ at water-air interfaces are considerably larger for PMMA than for A 74 and A 79, apparently indicating stronger mutual attractions between the PMMA molecules than between the polypropylene molecules.

For PMMA, the calculated ε_Δ values are nearly the same at the three temperatures. As already noted, previous research has indicated partial submersion of the polymer molecules in the subphase (water). The degree of submersion is doubtless a function of the temperature. The enthalpy changes associated with changes in the degree of submersion are not taken into account in the theoretical equations used here, except insofar as they affect the magnitude of ε_Δ. One might guess that these changes are responsible for the fact that the variation of ε_Δ with temperature is different for PMMA than for the polypropylene copolymers.

The water-water attraction energies referred to above are actually differences between the attraction energies between water molecules *in* the surface and between water molecules *below* the surface. If another liquid ("oil") is over the water, the difference between the attraction energies of the oil molecules *in* the surface and *above* the surface (and also the water-oil and oil-polymer attractions) are of course also involved, in obvious ways. Such reasoning can explain the much lower ε_Δ values for the monolayers at the water-oil interfaces than for those at the water-air interfaces.

The considerably larger ε_Δ values for the polypropylene copolymers at water-(petroleum ether)

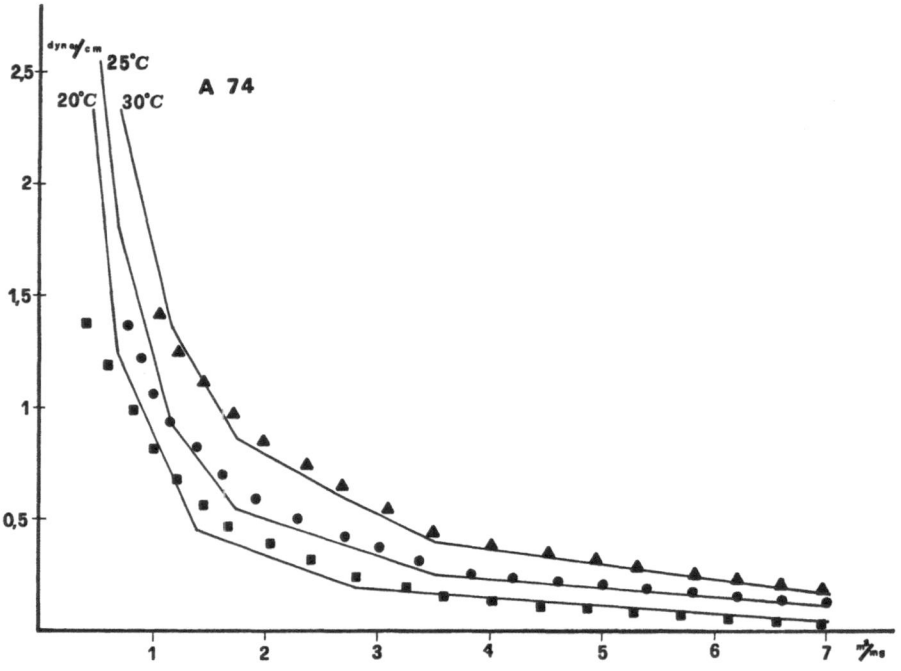

Fig. 7. Spreading iso-
therms of polymer A
74 at a water/(petro-
leum ether) interface

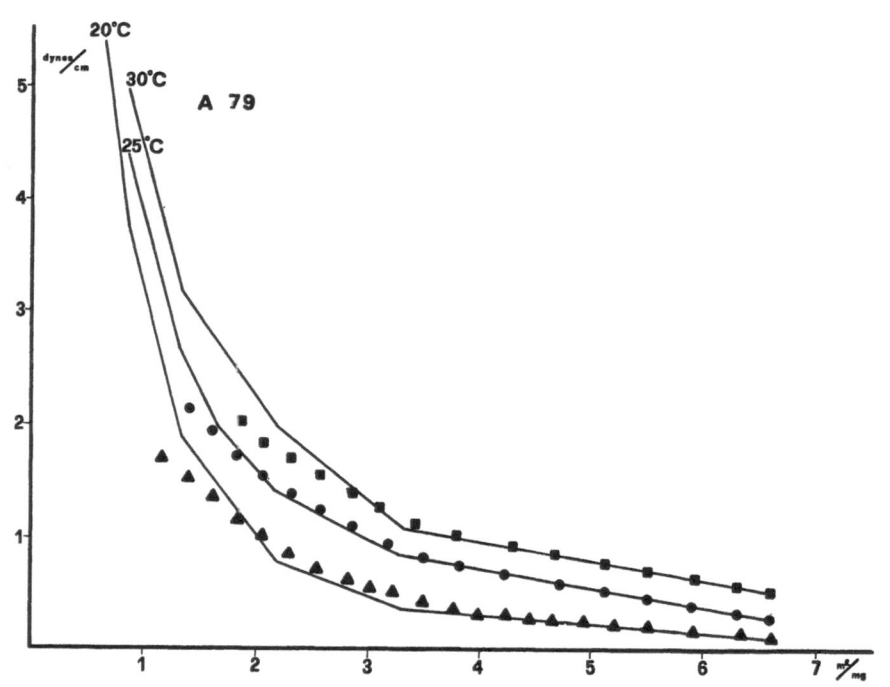

Fig. 8. Spreading iso-
therms of polymer A
79 at a water/(petro-
leum ether) interface

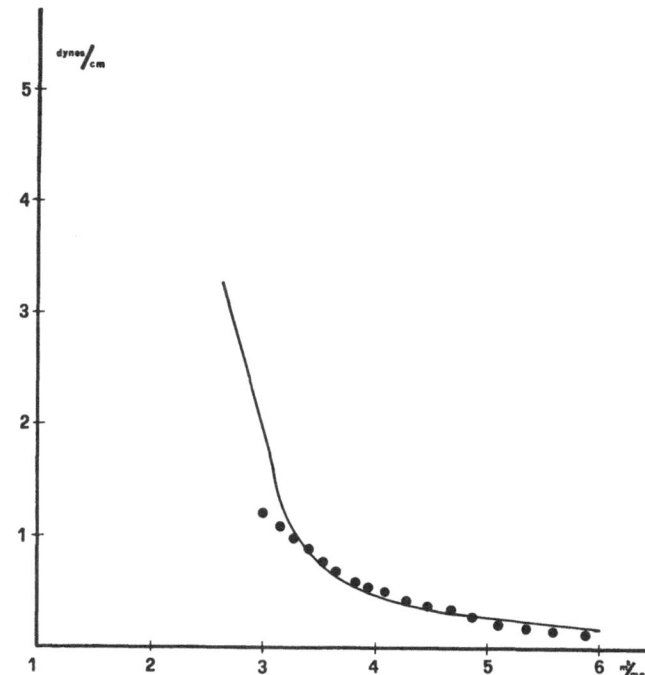

Fig. 9. Spreading isotherm of PMMA at a water/(petroleum ether) interface

interfaces than at water-hexane and water-toluene interfaces can be attributed to the lower solubilities of these polymers in petroleum ether than in hexane or toluene. This would result in larger $\varepsilon_{\alpha\beta}$ and larger ε_Δ for the system containing petroleum ether than for the others.

4.5. Entropy Parameter k_s'

The parameter k_s' measures the change in randomness (hence entropy) as the close-neighbor environments of the polymer segments and liquid molecules change. In part it is a measure of the change of polymer flexibility with changes of the close-neighbor environment of the segments. Further discussion of the k_s' values obtained in this study does not appear justifiable at this time. It seems better to wait until after data on more systems are available.

5. Conclusions

The theoretically deduced equations are satisfactorily consistent with the experimental pressure vs. area data, especially at pressures below that corresponding to the limiting area for monolayer stability.

Application of the theoretical equations to the experimental data yields parameters that are closely related to molecular and intermolecular dimensions and energies and to structural factors affecting the entropy. Comparison of the parameters for the different polymers, different temperatures, and different interfaces leads to a better understanding of the effect of each molecular property on monolayer properties.

Acknowledgement

The authors wish to thank Dr. *M. Puggelli* for her contribution in elaborating the computer-produced data.

Summary

A recently derived theoretical equation of state for linear polymers in monolayers is applied to experimental measurements of the dependence of surface area on surface pressure for monolayers of poly(methyl methacrylate) and polypropylene grafted with methyl acrylate, at water/air, water/n-hexane, water/toluene, and water/(petroleum ether) interfaces at 20°, 25° and 30°C. The experimental and theoretical curves are in satisfactory agreement for pressures up to those corresponding to the limiting area for stability of the monolayers. The theoretical parameters derived from the experimental data measure the molecular and intermolecular properties that determine the *Gibbs* free energies of the monolayers.

Zusammenfassung

Eine kürzlich hergeleitete theoretische Zustandsgleichung für lineare Polymere in zweidimensionalen Schichten wird hier mit experimentellen Messungen für ihre Abhängigkeit der Oberfläche vom Oberflächendruck, für Monoschichten auf Polymethylmethakrylat und Polypropylen, gepfropft mit Methylakrylat an Wasser/Luft, Wasser/n-Hexan, Wasser/Toluol, und Wasser/Petroläther Zwischenfläche angewandt.

Die experimentellen und theoretischen Kurven stehen in befriedigender Übereinstimmung, insbesondere für den genügend niedrigen Bereich der Oberflächendrucke bis zu jenen Werten, die den Grenzflächen für stabile Monoschichten entsprechen. Die hergeleiteten Parameter bestimmen die molekularen und zwischenmolekularen Eigenschaften, die die *Gibbs*sche Energie der Monofläche bestimmen.

References

1) *Gabrielli, G., M. Puggelli* and *E. Ferroni,* J. Colloid Interface Sci. **32**, 242 (1970).
2) *Gabrielli, G., M. Puggelli* and *E. Ferroni,* J. Colloid Interface Sci. **33**, 133 (1970).
3) *Gabrielli, G.* and *M. Puggelli,* J. Colloid Interface Sci. **35**, 460 (1971).
4) *Singer, S. I.,* J. Chem. Physics **16**, 872 (1948).
5) *Saraga, L. T.* and *I. Prigogine,* Mem. Ser. Chim. Etat (Paris) **38**, 109 (1953).
6) *Davies, J. T.* and *J. Llopis,* Proc. Roy. Soc. **A 227**, 537 (London, 1955).
7) *Frisch, H. L.* and *R. Simha,* J. Chem. Physics **27**, 702 (1957).
8) *Kawai, T.,* J. Polymer Sci. **35**, 401 (1959).
9) *Motomura, K.* and *R. Matuura,* J. Colloid Sci. **18**, 12 (1963).
10) *Huggins, M. L.,* J. Phys. Chem. **74**, 371 (1970).
11) *Huggins, M. L.,* Polymer **12**, 357 (1971); **13**, 554 (1972).
12) *Huggins, M. L.,* J. Phys. Chem. **75**, 1255 (1971).
13) *Huggins, M. L.,* Kolloid-Z. u. Z. Polymere **25**, 29 (1973).
14) *Huggins, M. L.,* Makromol. Chem. **87**, 119 (1965).
15) *Gabrielli, G., M. Puggelli* and *R. Faccioli,* J. Colloid Interface Sci. **37**, 213 (1971).
16) *Gabrielli, G., M. Puggelli* and *R. Faccioli,* J. Colloid Interface Sci. **41**, 63 (1972).
17) *Gabrielli, G., M. Puggelli* and *E. Ferroni,* Berichte vom VI. Internationalen Kongreß für grenzflächenaktive Stoffe, Band B II 329 (1973).
18) *Gabrielli, G., M. Puggelli* and *E. Ferroni,* J. Colloid Interface Sci. **47**, 145 (1974).
19) *Natta, G., F. Severini, M. Pegoraro* and *C. Tavazzini,* Makromol. Chem. **119**, 201 (1968).
20) *Gabrielli, G.* and *M. Puggelli,* J. Colloid Interface Sci. **45**, 217 (1973).
21) *Crisp, D. J.,* J. Colloid Sci. **1**, 49, 161 (1946).
22) *Isemura, T.* and *K. Hamaguchi,* Bull. Chem. Soc. Japan **25**, 49 (1952).
23) *Davies, J. T.* and *K. Rideal,* Interfacial Phenomena, p. 246 (New York, 1963).

Authors' addresses:

Prof. *G. Gabrielli* and Prof. *E. Ferroni*
Istituto di Chimica Fisica
Università di Firenze
Firenze, Italia

Dr. *M. L. Huggins*
Arcadia Institute for Scientific Research
135 Northridge Lane, Woodside
California 94062, U.S.A.

Progr. Colloid & Polymer Sci. **58**, 211—218 (1975)

Institut für chemisch technische Untersuchungen, 5357 Swisttal-Heimerzheim

Beschreibung der Transformation von Gläsern in Analogie zu den Telegraphengleichungen

C. O. Leiber

Mit 6 Abbildungen und 3 Tabellen

(Eingegangen am 25. September 1973)

Einleitung

In der elektrischen Nachrichtentechnik wird die Eigenschaft eines Kabels aus dem Vergleich zwischen einem Sende- und Empfangssignal ermittelt. Grundsätzlich gleich geht man bei Festkörpern vor. Findet man zwischen den elektrischen und mechanisch/akustischen Gesetzmäßigkeiten eine Isomorphie, so können Korrespondenzen geschaffen werden und die zahlreichen Lösungen aus der elektrischen Nachrichtentechnik direkt für den mechanisch/akustischen Fall umgeschrieben werden. Vielfach (1) wird bei *praktischen* Belangen so vorgegangen.

Mit dem Vorbehalt der Spekulation*) werden nachfolgend die Telegraphengleichungen zur Beschreibung des Transformationsverhaltens von Gläsern herangezogen. Als Resultat ergibt sich, daß der Transformationsbereich von Meßgeschwindigkeit und Probengröße abhängt, die obere und untere Kühltemperatur kann interpretiert werden. Schließlich kann gezeigt werden, daß eine monotone logarithmische Viskositäts-/Temperaturkurve durchaus mit der Existenz eines Transformationsbereiches verträglich ist und daß die Dämpfungskurve symmetrisch sein soll, wenn man Temperatureinflüsse vernachlässigt. Für Versprödungsmessungen eignet sich die Messung der Lebensdauer der Phos-

phoreszenz in Temperaturabhängigkeit und für die Abnahme der Versprödung die Rückprallhärte. Als weiteres Ergebnis sollte der reale Modul bei der Erweichung stärker abfallen als der Verlustanteil ($E''/E' > 1$). Dieses Ergebnis ist bisher nicht realisiert, zum einen deshalb, weil in diesem Fall meßtechnisch die Ankopplung versagt, möglicherweise ist dieses Ergebnis aber auch falsch, da beim Festkörper die Dämpfung frequenzproportional und bei Flüssigkeiten proportional mit dem Quadrat der Frequenz zunimmt, die Isomorphie also nicht mehr zutrifft.

Analogie zwischen elektrischen, mechanischen und akustischen Beziehungen

In Tabelle 1 sind die elektrische und mechanische Schwingungsgleichung einander gegenübergestellt, wobei sich die Koeffizienten und die Lösungen entsprechen.

Um akustische Analogien zu den mechanischen und elektrischen Beziehungen aufzufinden, muß berücksichtigt werden, daß der *Ohm*sche Widerstand und die geschwindigkeitsabhängige Reibung rein dissipativ und der akustische Widerstand konservativ sind. Macht man diesen aber verlustbehaftet, so behebt sich diese Schwierigkeit. Rechnet man bei der Dünndrahtnäherung der Schallgeschwindigkeit $c = \sqrt{E/\varrho}$ nicht mit dem Realteil E', sondern mit dem Verlustanteil E'', so ergibt sich eine echte formale Analogie zum *Ohm*schen Gesetz

$$\frac{p}{v} = \varrho\, c \sqrt{\operatorname{tg} \varphi} \quad \text{mit} \quad \operatorname{tg} \varphi = \frac{E''}{E'}. \qquad [1]$$

Die Masse und Federkonstante werden den mechanischen Werten gleichgesetzt, vgl. Tabelle 2.

---

*) Die Spekulation ist dann voll gerechtfertigt, wenn eine wirbelfreie Bewegung eines inkompressiblen Mediums betrachtet wird, welche ihrerseits von einem Geschwindigkeitspotential entsprechend der *Laplace*schen Gleichung abhängt. Dann besteht eine weitreichende Analogie zwischen mechanischem, elektrischem und Wärmefluß in einem einheitlichen Körper. Vgl. *J. W. S. Rayleigh*, The Theory of Sound II, § 243 (New York, 1945).

Bei Anwendung der Telegraphengleichungen auf Kabel geht man von Widerstands-, Kapazitäts- und Induktivitätsbelägen der Leitungen aus. Diese Werte werden also jeweils auf die Längeneinheit bezogen. In Tabelle 3 sind die mechanischen und akustischen Größen ebenfalls auf die Einheiten der Länge, $l = 1$, und der Fläche, $F = 1$, bezogen.

Wenn in Tabelle 3 Korrespondenzen tatsächlich vorliegen, müssen die einzelnen Lösungsschritte bei Auflösung der Telegraphengleichung sinnvolle akustische Resultate ergeben.

Tabelle 1: Analogien

Tabelle 3. Reduzierte Korrespondenzen gemäß Tabelle 2, bezogen auf die Längeneinheit $l = 1$ und die Flächeneinheit $F = 1$. Diese Korrespondenzen werden in die Telegraphengleichungen eingesetzt

elektrisch	mechanisch	akustisch
U	K	p
I	w	v
$R' = \dfrac{R}{l}$	$\dfrac{f}{l}$	$\dfrac{\varrho\, c\, \sqrt{\operatorname{tg}\varphi}}{l}$
$L' = \dfrac{L}{l}$	$\dfrac{m}{l}$	ϱ
$\dfrac{1}{C'} = \dfrac{l}{C}$	E	E
$C' = \dfrac{C}{l}$	$\dfrac{1}{E}$	$\dfrac{1}{E}$

Bezeichnungen:

U	Spannung
I	Strom
R	elektrischer Widerstand
L	Induktivität
C	Kapazität
K	mechanische Kraft
x	Weg
$w = \dot{x}$	Geschwindigkeit
f	Reibungskoeffizient der w-abhängigen Reibung
p	Druck
v	Schallschnelle bzw. Partikelgeschwindigkeit
c	Schallgeschwindigkeit bzw. Stoßwellengeschwindigkeit
$\operatorname{tg}\varphi$	Verlustwinkel des Elastizitätsmoduls $\operatorname{tg}\varphi = E''/E'$
E	Elastizitätsmodul
ϱ	Dichte
F	Fläche
l	Länge, z.B. der Feder oder des Stabes
S	Federkonstante

Signalausbreitung in einem Leiter

Die Signalausbreitung in einem elektrischen Kabel oder akustischen Leiter (Stab) kann durch die Telegraphengleichungen beschrieben werden.

Tabelle 2. Korrespondenzen gemäß Tabelle 1 und Text

elektrisch	mechanisch	akustisch
U	K	$F\,p$
I	$w = \dot{x}$	v
R	f	$F\,\varrho\, c\, \sqrt{\operatorname{tg}\varphi}$
L	m	$F\,l$
$\dfrac{1}{C}$	$S = \dfrac{F\,E}{l}$	$\dfrac{F\,E}{l}$
C	$\dfrac{1}{S} = \dfrac{l}{F\,E}$	$\dfrac{l}{F\,E}$

$$-\frac{\partial u}{\partial x} = R'\,i + L'\,\frac{\partial i}{\partial t} \qquad \text{(elektrisch)}$$

$$-\frac{\partial p}{\partial x} = \frac{\varrho}{l}\,c\,\sqrt{\operatorname{tg}\varphi}\;v + \varrho\,\frac{\partial v}{\partial t} \qquad \text{(akustisch)} \quad [2]$$

$$-\frac{\partial i}{\partial x} = C'\,\frac{\partial u}{\partial t} \qquad \text{(elektrisch)}$$

$$-\frac{\partial v}{\partial x} = \frac{1}{E}\,\frac{\partial p}{\partial t}\,. \qquad \text{(akustisch)} \quad [3]$$

Die Auflösung dieses Systems ist für den stationären Fall *H. Kaden*, Impulse und Schaltvorgänge in der Nachrichtentechnik (2) von Seite 70 bis 76 zu entnehmen.[1])

Daß die in Tabelle 3 angegebenen Korrespondenzen tatsächlich zutreffen, entnimmt man überprüfbaren Zwischenergebnissen, z.B. beim Wellenwiderstand, der Phasenlaufzeit und deren Frequenzabhängigkeit sowie der anomalen Dispersion. Damit können die aufgezeigten Korrespondenzen zur Betrachtung der interessanten instationären Probleme (Einschwingvorgänge) herangezogen werden, und Lehrbücher der Nachrichtentechnik als Nachschlagewerke erleichtern erheblich die Behandlung eines solchen Falles.

Betrachtung des angeschlagenen Stabes

Gibt man einem freihängenden Glasstab einen sehr lange dauernden Druckstoß, so entspricht dieser Fall dem Einschalten eines elektrischen Kabels ohne Belastung. Zur Zeit $t = 0$ wird an den Eingang des Kabels die Spannung E angelegt, und es soll der Spannungsverlauf am Kabelende ermittelt werden. Im akustischen Fall entspricht dem Anlegen der Spannung E das Einwirken eines langdauernden Druckes P_0, und es wird die Druck/Zeitgeschichte am offenen Ende des Stabes betrachtet. Die sich bildende elastische Stoßwelle kann aus dem Stab nicht austreten, da sie am Ende reflektiert wird.

Den Spannungsverlauf im Kabel findet man durch Einsetzen der dem Kabel eigenen „Stammfunktion" in die *Heaviside*schen Formeln. Führt man die Dämpfung a_∞ sowie die Laufzeit t_∞ des Gesamtkabels der Länge l ein, so erhält man für den elektrischen und akustischen Fall

$$a_\infty = \alpha_\infty l \,, \qquad\qquad [4]$$

$$t_\infty = \tau_\infty l \,, \qquad\qquad [5]$$

wobei α_∞ und τ_∞ die auf die Längeneinheit bezogene Dämpfungskonstante und Phasenlaufzeit für $\omega \to \infty$ sind. Aus (2) entnimmt man

Diese Gleichung konvergiert nur schwach und muß elektronisch gerechnet werden. Die Lösung des elektrischen Falles wird *Kaden* (2) entnommen.

Für den akustischen Fall wird lediglich $u(t)/E$ und a_∞ (elektrisch) gegen $p(t)/P_0$ und a_∞ (akustisch) ausgetauscht.

Für die Dämpfung $a_\infty = 0$ ist der Druckverlauf $p(x, t)$ in Zeit- und Ortsabhängigkeit in Abb. 1

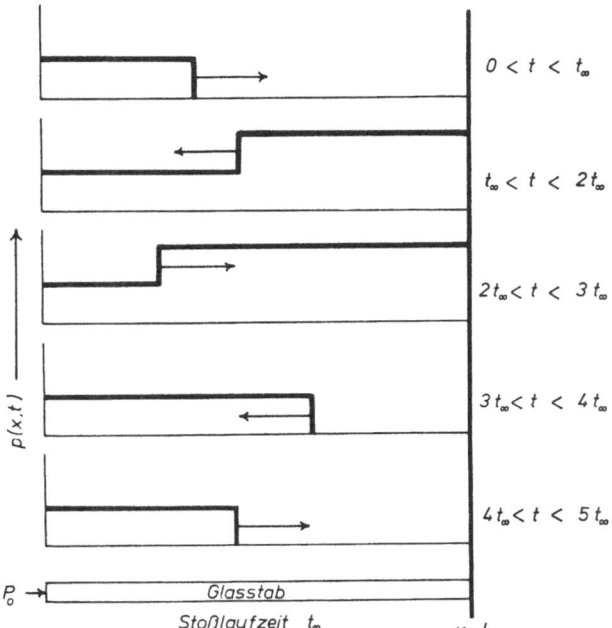

Abb. 1. Örtliche und zeitliche Ausbreitung einer Stoßfront in einem Glasstab für die Dämpfung $a_\infty = 0$. Aus *Kaden* (2), Bild 49, mit freundlicher Genehmigung vom Autor und Verlag

dargestellt. Der Verlauf ist unabhängig von der primären Druckamplitude, solange nur $2P_0$ unterhalb der Materialfestigkeit bleibt. Den Einfluß der reflektierten Welle bemerkt man augenfällig wenn $2P_0$ die Eigenfestigkeit überschreitet; vgl. Abb. 2, wo nach Untersuchungen von *J. E. Field* (3) die Bruchausbreitung in einer Glasprobe, welche durch eine Luftgewehr-

$$\frac{u(t)}{E} = 1 - \frac{4}{\pi}\, e^{-a_\infty \frac{t}{t_\infty}} \sum_{\nu=1,3,5\ldots}^{\nu=\infty} \frac{(-1)^{(\nu-1)/2}}{\nu} \left\{ \frac{a_\infty \sin \dfrac{t}{t_\infty} \sqrt{\left(\dfrac{\pi}{2}\nu\right)^2 - a_\infty^2}}{\sqrt{\left(\dfrac{\pi}{2}\nu\right)^2 - a_\infty^2}} + \cos \dfrac{t}{t_\infty} \sqrt{\left(\dfrac{\pi}{2}\nu\right)^2 - a_\infty^2} \right\}. \quad [6]$$

---

[1]) Herr Prof. *Kaden* wies darauf hin, daß in Gl. (58) $Z = Z_1 - iZ_2$ zu setzen ist.

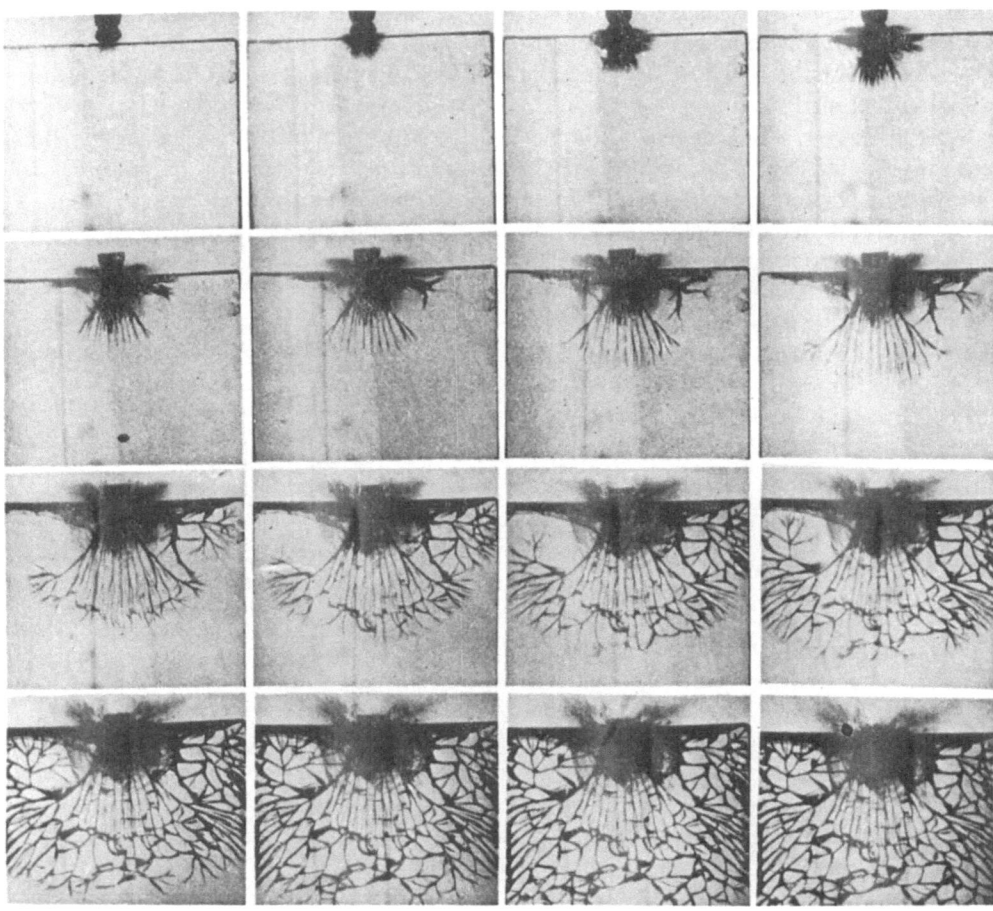

Abb. 2. Hochfrequenzkinematographische Aufnahme der Rißausbreitung in Glas, welches durch ein kleines Projektil beschossen wurde. Bildabstand 2 μs. Ab dem 8. Bild (von oben links nach rechts) ändert sich der Bruchcharakter durch das Eintreffen der Reflexionswelle. Dr. *J. E. Field* sei für das Überlassen der Aufnahme gedankt

kugel beaufschlagt wurde, hochfrequenzkinematographiert wurde. Das Bruchverhalten ändert sich wesentlich nach dem Eintreffen der reflektierten Welle.

In Abb. 3 sieht man einige Glasstäbe, welche stirnseitig einer Bleiazid-Detonation ausgesetzt waren. Dabei war der erregende Druckstoß gerade so bemessen, daß die Eigenfestigkeit des

Glases zwischen P_0 und $2 P_0$ lag. An Stellen, wo die dynamische Eigenfestigkeit überschritten wurde, liegen die Zerstörungsstrukturen. Die Zerstörungszonen finden sich in periodischer Anordnung, wobei der heile Bereich etwa gleich lang wie der zerstörte ist. Diese Bilder wurden von *E. Wehner* aufgenommen und von *E. Rexer* (4) publiziert.

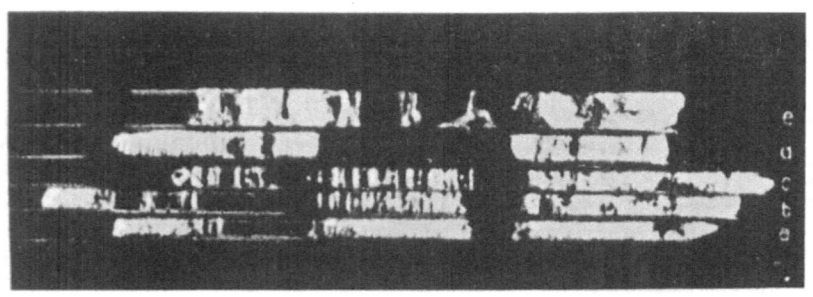

Abb. 3. Glasstäbe, welche stirnseitig Bleiazid-Detonationen ausgesetzt waren; nach *E. Wehner.* Man beachte die periodische Zerstörungsstruktur

Für das Stabende erhält man aus Gl. [6] für kleine Dämpfungen eine periodische Struktur der Stoßantwort, welche in Abb. 4 dargestellt ist. Bei großer Dämpfung, $a_\infty > 5$, erhält man nur ein viskoses Ansteigen, zwischen $5 > a_\infty > \pi/2$ kommt die Stoßsteilheit der Sendefunktion zunehmend in der Antwortfunktion zum Vorschein und für $a_\infty < \pi/2$ auch in Reflexion.

Will man die Antwortfunktion eines kurzen Druckstoßes als Sendeimpuls, so bildet man diese durch

$$P_\text{Impuls} = p(t) - p(t - T),\qquad [7]$$

wobei T die Pulsdauer des Sendeimpulses ist.

Multipliziert man die relative Zeit der Abb. 4 mit einer relativen Geschwindigkeit, so erhält man ein Spannungs-Dehnungs-Diagramm. Die Kurve für $a_\infty = 5$ entspräche rein plastischer Verformung, für $5 > a_\infty > \pi/2$ bildet sich sukzessive ein elastischer Bereich aus, für $a_\infty < \pi/2$ findet man eine Verfestigung auch in Reflexion. Schließlich wird das Verhalten für $a_\infty \to 0$ rein elastisch spröd.

Deutung der Dämpfung a_∞

Ein idealelastischer Körper besitzt die Dämpfung 0. Der reale Körper ist stets verlustbehaftet, und die einfachste Modellbeschreibung seines Verhaltens gibt der *Kelvin-Voigt*-Körper wieder, bei welchem Feder und Dämpfer parallel geschaltet sind. Im Gegensatz zum linearen Standardkörper treten nur zwei Parameter auf, wodurch die Verhältnisse zwar übersichtlich bleiben, die Ergebnisse jedoch nur mäßig genau beschrieben werden können. Phänomenologisch macht sich die Dämpfung dadurch bemerkbar, daß bei zyklischer Beanspruchung eines Körpers die aufgebrachte Spannung σ und die entsprechende Dehnung ε nicht mehr in Phase sind. Je nachdem, ob Normal- oder Schubspannungen an den Körper gelegt werden, erhält man

$$\frac{\sigma = E' + iE''}{\varepsilon = G' + iG''}\quad \text{mit}\ \ \begin{matrix}E'' = E'\,\text{tg}\,\varphi_E\\ G'' = G'\,\text{tg}\,\varphi_G,\end{matrix}\qquad [8]$$

wobei E bzw. G die komplexen Elastizitäts- bzw. Schubmoduln sind.

Die *Newton*sche Viskosität y_N^* bzw. die *Trouton*sche Viskosität y_T^* ergeben sich entsprechend Gl. [8] zu

$$y_N^* = \frac{\sigma}{\varepsilon} = -\frac{i}{\omega}\frac{\sigma}{\varepsilon} = \frac{G''}{\omega} - i\frac{G'}{\omega} = y_N' - i\,y_N'',$$

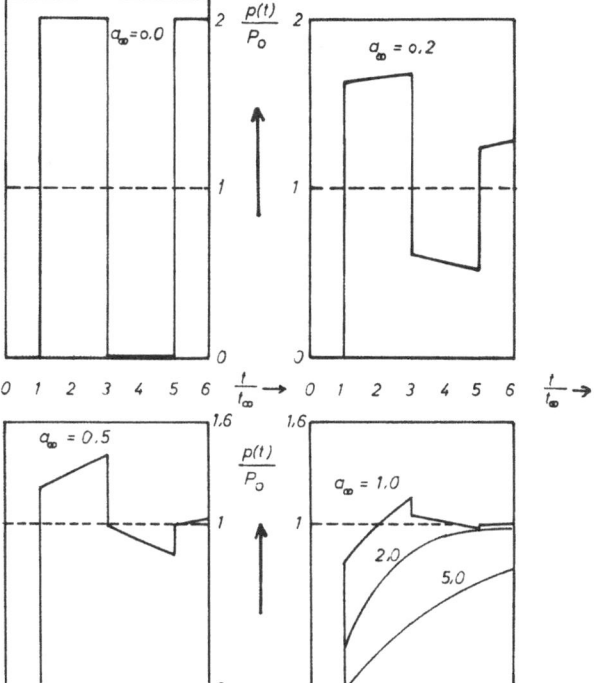

Abb. 4. Druckverlauf $p(t)$ an einem Stabende aus einem Material unterschiedlicher Dämpfung a_∞. Der einwirkende Druckstoß P_0 ist elastisch und über $6\ t/t_\infty$ lang. Aus *Kaden* (2), Bild 50, mit freundlicher Genehmigung vom Autor und Verlag

$$y_T^* = \frac{\sigma}{\varepsilon} = -\frac{i}{\omega}\frac{\sigma}{\varepsilon} = \frac{E''}{\omega} - i\frac{E'}{\omega} = y_T' - i\,y_T''.\qquad [9]$$

Der Realteil der *Trouton*-Viskosität y_T' unterscheidet sich von der *Newton*-Viskosität y_N'. Nimmt man aber gleiche Kreisfrequenzen ω an, so erhält man mit

$$G = \frac{1}{2}\,\frac{E}{1+v}\qquad [10]$$

und üblichen Querkontraktionszahlen v

$$y_T' = 2{,}4\ldots 2{,}6\,y_N'.\qquad [11]$$

Für $\omega \to 0$ gehen die Viskositätswerte in die statisch meßbaren über.

Aus den Gln. [8] und [9] erhält man

$$\sigma = E'\varepsilon + \dot\varepsilon\,y_T'.\qquad [12]$$

Für einen Spannungssprung σ_0 erhält man den Dehnungsverlauf

$$\varepsilon = \frac{\sigma_0}{E'}\left(1 - e^{-\frac{E't}{y_T'}}\right),\qquad [13]$$

womit die Relaxationszeit

$$\tau = \frac{y_T'}{E'} = \frac{E''}{\omega E'} = \frac{\text{tg}\,\varphi}{\omega} \qquad [14]$$

eingeführt wird.

Für den allgemeinen Fall hängt tg φ in komplizierter Weise von der Dämpfung a_∞ ab, bei Verwendung der Dünndrahtnäherung erhält man mit dem Ausdruck für α_∞ in Dünndrahtnäherung (Gl. [60] aus *Kaden* (2) und den Korrespondenzen der Tabelle 3)

$$\alpha_\infty = \frac{\sqrt{\text{tg}\,\varphi}}{2\,l} \qquad [15]$$

und Gl. [14]

$$\text{tg}\,\varphi = 4 a_\infty^2 = \tau\,\omega . \qquad [16]$$

Gemäß dem vorhergehenden Abschnitt und Abb. 4 erhält man mit Gl. [16]

tg $\varphi = \omega\,\tau \geqq 100$ für plastisches Fließen,

bis tg $\varphi = \omega\,\tau > 9,9$ Ausbildung einer elastischen Eigenfestigkeit.

Bei tg $\varphi = \omega\,\tau < 9,9$ macht sich die Eigenfestigkeit auch in der Stoßreflexion bemerkbar.

tg $\varphi = \omega\,\tau \to 0$ Vollausgeprägtes elastisch sprödes Verhalten.

Diskussion der Gl. [16] und Meßmethode zur Bestimmung der Versprödung

Das Produkt $\omega\,\tau$ läßt sich sowohl für $\omega > \tau$ als auch für $\tau < \omega$ bilden. Im einen Fall gelangt man vom Festkörper zur Erweichung bzw. umgekehrt. Wenn die Relaxationszeiten konstant blieben, sollte die tg φ-Kurve also symmetrisch sein, wobei nur der Ast vom Festkörper zur Erweichung durch eine Erhöhung der Beweglichkeit gekennzeichnet ist, in diesem Bereich nimmt auch die Viskosität dementsprechend stark ab.

Dieses Verhalten kann direkt zur Ermittlung der Versprödung bzw. Erweichung ohne Messung der mechanischen Moduln ausgenutzt werden.

Bekanntlich rührt die Phosphoreszenz von einem metastabilen, an sich verbotenen und deshalb sehr unwahrscheinlichen Triplett-Singlett-Übergang her, wobei die thermischen Des-

aktivierungsvorgänge weitestgehend vermieden sein müssen (5). Da die Triplett-Singlett-Übergangswahrscheinlichkeiten mit den strahlungslosen thermischen Desaktivierungsvorgängen konkurrieren, muß sich eine Zunahme der Desaktivierung — welche z.B. durch eine zunehmende Fluidisierung der Matrix erfolgt — durch Abfall der mittleren Phosphoreszenzlebensdauer bemerkbar machen.

Zum Nachweis dieses Verhaltens wurde Phenanthren in einem Zuckerglas eingeschmolzen (6). In Abb. 5 sind die Moduln des Zuckerglases sowie die Lebensdauern der Phenanthren-Phosphoreszenz in Temperaturabhängigkeit wiedergegeben. Wie man sieht, nimmt die Lebensdauer der Phosphoreszenz tatsächlich erst rechts vom G''-Maximum stark ab. Bei tiefen Temperaturen war das Zuckerglas sehr spröde. Da es klebrig wurde und die Kopplung abriß, konnten die Moduln leider nur bis ca. 30 °C gemessen werden.

Grundsätzlich ist diese Methode auf alle Gläser, welche dotiert sind, anwendbar, auch auf Kunststoffe, da diese meist ohnehin phosphoreszenzfähige Stoffe enthalten, leider aber mitunter mehrere, so daß eine Trennung meist schwierig ist.

Die Transformation des Glases

Nimmt man an, daß die Glastransformation gerade dann einsetzt, wenn eine gerade beginnende Eigenfestigkeit einsetzt, so ergibt sich mit $a_\infty = 5$ und Gl. [16] tg $\varphi =$ ca. 100. Nach *M. Coenen* und *E.-M. Amrhein* (7) beträgt der Schubmodul eines Natriumdisilikatglases beim Transformationspunkt ca. 10^{11} dyn/cm^2, nimmt man ferner an, daß die Meßgeschwindigkeit etwa $\omega \approx 1$/sec beträgt, so erhält man mit diesen Werten zusammen tatsächlich die Transformationsviskosität von 10^{13} Poise. Aus Gl. [16] wird nun auch verständlich, daß die Transformation in einem Bereich erfolgt, welcher von der Probengeometrie ($a_\infty = \alpha_\infty l$) und der Geschwindigkeit der Versuchsdurchführung (ω) abhängt. Bei rascher Versuchsdurchführung steigt also die Transformationstemperatur an. Es ist plausibel, daß die obere Kühltemperatur auch der Transformationstemperatur entspricht, da sich hier Verspannungen und Störungen noch nicht auswirken können. Kühlt man langsam ab, so bildet sich zunehmend eine Eigenfestigkeit aus,

Rißentstehung wäre möglich ohne daß ein Ausheilen mehr erfolgt, dennoch sind noch Transportmöglichkeiten vorhanden, um geringere Störungen auszugleichen. Schließlich wird die untere Kühltemperatur dann erreicht, wenn die Eigenfestigkeit so hoch ist, daß thermische Verspannungen elastisch aufgenommen werden können. Spekulativ entspricht der unteren Kühltemperatur die Dämpfung $a_\infty = \pi/2$. Darunter ist zügiges Abkühlen möglich. Die Dämpfung kann also durchaus monoton verlaufen, trotzdem sind relativ rasche Einfriervorgänge möglich, welche allerdings stark von der Meßgeschwindigkeit bezüglich der Temperatur abhängen.

Experimentell einfache Ermittlungsmöglichkeit kleiner Dämpfungen

Schlägt man den Glasstab der Abb. 1 mit einer gleichen Glaskugel an einem Pendel und befindet sich am anderen Ende des Stabes in direktem Kontakt ein ähnliches Pendel, so wird dessen Auslenkung von der Dämpfung des Glasstabes abhängen, aus dem Vergleich der Auslenkung kann dann auf die Dämpfung geschlossen werden.

Der elektrische Analogiefall entspricht dem Spannungsverlauf am Ende eines mit seinem Wellenwiderstand abgeschlossenen Kabels nach dem Anlegen einer Gleichspannung am Kabelanfang zur Zeit $t = 0$ und ist in (2), Bild 57 dargestellt. In Abb. 6 ist der elastische Anteil der Antwortfunktion in Abhängigkeit der Dämpfung dargestellt. Aus dem Verhältnis der Pendelausschläge kann also auf die Dämpfung a_∞ bzw. nach Umrechnung über den *Kelvin-Voigt*-Modellkörper auf tg φ geschlossen werden. Für $a_\infty = 5$ wird der Ausschlag verschwinden.

Lagert man die Probe so, daß der eintretende Stoß am anderen Ende nicht austreten kann, so wird er reflektiert und wirft das erregende Pendel selbst zurück. Damit hat man aber eine Messung der Rückprallelastizität. Bei käuflichen Instrumenten ist der Hammer aus einem definierten vollelastischen Material, und das zu untersuchende Gut ändert sich, so daß sich die Stoßanpassung (Wellenwiderstand) ebenfalls verändert. Bei ein und demselben Material können aber Dämpfungsänderungen — z.B. in Temperaturabhängigkeit — im unteren Bereich *relativ* empfindlich festgestellt werden.

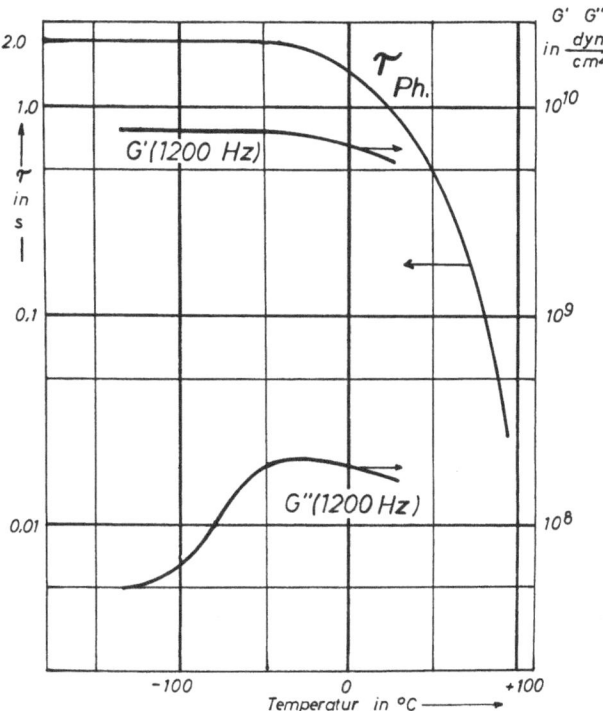

Abb. 5. Lebensdauer τ der Phosphoreszenz des Phenanthrens in Zuckerglas im Vergleich zum komplexen Schubmodel in Temperaturabhängigkeit. Die Modulmessungen verdanke ich Herrn *S. Blasenbrey*, damals I. Physikalisches Institut, TH Stuttgart

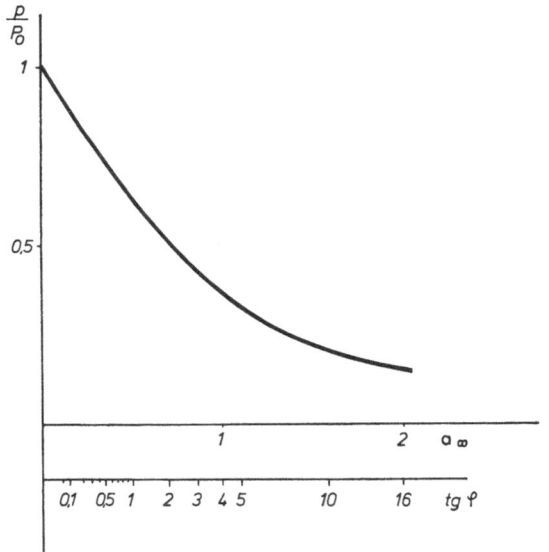

Abb. 6. Relative Rückprallhöhe in Abhängigkeit der Dämpfung bei voller Stoßanpassung zwischen Material und Hammer

Zusammenfassung

Durch analoge Gesetzmäßigkeiten elektrischer und akustischer Größen im Festkörper werden Korrespondenzen vorgeschlagen, welche in die Telegraphengleichungen eingesetzt werden. In Analogie zum freilaufenden Kabel werden der Transformationsbereich sowie die Kühltemperaturen bei Gläsern interpretiert. Es werden auch meßtechnisch ausnutzbare Methoden zur Bestimmung der Versprödung skizziert.

Summary

By analogies in mathematical interrelationships between electrical and mechanical properties of a solid it is possible to use the still solved solutions of the telegraphic equations for finding out conveniently the solution of the corresponding mechanical case. Doing so, for example the vitreous/liquid transition of glasses, and the annealing points are tractable in analogy to a free running electrical transmission line. Some possible tests for determination of the embrittlement are outlined.

Literatur

1) *Naslin, P.*, Dynamik linearer und nichtlinearer Systeme (München/Wien, 1968).
2) *Kaden, H.*, Impulse und Schaltvorgänge in der Nachrichtentechnik (München, 1957).
3) *Field, J. E.*, Contemp. Phys. 12, 1—31 (1971).
4) *Rexer, E.*, Glastechn. Ber. 17, 33—38 (1939).
5) *Förster, Th.*, Fluoreszenz organischer Verbindungen (Göttingen, 1951).
6) *Leiber, C. O.*, Der Druckeinfluß bei Reaktionen elektronisch angeregter Moleküle in Lösung (Stuttgart, 1963).
7) *Coenen*, M. und *E.-M. Amrhein*, C. R. Symp. Res. Mec. du Verre, 529—550 (Charleroi, 1962).

Adresse des Autors:

Dr. *C. O. Leiber*
Institut für chemisch-technische
Untersuchungen
D-5357 Swissttal-Heimerzheim

Für die Schriftleitung verantwortlich: Für Originalarbeiten Prof. Dr. F. H. Müller, 3550 Marbach b. Marburg/L.
und Prof. Dr. Armin Weiss, 8000 München 2
Dr. Dietrich Steinkopff Verlag, 6100 Darmstadt, Saalbaustraße 12
Herstellung: Konrad Triltsch, Graphischer Betrieb, 8700 Würzburg

Neu:

Konzepte der Kolloidchemie

Aussagen aus fünf Jahrzehnten

Ausgewählt von **Jürgen Steinkopff,** Darmstadt
VIII, 145 Seiten mit 16 Abb., 2 Schemata und 8 Tab.
Kunststoffeinband DM 38,—

Inhalt:

Wolfgang Ostwald, Zur Gründung der Kolloid-Gesellschaft (1922)
Alfred Lottermoser, Zur Geschichte der Kolloidchemie (1930)
Wolfgang Ostwald, Zur Topographie und Nomenklatur kolloider Systeme (1923)
Peter von Weimarn, Der kolloide Zustand als allgemeiner Zustand (1925)
Raphael Eduard Liesegang, Über die Gestalt der Kolloide (1926)
Heinrich Bechhold, Über Biokolloide (1929)
Raphael Eduard Liesegang, Kolloide in Biologie und Medizin (1935)
Aladar von Buzágh, Grundlagen der Kolloidik (1936)
Heinrich Thiele, Abgrenzung und Darstellung der Kolloide (1950)
Martin H. Fischer, Kolloidchemie und Biologie (1951)
Alfred Lottermoser, Zur Nomenklatur und Systematik in der Kolloidchemie (1944)
Erich Manegold, Kolloide Systemzusammenhänge (1959)
Joachim Stauff, Grenzen und Aufgaben der Koloidchemie (1960)
Erhard Ühlein, Lexikalisches zum Stichwort Kolloidchemie (1966)
Egon Matijević, Kolloide: Die Welt der vernachlässigten Dimensionen (1974)
Jürgen Steinkopff, Der Steinkopff Verlag und die Kolloidchemie (1975)
Hans Wolfgang Kohlschütter, Kolloidchemie im Verbundsystem der Naturwissenschaften (1975)

Kolloidchemie

Von Dr. **Kurt Edelmann,** Domat/Ems, Graubünden
(UTB Uni-Taschenbücher 512)
VII, 139 Seiten mit 29 Abb. und 10 Tab. Kunststoffeinband DM 18,80

Inhalt:

Einführung
Anorganische Kolloidchemie
Organische Kolloidchemie
Literatur — Sachverzeichnis

DR. DIETRICH STEINKOPFF VERLAG · DARMSTADT